Structure and Reactivity of Coal

Ke-Chang Xie

Structure and Reactivity of Coal

A Survey of Selected Chinese Coals

Springer

Ke-Chang Xie
Taiyuan University of Technology
Taiyuan
China

and

Tsinghua University
Beijing
China

ISBN 978-3-662-47336-8 ISBN 978-3-662-47337-5 (eBook)
DOI 10.1007/978-3-662-47337-5

Library of Congress Control Number: 2015941120

Springer Heidelberg New York Dordrecht London

Printed on acid-free paper

Springer-Verlag GmbH Berlin Heidelberg is part of Springer Science+Business Media
(www.springer.com)

Contents

Chapter 1
Geological Characteristics of Coal

Abstract Coal is a complex and diverse material. It is important to consider the geological origins and history of coal as these considerations allow for the categorization of coal into various types. The classification of coal into various types is itself a valuable step towards using coal resources effectively. It is also essential for understanding the various aspects of coal structure and reactivity. In this chapter, we discuss the plant materials that are the precursors of coal in terms of their organic constituents. The chemical structures of plant sugars, proteins, lipids, and other organic components are related to their stabilities and the geochemical processes by which these organic raw materials are eventually converted into various types of coal. We review the basic generic types of coal in terms of precursor material, geochemical history and relate this to appearance and basic physical properties. We introduce the categorization concepts of coal maceral, rank, and lithotype. The qualities and factors that are used to determine coal type are discussed in terms of the distribution and type of coal resources in China. The chapter forms the necessary background to understanding the origin of materials discussed in later chapters and provides a context for the studies we present.

1.1 Coal Formation

Coal is an organic biological rock, gradually formed from dead plant remains that have accumulated to a certain thickness and become covered by water or silt under physical, chemical, and biological processes in an appropriate geological environment over a long geological age. A great variety of materials can be converted into coal, and the external conditions and historical era of coal formation vary. Coal has some features in common with other minerals, but is generally more diverse and has a more complex structure in terms of mineralogical, petrographic, and basic physical and chemical characteristics.

K.-C. Xie, *Structure and Reactivity of Coal*, DOI 10.1007/978-3-662-47337-5_1

1.1.1 Components of Coal Formation

Plant is the principal precursor of coal. Plant cells consist of cell walls made up of celluloses, hemicelluloses and lignins, and cytoplasm, mainly composed of proteins and fat. The amount of cytoplasm in the cells of higher plants is less than that in lower plants. There are four types of organic components in plants: saccharides and their derivatives, lignins, proteins, and lipids.

1.1.1.1 Saccharides and Derivatives

Saccharides and their derivatives include celluloses, hemicelluloses, and pectins. Celluloses are high-molecular-weight carbohydrates, known as polysaccharides, with chain structures of general formula $(C_6H_{10}O_5)_n$ (Fig. 1.1). Celluloses are extremely stable in living plants; however, in dead plant material, aerobic bacteria can hydrolyze the celluloses to monosaccharides, catalyzed by cellulose hydrolytic enzymes, and the monosaccharides can be further be oxidized to CO_2 and H_2O, as shown below:

$$(C_6H_{10}O_5)_n + nH_2O \xrightarrow{\text{Bacterial action}} nC_6H_{12}O_6$$
$$C_6H_{12}O_6 + 6O_2 \rightarrow 6CO_2 \uparrow + 6H_2O + \text{heat}$$

When the environment gradually becomes anoxic, anaerobic bacteria can ferment the celluloses to generate intermediate products such as CH_4, CO_2, C_3H_7COOH and CH_3COOH. These hydrolysis products and fermentation products may react with other plant decomposition products to form more complex substances and participate in coalification.

Hemicelluloses are also polysaccharides, with various structures, e.g., pentosan $(C_5H_8O_4)_n$. Hemicelluloses can be decomposed to monosaccharides under the action of microorganisms. Pectins, which consist mainly of condensed galacturonic acid and galacturonic acid methyl ester units, are gel-like saccharide derivatives, present in fruit and plant xylem. Pectin molecules contain galacturonic acid, so pectins are acidic. Pectins are relatively unstable and can be hydrolyzed to generate a series of monosaccharides and uronic acids by biochemical action in the initial

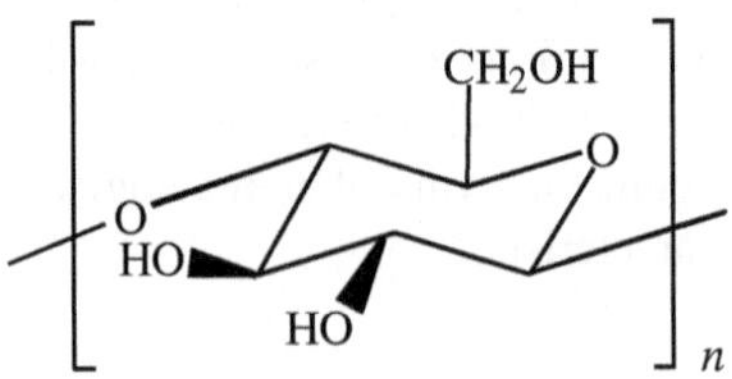

Fig. 1.1 Molecular structure of cellulose

stage of peat formation. Dead plant also contains glucosides, which are saccharides condensed through reducing groups and other hydroxyl-containing substances such as alcohols, phenols, and sterols.

1.1.1.2 Lignins

Lignins are the main organic components in coal-forming materials. They are mainly distributed in the cell walls of higher plants, the surrounding celluloses and fill space to increase the stem firmness. The lignin composition varies depending on the plant variety, but all contain an aromatic nucleus with side chains, such as $-OCH_3$, –OH, –O–, and other functional groups. See Table 1.1 for three types of lignin monomer. These lignin monomers connect in different ways to form three-dimensional macromolecules, making lignins more stable than celluloses and more difficult to hydrolyze. However, under oxygen-rich conditions, lignins are easily oxidized to form aromatic acids and fatty acids under the action of microorganisms.

1.1.1.3 Proteins

Proteins are the main component of plant cell protoplasm, and are the most important organic substances in terms of the origin of life. Proteins are macromolecular compounds with complex multilevel structures, consisting of amino acid molecules condensed according to certain arrangement rules (Fig. 1.2). The –COOH in an amino acid molecule and the $-NH_2$ in another amino acid molecule form an amido bond; the –CO–NH– group in protein molecules is called a peptide bond. Proteins are natural polypeptides, with molecular weights above 10,000, containing groups such as carboxyl, amido, hydroxyl, and disulfide bonds. The N and S in coal are derived from plant proteins. When the plant is dead, proteins can decompose to form gaseous products under oxidizing conditions. In a peat marsh, proteins can be hydrolyzed to generate nitrogen-containing compounds

Table 1.1 Three types of lignin monomer

Type of plant	Conifer	Broad-leaved tree	Grass
Monomer	Coniferyl alcohol	Sinapyl alcohol	γ-Tonquinol
Structural formula	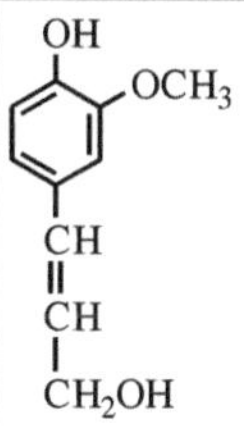	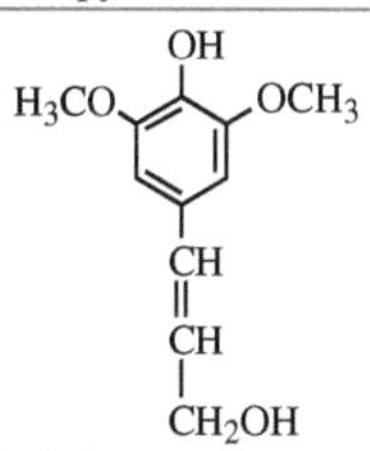	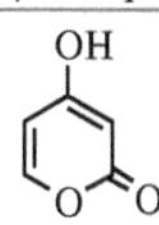

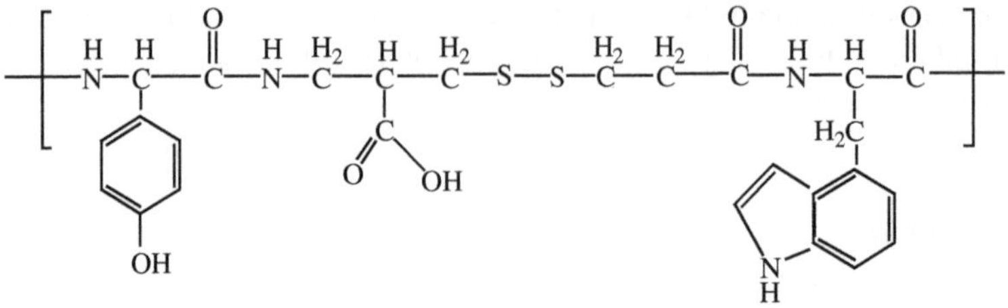

Fig. 1.2 Typical chemical structure of protein segment

such as amino acids and porphyrins, which are involved in coalification; for example, amino acids and saccharides can be condensed to produce more complex humic substances.

1.1.1.4 Lipids

The term lipid typically refers to organic compounds that are insoluble in water but soluble in organic solvents, including fats, resins, waxes, cutins, suberins, and sporopollenin. One common feature of lipids is a high chemical stability, which leads to the preservation of these molecules in coal.

Fats are glycerides of long-chain fatty acids, e.g., palmitin (Fig. 1.3). Lower plants generally contain relatively large amounts of fats; algae, for example, have fat contents as high as 20 %. In general, higher plants only contain 1–2 % fat, mostly concentrated in the plant spores or seeds. Fats can be hydrolyzed under biochemical action to form fatty acids and glycerin; the former are involved in coalification.

Resins are secreted during plant growth. In higher plants, acicular-leaved plants contain the largest amounts of resins. Resins are mixtures of mainly diterpene and triterpene derivatives. Typical acids present in resins include abietic acid and dextropimaric acid, which are non-saturated and can be involved in polymerization. Resins are extremely chemically stable and unaffected by microorganisms. They are also insoluble in organic acids; resins therefore remain relatively intact in coal.

Waxes are similar to fats in terms of chemical properties, but are more stable. The leaves, stems, and fruit of plants are normally covered with thin layers of wax. Wax components are complex consisting of long-chain fatty acids and esters formed from monohydric alcohols containing 24–36 (or more) carbon atoms (e.g., glycerin stearic acid). Waxes are chemically stable and do not decompose easily. Waxes are often found in peat and lignite.

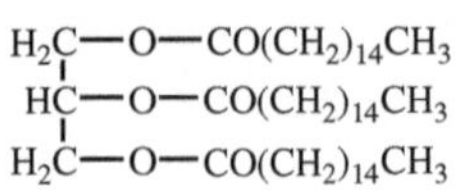

Fig. 1.3 Structural formula of palmitin

Cutins are the main components of cutin membranes, and the surfaces of leaves, twigs, germs, and fruit of plants are often covered with cutin membranes. Cutins are produced from fatty acids by dehydration or polymerization, and mainly consist of cutic acids containing 16–18 carbon atoms.

The main components of suberins are ω fatty alcohol acids, dicarboxylic acids, long-chain carboxylic acids with more than 20 carbon atoms, and alcohols.

Sporopollenin is the main organic component of the external walls of plant reproductive organs, i.e., spores and pollen, and has an aliphatic–aromatic reticular structure. It is chemically stable, has good resistance to acid and alkali attacks, is insoluble in organic solvents, and does not decompose at high temperatures.

In addition to the stated four categories of organic compounds, small amounts of tannins, pigments, etc. are present in plants.

1.1.1.5 Coal Types

Generally, coal is classified into two types, according to the coal-forming plants: humic coal, mainly formed from higher plants, and sapropelic coal, mainly formed from lower plants. The majority of humic coal consists largely of plant lignins and celluloses; it is the most widespread coal in nature, with the largest reserves. Sapropelic coal includes boghead coal and saprocollite. Boghead coal is mainly formed from algae, and saprocollite, a type of structureless sapropelic coal, consists almost completely of matrix materials, because the plants are thoroughly decomposed. In addition, there is one type of composite coal, a mixture of humic coal and sapropelic coal, called humic–sapropelic coal, which is sometimes classified separately.

1.1.2 Formation of Humic Coal

There are well-preserved paleophyte fossils and carbonized tree trunks in coal seams, and sometimes even the original shapes of fractured trunks are preserved. The bottom of a coal bed is rich in plant root fossils, proving that it was the soil for plant growth. The remains of the cell structures of the originals plant and other tissues such as spores, pollen, resins, cutin layers, and suberins can be observed by examining a thin coal sheet under a microscope. Simulation of the coal-forming process shows that materials with appearances and properties similar to those of natural coal can be formed using trees as the starting material. This proves that coal is produced from plants, and mainly from higher plants.

Plant evolution has a significant impact on coal formation, and coalification only occurs when plants are widely distributed and grow abundantly. A coal seam of industrial significance can only be formed under warm, humid paleoclimatic conditions, with mass propagation of woody plants and mass accumulation of dead plants. There are three large coal-accumulating periods in the entire geological age:

the Carboniferous and Permian periods in the Paleozoic era, when the main coal-forming plants were cryptogams; the Jurassic and Cretaceous periods in the Mesozoic era, when the main coal-forming plants were gymnosperms; and the Tertiary period of the Cainozoic era, when the main coal-forming plants were angiosperms. The plant growth and coal-forming processes in different geological ages are listed in Table 1.2.

Although coal-accumulating periods are closely linked to plant growth, appropriate geological conditions are also needed, i.e., vertical movement of the Earth's crust, either rising or descending. As a result, large amounts of dead plant can form peat layers and be buried underground, the conditions for coal formation such as the required pressure and temperature are met. Coal formation is an extremely long and complex process. The causes of coal formation (such as the types of coal-forming plants, the environment and methods of accumulation of dead plant, and

Table 1.2 Stratigraphic systems, geological ages, coal-forming plants, and major coal types

Era (erathem)		**Cainozoic**			**Mesozoic**		
Period (system)		Quaternary	Neogene[a]	Eogene[a]	Cretaceous[a]	Jurassic[a]	Triassic
Age (million years)		1.6	23	65	135	205	250
Organic Evolution	Plants	Angiosperm			Gymnosperms		
	Animals	Fossil humans	Mammals		Reptiles		
Coal Type		Peat	Large amount of lignite and small amount of bituminous coal		Lignite, bituminous coal, and small amount of anthracite		
Distribution in China			Yunnan, Guangxi, Guangdong, Liaoning, Taiwan		Xinjiang, Inner Mongolia, Liaoning Fuxin, Shanxi Datong, Jiangxi Pingxiang		
Era (erathem)		**Paleozoic**					
		Neopaleozoic			**Early Paleozoic**		
Period (system)		Permian[a]	Carboniferous[a]	Devonian	Silurian	Ordovician	Cambrian
Age (million years)		290	355	410	438	510	570
Organic Evolution	Plants	Ferns		Gymnosperms	Thallophytes		
	Animals	Amphibians			Fish	Invertebrates	
Coal Type		Bituminous coal and anthracite			Stone coal		
		Yunnan Luquan, Guangdong Taishan, Liaoning Benxi, Anhui Huainan, Henan			Southern provinces		
		Kailuan, Shanxi					

[a]Main coal-accumulating period.

biochemical actions during the paludification stage) define the coal type. This essentially determines the various macerals of coal that are seen under a microscope. The metamorphic factors of coal (such as coal metamorphic type, temperature, pressure, and time after peat became rock) determine the coal rank, i.e., the chemical maturity of coal, which is also known as the coalification extent.

The generation of humic coal is a coal-forming process, in which higher plants underwent continuous growth and death in peat swamps. Their remains accumulated and underwent long and complex biochemical, geochemical, and physical reactions before gradually evolving into peat, lignite, bituminous coal, and anthracite. The whole transformation process from plants to coal is called coalification. The coal-forming process can be divided into two stages, i.e., paludification and coalification.

1.1.2.1 Paludification Stage

The paludification stage is the process in which the remains of higher plants underwent biochemical and geochemical reactions in swamps and evolved into peat. Plant remains were the source material of coal, and they could only accumulate and be transformed into peat under certain external conditions. This transformation needed a high level of plant reproduction, and an appropriate environment to preserve plant remains. Swamps provided such conditions. Swamps are low, flat terrains, with poor drainage, and plenty of plants. Incompletely decomposed plant remains accumulated in the swamps year by year, and paludification eventually started. During this process, the organic components of the plants and swamp microorganisms both participated in coalification.

Initially, the plant remains were exposed to air or were in the shallow parts of the swamps, and experienced oxidative decomposition and hydrolysis reactions, which were caused by microorganisms such as aerobic bacteria and fungi. Part of the plant remains was completely destroyed, and the other part decomposed to simple organic components, some of which were further transformed to humic acids under certain conditions, and some of which were stable and therefore retained. As the plant remains piled up to a certain height and were completely covered by swamp water, the decomposing plant remains were gradually isolated from the air. At the same time, the accumulation of products from the decomposition of plant remains such as hydrogen sulfide, organic acids, and phenols during the transformation inhibited the presence and activities of aerobic bacteria and fungi. The system became a weak oxidative or even reducing environment. At this stage, the microorganisms were replaced by anaerobic bacteria, and hydrogen-rich products were generated by oxygen-deficient reduction reactions. These products were then transformed into stable new materials such as humic acids and asphaltene. If the plant remains were kept in an aerobic environment or one with an adequate oxygen supply, they were strongly oxidized and decomposed, and underwent full decay or partial decay that did not lead to peat formation.

In the paludification process, the plant remains underwent significant physical and chemical changes. The different degrees of oxidative decomposition and the differences in timing of the change to reduction reactions, resulted in the plant remains undergoing two types of reaction in the paludification stage: gelatinization, which occurred in weakly oxidative or reducing environments and led to the formation of gelated materials such as humic acids and asphaltine; and fusinization, which occurred under strongly oxidizing conditions and led to the formation of hydrogen-lean and carbon-rich fusinized materials, collectively called fusain. When gelatinization was weak, the expansion of the plant cell walls was low. The plant cell tissue retained its original array patterns, and the cell cavity was obvious. When gelatinization was extremely strong, the plant cell structure disappeared completely, forming a uniform gel.

The accumulation environment for peat has a significant impact on the lithofacies composition, sulfur content, and reduction degree of coal. Physical conditions such as the water depth and mobility influence the chemical conditions of a peat swamp. Chemical conditions such as the proton concentration (pH) and oxidation/reduction potential affect microorganism activity. The interrelated physical, chemical, and microorganism conditions influence the coal-forming plant materials and give peat's special characteristics. For example, coal from an offshore coalfield is rich in vitrinite and its sulfur content is high—sometimes as high as 8–12 %—whereas coal from an inland coalfield is rich in resinite and inertinite, and the sulfur content is low.

1.1.2.2 Coalification Stage

The end of the paludification stage occurs when peat becomes covered with inorganic sediments, at which point the biochemical effect was weakened and gradually stopped. Transformation of peat into lignite, bituminous coal, and anthracite occurred as a result of physical and chemical reactions at certain temperatures and pressures. This process is called the coalification stage. Based on the different impacts and results, the coalification stage can be divided into diagenetic and metamorphic stages.

In the diagenetic stage, the amorphous peat was subjected to high pressure from overlying silt and other inorganic sediments, experienced a series of physical and chemical changes such as compacting, dehydration, an increase in the amount of carbon, colloidal aging, indurations, and voidage reduction, and was eventually transformed into young lignite with biogenetic rock characteristics. During the diagenetic process, the plant components that remained in the peat (such as celluloses, hemicelluloses, and lignins) gradually disappeared, and humic acids, hydrogen, oxygen, and carbon underwent significant changes.

In the metamorphic stage, the lignite descended to the bottom of the Earth's crust, underwent chemical reactions at relatively high temperatures and high pressures over a long time, and was transformed into bituminous coal and anthracite after its components, structures, and properties had changed. During this transformation the coal seam is subjected to pressures of tens to hundreds of megapascals, and the temperature generally remains below 200 °C. The main factors causing coal metamorphism are temperature, time, and pressure. Temperature is the main factor in coal metamorphism, and there appears to be a critical temperature for this process. The geothermal gradient is always positive, i.e., the temperature increases gradually with depth underground, but its variation range is 0.5–25 °C/100 m. The temperatures required for different stages of coalification are as follows: lignite 40–50 °C, long-flame coal 100 °C, and typical bituminous coal no higher than 350 °C. Time is also an important factor in coal metamorphism. The impacts of temperature and pressure on coal metamorphism change with heating and pressure durations. For the same temperature, if the heating time is short, the coal metamorphism extent is low; if the heating time is long, the coal metamorphism extent is relatively high. Different conditions, i.e., relatively short times and high temperatures, or relatively long times and low temperatures, can therefore give the same degree of metamorphism. Pressure is another important factor in coal metamorphism. The pressure can change the shapes of coal-forming materials; it makes coal compact, reduces the porosity and moisture content, and arranges the coal lithofacies components along the direction of the vertical pressure, and arranges rings of aromatic clusters in parallel planes. Pressure is generally considered to be a secondary factor in coal metamorphism.

Coal metamorphism can be divided into three types, according to the coal metamorphic conditions and characteristics: plutonic metamorphism, magma metamorphism, and dynamic metamorphism. Plutonic metamorphism occurs when the coal is buried relatively deep in the ground; geothermal effects and the static pressure of the overlying rock stratum cause metamorphism. This metamorphism is directly related to the vastly lifting motions of the Earth's crust. Coal metamorphism is distributed vertically. In the same coalfield under similar tectonic conditions, the degree of metamorphism increases along the burial depth. For a depth increase of 100 m, the volatile content on a dry and ash free basis (V_{dat}) of the coal decreases by about 2.3 %. This is called Hilt's law [1]. Coal metamorphism is also distributed horizontally. In the same coalfield, the settlement range of the same coal seam deposition may differ. According to Hilt's law, the metamorphic degrees of such a coal seam differ, resulting in metamorphism being distributed in stripes or circles. Magma metamorphism occurs when the coal seam is affected by high temperature, volatile gases, and pressure from magma, causing abnormal coal metamorphism; this is a regional metamorphic phenomenon. Coal metamorphism is enhanced by the direct intrusion, passing, or approaching of magma to the coal, where the magma is mainly from a shallow rock layer. This effect is called contact metamorphism. The coal metamorphism was increased by the intrusion of magma at the bottom of the coal seam is called regional thermal metamorphism. Dynamic metamorphism is caused by the dynamic pressure and heat produced by crust structural changes; this is also a regional metamorphic phenomenon.

1.2 Classification of Coal

In terms of reserves, the substance commonly referred to as coal is humic coal, mainly formed from lignins and celluloses. Humic coal is the main coal type used in modern times, and also the focus of coal science and technology research. See Table 1.3 for the main characteristics of humic coal and sapropelic coal.

Humic coal can be classified into peat, lignite, bituminous coal, and anthracitic coal, depending on the coalification process. Coals differ from each other in appearance and properties. In general, the major types can be distinguished visually.

1.2.1 Peat

Peat, which is a non-uniform deep brown color, is a transitional product between dead plant and coal. Peat is formed in marshes, and has an extremely high water content because of the large amount of preserved undecomposed roots, stems, and leaves of plants. This water content is generally as high as 85–95 %. The water content in the exploited peat is reduced to 25–35 % by natural drying. Dry peat is brownish-black or blackish-brown clods, with a true density of 1.29–1.61 g cm^{-3}.

The main organic matter in peat includes humic acids, asphaltine, and plant exinite, as well as undecomposed and incompletely decomposed plant cluster components. Humic acids, the main organic component of peat, are complex compounds composed of macromolecular hydroxycarboxylic acids. Humic acids are soluble in alkaline solution, and flocculent sediments are separated out when the pH of the solution is adjusted to be acidic. Asphaltine is organic matter that can be extracted by organic solvents such as benzene and methanol. Plant exinites are components that have hardly changed compared with the original plant components, i.e., cutins and resins; undecomposed or incompletely decomposed plant clusters consist mainly of celluloses, hemicelluloses, and lignins.

Table 1.3 Main characteristics of humic and sapropelic coals

Characteristic	Humic coal	Sapropelic coal
Color	Brown and black, mostly black	Mostly brown
Luster	Mostly bright	Dull
Ignition with match	Non-combustion	Combustion, with asphalt smell
Hydrogen content (%)	Generally <6	Generally >6
Low-temperature-carbonization tar yield (%)	Generally <20	Generally >25

1.2.2 Lignite

Most lignite is brown or darkish brown and lackluster, so it is also called brown coal. Lignite is the initial product from settled peat on dehydration, compression, and other coalification actions. Lignite has a high water content, ranging from 30 to 60 %. After air drying, it still contains 10–30 % moisture, and has a true density of 1.10–1.40 g cm^{-3}. It is easily weathered and broken into pieces. China has large reserves of lignite, approximately 350 billion tons, accounting for about 20 % of the proven coal reserves [2].

In terms of appearance, the main difference between lignite and peat is that lignite contains no undecomposed dead plant and forms layers geologically. Compared with peat, the aromatic nucleus condensation degree of humic acids in lignite is higher, the number of oxygen-containing functional groups is smaller, the side chains are shorter, and the number of side chains is smaller. Moreover, some conversions of humic acids to neutral humic species occur.

1.2.3 Bituminous Coal

Bituminous coal is the most important and the most widely spread coal; it is also the most diverse with the largest natural abundance. The coal rank of bituminous coal is lower than that of anthracite and higher than that of lignite. It produces large amounts of smoke on combustion. The humic acids in bituminous coal are already completely converted to neutral humus, which is more complex; therefore bituminous coal does not dye acid and alkaline solutions. Common bituminous coal has a luster of varying brightness. The majority has alternate bright–dull stripes, is relatively dense, with a true density of 1.20–1.45 g cm^{-3}, and relatively hard.

Based on coalification, bituminous coal can be roughly classified as long-flame coal, non-caking coal, weakly caking coal, gas coal, fat coal, coking coal, lean coal, and meager coal. Gas coal, fat coal, coking coal, and lean coal can be softened and melted to a plastic phase to some extent during high-temperature carbonization after being crushed, and then solidified to coke chunks. These are therefore also called coke-making coals.

1.2.4 Anthracite

Anthracite is the oldest type of humic coal, and it produces no smoke during combustion. Anthracite is grayish black with a metallic luster, and has no obvious stripes. Anthracite has the lowest amount of volatiles, highest true density (1.35–1.90 g cm^{-3}), and highest hardness of all coals. Its ignition point is as high as 360–410 °C.

Anthracite is mainly used as fuel for heating, power generation, and coal gas production as a raw material for synthetic ammonia production, and in carbon electrodes, electrode pastes, and activated carbons.

Table 1.4 lists the main characteristics of anthracite, and the differences between it and the other humic coals described above.

1.2.5 Technical Classification of Coal in China

The Chinese *Technical Classification Scheme of Chinese Coal*, which was subject to trial implementation in 2009, took effect in 2010 as the national standard for the *Chinese Classification of Coals* (GB/T 5751-2009). The classification scheme in this national standard consists of five tables (*Anthracite, Bituminous Coal and Lignite Classification*; *Anthracite Subcategory*; *Bituminous Coal Classification*; *Lignite Subcategory*; and *Chinese Summary Classification of Coals*) and one figure (*Chinese Coal Classification*). Tables 1.5 and 1.6 show the most important details of the classification scheme.

As shown in Table 1.5 (*Anthracite, Bituminous Coal and Lignite Classification*), the main difference among anthracite, bituminous coal, and lignite is the volatile content on a dry and ash-free basis, V_{daf}. If V_{daf} is above 37 % and the caking index, *G*, is less than or equal to 5, the index of transmittance, P_M, is used to distinguish bituminous coal and lignite. All the coal types in the table are represented by double Arabic numerals. The tens digit refers to the volatile matter in the coal, specifically,

Table 1.4 Characteristics of and differences among four types of humic coal

Characteristic	Peat	Lignite	Bituminous coal	Anthracite
Color	Dark brown	Brown and blackish brown	Black	Grayish black
Luster	No	Mostly no	Certain luster	Metallic luster
Appearance	With dead original plant, earthy	Without dead original plant or obvious stripes	Striped	Without obvious stripes
Color in boiling KOH solution	Brownish red–brownish black	Brown	Colorless	Colorless
Color in dilute HNO_3 solution	Brownish red	Red	Colorless	Colorless
Natural moisture	High	Relatively high	Relatively low	Low
Humic acids	Plenty	Part	No	No
Density (g cm^{-3})	–	1.10–1.40	1.20–1.45	1.35–1.90
Hardness	Extremely low	Low	Relatively high	High
Upon burning	Some smoke	Some smoke	Large amount of smoke	Smokeless

Table 1.5 Anthracite, bituminous coal, and lignite classification

Type	Symbol	Numeral	Classification indicator	
			V_{daf} (%)	P_M (%)
Anthracite	WY	01, 02, 03	≤10.0	–
Bituminous coal	YM	11, 12, 13, 14, 15, 16	>10.0–20.0	–
		21, 22, 23, 24, 25, 26	>20.0–28.0	–
		31, 32, 33, 34, 35, 36	>28.0–37.0	–
		41, 42, 43, 44, 45, 46	>37.0	–
Lignite	HM	51, 52	>37.0[a]	≤50[b]

[a]Only under the conditions $V_{daf} > 37.0$ % and $G \leq 5$, P_M is used to distinguish bituminous coal and lignite
[b]If $V_{daf} > 37.0$ % and $P_M > 50$ %, it is classified as bituminous coal; if P_M is 30–50 % and the gross calorific value on a moist ash-free basis $Q_{gr,\ maf} > 24$ MJ kg^{-1}, it is classified as long-flame coal; otherwise, it is classified as lignite

0 for anthracite, 1–4 for bituminous coal, and 5 for lignite. The single digit 1–3 for anthracite represents the coalification, 1–6 for bituminous coal represents the caking property, and 1 or 2 for lignite represents the coalification. For the bituminous coal, it is classified into lean coal, lean meager coal, meager coal, coking coal, fat coal, 1/3 coking coal, gas-fat coal, gas coal, 1/2 medium caking coal, weakly caking coal, non-caking coal, long flame coal, according to V_{daf}, G, $\underline{Y}$ (maximum thickness of plastic layer) and b (Audiberts-Arnu dilation) as shown in Table 1.6.

1.2.6 Coal Resources and Distribution in China

China has rich coal resources; however, these are unevenly distributed. A large proportion of coal resources are distributed in the middle and west of China, including large reserves of high-quality coal, and a wide variety of coal types, accounting for more than 85 % of the total available coal reserves. In contrast, coal resources are lower in eastern and southern areas. Xinjiang has the largest share of coal total resources, accounting for about 40 %, followed by Inner Mongolia of about 26 %. According to available coal reserves [2], Inner Mongolia has the largest share of coal resources, accounting for about 44 %, followed by Shanxi and Xinjiang, accounting for about 15 and 11 %, respectively. See Table 1.7 for the distributions and proportions of coal resources in the main Chinese coal-producing regions.

Table 1.6 Classification of bituminous coal

Type	Symbol	Numeral	Classification Indicator			
			V_{daf} (%)	G	Y (mm)	b^{b} (%)
Lean coal	PM	11	>10.0–20.0	≤5		
Lean meager coal	PS	12	>10.0–20.0	>5–20		
Meager coal	SM	13	>10.0–20.0	>20–50		
		14	>10.0–20.0	>50–65		
Coking coal	JM	15	>10.0–20.0	>65[a]	≤25.0	≤150
		24	>20.0–28.0	>50–65	≤25.0	≤150
		25	>20.0–28.0	>65[a]		
Fat coal	FM	16	>10.0–20.0	(> 85)[a]	>25.0	>150
		26	>20.0–28.0	(>85)[a]	>25.0	>150
		36	>28.0–37.0	(>85)[a]	>25.0	>150
1/3 Coking coal	1/3JM	35	>28.0–37.0	>65[a]	≤25.00	≤220
Gas–fat coal	QF	46	>37.0	(>85)[a]	>25.0	>220
Gas coal	QM	34	>28.0–37.0	>50–65	≤25.0	≤220
		43	>37.0	>35–50		
		44	>37.0	>50–65		
		45	>37.0	>65[a]		
1/2 Medium caking coal	1/2ZN	23	>20.0–28.0	>30–50		
		33	>28.0–37.0	>30–50		
Weakly caking coal	RN	22	>20.0–28.0	>5–30		
		32	>28.0–37.0	>5–30		
Non-caking coal	BN	21	>20.0–28.0	≤5		
		31	>28.0–37.0	≤5		
Long flame coal	CY	41	>37.0	≤5		
		42	>37.0	>5–35		

[a]When the measured caking index of bituminous coal, G, is less than or equal to 85, the dry and ash-free basis volatile matter, V_{daf}, and caking index G are used for coal classification. If the measured caking index G is greater than 85, V_{daf} and the maximal thickness of plastic layer, Y, or V_{daf} and the Audibert–Arnu dilatation, b, are used for coal classification. When $G > 85$ and $Y > 25.00$ mm, V_{daf} is used to classify fat coal or gas-fat coal; When $G > 85$ and $Y \leq 25.0$ mm, V_{daf} is used to classify coking coal, 1/3 coking coal or gas coal

[b]When $G > 85$, Y and b are used in parallel as classification indicators. When $V_{daf} \leq 28.0$ % and $b > 150$ %, it is classified as fat coal; when $V_{daf} > 28.0$ % and $b > 220$ %, it is classified as fat coal or gas-fat coal. If the value of b contradicts that of Y, Y is used for coal classification

1.3 Coal Macerals

Coal is a type of organic rock. The study of coal from a geological perspective with a particular focus on the microscopic properties of rocks is called coal petrology. In general, there are two approaches to study coal, i.e., macro and micro approaches. The purpose of such research is to determine the origins of coal and the impact of lithotype changes on properties during formation. Identifying the coal lithotype can

Table 1.7 Distribution and proportions of coal resources in Chinese main coal-producing region

Region	Total reserves (Gt)	Proportion (%)	Available reserves (Gt)	Proportion (%)
China (total)	5569.7	100	1821.8	100
Xinjiang	2200.0	39.5	190.4	10.5
Inner Mongolia	1447.6	26.0	798.8	43.8
Shanxi	655.2	11.8	271.8	14.9
Shaanxi	356.8	6.4	188.1	10.3
Guizhou	240.5	4.3	67.5	3.7
Ningxia	203.0	3.6	34.0	1.9
Henan	115.8	2.1	61.8	3.4
Hebei	78.7	1.4	34.6	1.90
Yunnan	62.9	1.1	29.0	1.6
Shandong	67.2	1.2	22.8	1.3
Other	142.0	2.6	123.0	6.7

help to classify coal types more rationally, and to understand the physical, chemical, and process properties, to enable more rational use of coal. Macroscopic research is performed to identify the petrographic composition of the coal seam, based on the apparent properties such as color, luster, fracture, fissure, and hardness, which can be identified visually. This method, which identifies four macroscopic lithotypes of coal, is simple but rough. In microscopic research, a microscope is used to observe and identify the coal maceral. The primary indexes are color (transmitted light and reflective light), shape, and physical structure and protuberances. As microscope technology develops, new methods of microscopic research are being introduced, and new findings are being made.

1.3.1 Coal Lithotype

The coal in a coal bed can be visually divided into four lithotypes based on the color, luster, fractures, fissure, and hardness, i.e., vitrain, clarain, durain, and fusain, which are the basic visible units of coal.

Vitrain is a black and lusterous, uniform lithotype, with developed endokinetic fissures. It is derived from the woody fibrous tissues of plants through gelification.

Clarain is black and complex non-uniform lithotype. It is inferior to vitrain in terms of luster, brittleness, density, structural homogeneity, and endokinetic fissure development. It usually forms a relatively thick layer or even the whole bed.

Durain is gray–black and hard, with a dull luster, rough surface, and undeveloped endokinetic fissures. It is a complex and non-uniform lithotype.

Fusain is gray–black and charcoal-like in appearance, with a silky luster and fibrous structure. It is a fragile, homogeneous lithotype. Fusain is derived from the woody fibrous tissues of coal-forming plants through fusinization. Microscopic observations show that fusain contains clear cell structures and even annual growth ring structures.

The macro lithotype of the coal is used to distinguish coal beds that have similar lusters. Usually, there are four types of coal. Bituminous coal and anthracite contain bright coal, semi-bright coal, semi-dull coal, and dull coal. The four types of coal represent the natural symbiotic association of the four lithotypes in a coal bed.

Bright coal has the brightest relative luster. It has a homogeneous composition, immature striped structure, and shell-like fracture, with developed endokinetic fissures, and is fragile. Bright coal is mainly composed of vitrain and clarain (together accounting for more than 75 %) with small amounts of durain and fusain.

Semi-bright coal is the most common coal type, with a relatively bright luster, an obvious striped structure, developed endokinetic fissures, and angular or terraced fractures. It consists mainly of vitrain and clarain (together accounting for above 50–75 %) with a certain amount of durain, and sometimes fusain.

Semi-dull coal has a low luster, and high hardness and toughness. It consists of 25–50 % vitrain and clarain, with the rest being durain and a small amount of fusain.

Dull coal has the lowest luster, and usually has a block structure. It has no obvious stratification, high hardness and toughness, a high density, and undeveloped endokinetic fissures. It consists of less than 25 % vitrain and clarain, with the rest being durain and a small amount of fusain.

1.3.2 Organic Macerals of Coal

A maceral is the basic coal composition that can be distinguished and recognized under a microscope. Macerals are organic or inorganic depending on their compositions and properties. Organic macerals are derived from organic plant, whereas inorganic macerals are mineral substances that can be observed under a microscope.

1.3.2.1 General Description of Various Organic Macerals

The macerals of humic coal are divided into four types, namely gelified microcomponent (vitrinite), fusinized microcomponent (inertinite), chitin microcomponent (exinite), and a transitional microcomponent between gelatinous and fusinized macerals (e.g., semivitrinite). Based on its microscopic characteristics, each maceral can be further divided into several macerals or submacerals.

The gelatinous microcomponent is the main maceral of coal, with a content of about 60–80 % in many Chinese coalfields. Vitrinite is mainly formed from woody fibrous tissues such as the stems and leaves of plant. Various gelatinous materials are formed through gelification during paludification, so it is called gelatinous maceral and vitrinite for classification. Vitrinite is orange–red to brown–red under diascoptic lighting, and the color deepens with increasing metamorphic degree. Under an oil-immersion reflector, vitrinite is gray to light gray, and the color fades with increasing metamorphic degree; no protrusions are observed. When it is close to anthracite stage, vitrinite is opaque under a transmission microscope and brilliant white under an incident-light microscope. Inhomogeneity gradually increases with increasing metamorphic degree. Based on different degrees of gelification and properties such as cell structures and color under a microscope, vitrinite is divided into seven submacerals, namely telinite 1, telinite 2, telocollinite, gelocollinite, desmocollinite, corpocollinite, and vitrodetrinite (in the order of increasing degree of gelification).

The fusinized microcomponent is a common maceral of coal, but less common than vitrinite The inertinite content is about 10–20 % in many Chinese coalfields. It is also derived from the woody fibrous tissues of plants by paludification. Fusinization affects the macerals subjected to gelification to different degrees, generating various series of materials corresponding to gelatinous products; these are usually called inertinite. It is black and opaque under diascoptic lighting and brilliant white to yellowish white, with large protrusions, under reflected light. Inertinite does not change significantly with increasing metamorphic degree. Based on the intactness and morphological features of the cell structures, it is divided into fusinite, sclerotinite, macrinite, and micrinite.

The chitin microcomponent is derived from plant skin and shell tissues, and secretions, as well as associated secondary compounds such as spores, horny materials, bark, resins, and oozed asphalt. It is classified as exinite and has recognizable specific morphological features. It is gray–black to black–gray with medium-high protrusions under an oil-immersion reflector, and has the lowest reflectivity of similar metamorphosed coals. Exinite is lemon yellow, orange, or red under a transmission microscope, with an obvious outline, particular shape, and clear fluorescent effect. If excited by blue light, the reflected fluorescence color ranges from greenish-yellow to orange brown, and the fluorescence intensity differs with the composition and metamorphic degree. The color of exinite changes greatly under different lenses. At a low metamorphic degree, it is gray–black under an oil-immersion reflector; at a medium metamorphic degree, when the volatile content is about 28 %, it is dull gray, and when the volatile content is about 22 %, it is pale gray. It is difficult to distinguish it from vitrinite and the protrusions also gradually become the same as for vitrinite. With diascoptic lighting, it is golden yellow to golden brown at low metamorphic degrees, but becomes light red with increasing metamorphic degree. At medium metamorphic degrees, it is red, which is similar to vitrinite. Moreover, the fluorescence fades with increasing metamorphic degree. Based on the composition source and morphological features, it is divided into sporinite, cutinite, barkinite, resinite, and bituminite.

Transitional microcomponents are macerals in a state between a gelatinous and fusinized macerals, such as semivitrinite and semifusinite. They are all derived from the woody fibrous tissues of plants by gelification and fusinization during paludification. However, they are less affected by fusinization than inertinite. Among the transitional macerals, the maceral that is only slightly affected by fusinization is called semivitrinite; its microscopic characteristics and properties are similar to those of vitrinite. The maceral that is most influenced by fusinization is called semifusinite; its microscopic characteristics and properties are similar to those of inertinite.

1.3.2.2 Classification and Nomenclature of Organic Macerals

There are many classification schemes both in China and elsewhere, which use different terms in the classification of organic macerals. Classification schemes can be divided into two types. One type focuses on the coal origin, and is a detailed classification, with transmission microscopy commonly being used. The other focuses on process properties and applications; it is a simple classification, in which incident light microscopy is commonly used.

Classification of Macerals for Bituminous Coal in China (GB/T 15588-2013) (Table 1.8) gives priority to the characteristics observed using a microscope oil-immersion reflector, combined with the transmitted light and fluorescence features. It includes three groups, twenty macerals, and fourteen submacerals, based on two aspects, namely coal origin and process properties.

The *International Classification* of *Macerals of Hard Coal* (including bituminous coal and anthracite) put forward by the International Committee for Coal Petrology (ICCP) focuses on chemical process properties. In this classification (Table 1.9), the macerals are divided into three groups, vitrinite, exinite, and inertinite, including maceral, submaceral, and maceral variety. The classification of macerals and submacerals is relatively simple, and the maceral variety is determined according to the plant class and parts.

1.3.2.3 Genesis of Organic Macerals

Gelification and fusinization are two typical processes during paludification. They occur not only in paludification but also for a long period in the diagenetic process. After diagenesis and metamorphism, the fragments in dead plants subjected to gelification and fusinization are respectively converted to two typical organic macerals: vitrinite, which is uniform, with a low carbon content (78–80 %) and high hydrogen content, and inertinite, which contains a large amount of carbon (92–94 %), a small amount of hydrogen, and a large amount of oxygen. Asphaltization is another type of conversion, which stimulates exinite formation from lipoids in dead plants. The conversions of different parts in dead plants to the three main macerals through a series of evolutions and changes (mainly the three actions mentioned above) are illustrated in Fig. 1.4.

The conversions of the three macerals during the coal-forming process are very different from each other, as shown in Fig. 1.5. Fusinized microcomponents change

Table 1.8 Classification of macerals for bituminous coal in China

Maceral group	Symbol	Maceral	Symbol	Submaceral	Symbol
Vitrinite	V	Telinite	T	Telinite 1	T_1
				Telinite 2	T_2
		Collinite	C	Telocollinite	TC
				Desmocollinite	DC
				Corpocollinite	CC
				Gelocollinite	GC
		Vitrodetrinite	VD		
Inertinite	I	Fusinite	F	Pyrofusinite Oxyfusinite	PF OF
		Semifusinite	SF		
		Funginite	Fu		
		Secrerinite	Se		
		Macrinite	Ma	Macrinite 1 Macrinite 2	Ma1 Ma2
		Micrinite	Mi		
		Inertodetrinite	ID		
Exinite	E	Sporinite	Sp	Macrosporinite Microsporinite	MaS MiS
		Cutinite	Cu		
		Resinite	Re		
		Suberinite	Sub		
		Barkinite	Ba		
		Bituminite	Bt		
		Exsudatinite	Ex		
		Fluorophor	Fl		
		Alginite	Alg	Telalginite Lamalginite	TA LA
		Liptodetrinite	LD		

a great deal during paludification, but changes less in the subsequent coalification stages. Chitin microcomponents are stable under biochemical actions, so it seldom changes at the paludification stage, but changes greatly during deep metamorphism. Only the gelatinous microcomponents change gradually and regularly throughout the coal-forming process. To sum up, the three macerals become more similar with increasing metamorphic degree.

The relationships among three macerals and four lithotypes are illustrated in Fig. 1.6. It is generally believed that vitrain and fusain consist of gelatinous and fusinized microcomponents, respectively. In fact, vitrain also contains resinite (chitin maceral) and small amounts of fusinite and semifusinite (fusinite).

Table 1.9 International classification of macerals of hard coal

Group maceral	Maceral	Submaceral	Maceral variety
Vitrinite	Telinite	Telinite 1 Telinite 2	Cordaitotelinite Fungotelinite Xylotelinite Lepidophytotelinite Sigillariotelinite
	Collinite	Telocollinite Gelocollinite Desmocollinite Corpocollinite	
	Vitrodetrinite		
Exinite	Sporinite		Tenuisorinite Crassisporinite Microsporinite Macrosporinite
	Cutinite Resinite		
	Alginite		Pila-Alginite Reinschia-Alginite
	Liptodetrinite		
Inertinite	Micrinite Macrinite Semifusinite		
	Fusinite	Pyrofusinite Degradofusinite	
	Sclerotinite	Fungosclerotinite	Plectenchyminite Corposclerotinite Pseudocorposclerotinite
	Inertodetrinite		

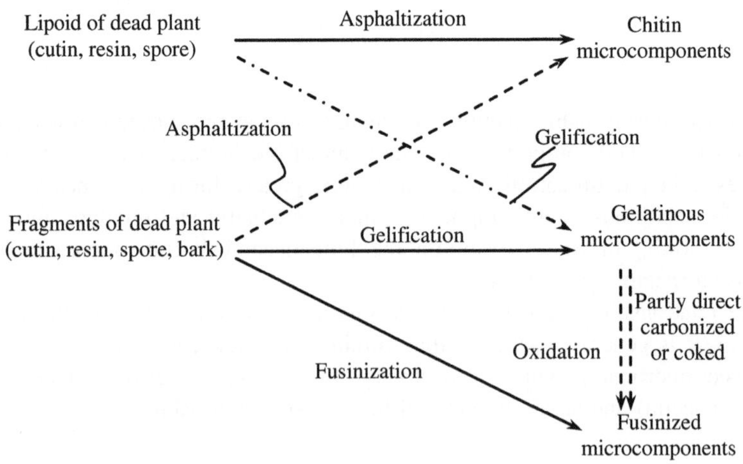

Fig. 1.4 Evolutionary mode for formation of three main macerals

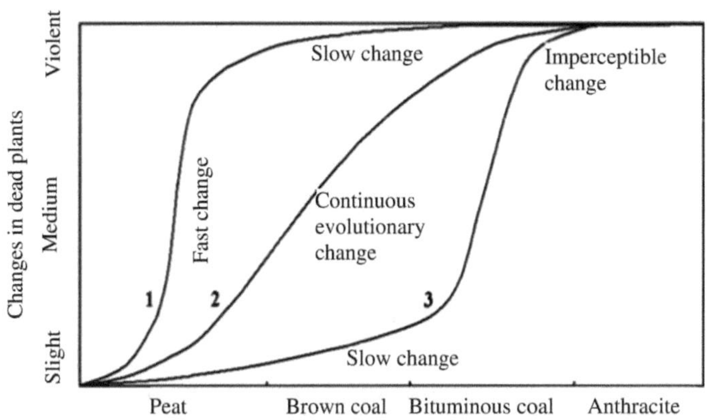

Fig. 1.5 Changes of macerals during coal-forming process: *1* fusinization of lignins and celluloses; *2* gelification of lignins and celluloses; *3* asphaltization of wax resins

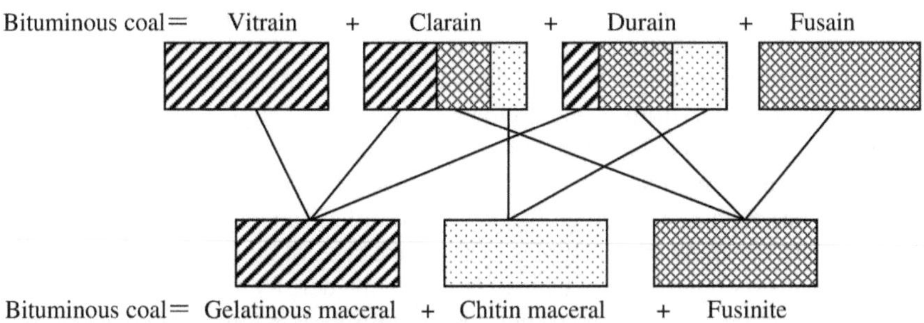

Fig. 1.6 Relationship between three macerals and four lithotypes

1.3.3 Inorganic Macerals

Inorganic macerals are mineral substances in coal. They are generated from the inorganic constituents of coal-forming plants and mineral substances mixed during the coal-forming process, with the latter being the main source. Common mineral substances include clay minerals, sulfides, oxides, and carbonates.

Clay minerals, the main components of mineral substances, mainly include kaolin and hydromica. They are present in coal as thin layers, lenses, blocks, and impregnated and irregular shapes. They are commonly found in the cell cavities of telinite, organized semifusinite and fusinite, or scattered in collinite.

Sulfide minerals include mainly iron pyrites and binarite, which are present in coal as nodular and impregnated shapes, and coccus aggregations, or filling in fissures and holes. Sometimes, these may even fill the cell cavities, or be directly embedded in organic macerals.

Carbonate minerals are mainly calcite and siderite. They usually fill the cell cavities of organic macerals or small fissures. Siderite is usually present in nodular shapes or spherulitic aggregations.

Oxide minerals include mainly quartz, chalcedonite, and opal. It is difficult to distinguish oxide minerals from clay minerals in coal. Sometimes, quartz veins can be found filling in fissures.

1.3.4 Separation and Enrichment of Coal Macerals

The study of macerals provides essential information on coal characteristics, and identifies the properties shared by, and reasons, for the differences between different types of coal. It is therefore important to obtain macerals that are as pure as possible, which requires maceral separation and enrichment. It has been confirmed experimentally that different coal macerals have different reactivities, and the maceral composition of coal has a significant impact on the reactivity and processing of coal. Maceral separation is therefore very important. In principle, maceral separation is simple, but it is difficult in reality. In particular, it is difficult to separate the macerals from a single type of coal simultaneously. The separation of macerals consists of hand sorting, rescreening, and selection based on density. This procedure enables separation of high-purity macerals. Much research on maceral separation has been performed globally; the author has separated and enriched three basic organic macerals from samples of the same type of coal, with purities higher than the levels reported outside China [3]. Isopycnic density gradient centrifugation is more effective than hand sorting and flotation for the separation of coal macerals.

Maceral separation involves two main steps, i.e., preliminary separation (hand sorting, screening, zinc chloride or cesium chloride liquid density separation) and fine separation (organic liquid density natural sedimentation and centrifugal separation).

Hand sorting is used to select macerals based primarily on the luster and other physical characteristics of coal components; for example, exinite is more concentrated in dull coal, which is hard and less dense. Hand sorting based on the appearance can initially enrich a certain maceral. For inhomogeneous coal, coal samples can be screened based on the anti-crushing performances of coal macerals. In general, soft fusain is the most brittle, and is concentrated in the fraction with the smallest particles. Vitrinite has weak crush resistance and is enriched at small sieve meshes. Dull coal, which has great toughness and crush resistance, is concentrated in the coarse fraction. The appropriate sieved sample is separated using zinc chloride solution, and a high degree of enrichment of a maceral is achieved. The separation process is shown in Fig. 1.7. It is difficult to adjust the density of zinc chloride solution, and a residue remains after cleaning that may affect the determination of the reactivity. Use of a more expensive cesium chloride solution is recommended to avoid these issues.

Fig. 1.7 Coal maceral density separation process: *d* is density (g cm^{-3})

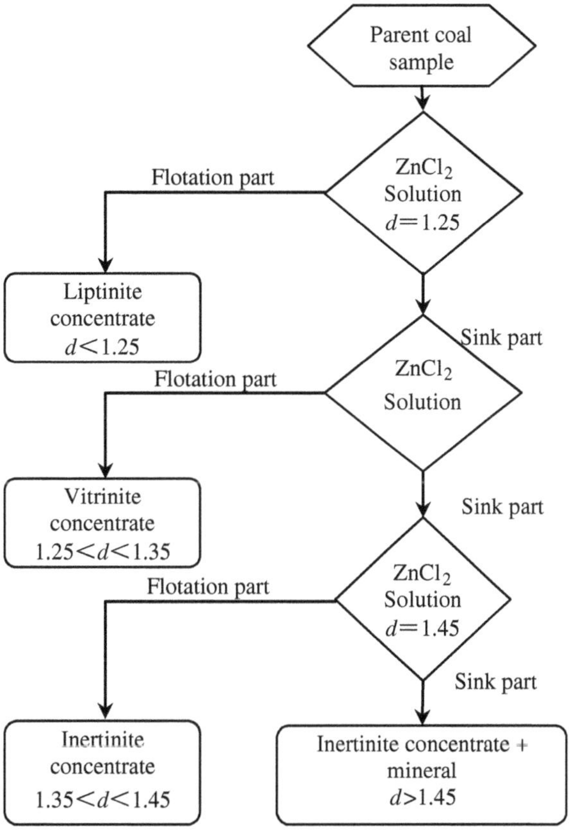

Fine separation refers to the natural sedimentation or centrifugation of coal samples in organic density liquid. This step gives maceral samples of the required purity. The organic density liquid should have properties such as low viscosity, strong wetting ability, fast layering, and volatility, and should leave no residue after drying and have little effect on the coal properties. Generally, a mixture of benzene and carbon tetrachloride is used for separation.

1.3.5 Quantitative Analysis of Coal Macerals

In maceral content determination using a microscope, the area percentages of the respective components, which are proportional to the volume percentages, and the component densities can be used to calculate the mass percentages.

A common method for maceral determination is the point-count method, which has high efficiency and a certain precision; an electrical point-count meter is used. Components with high contents have a higher probability of appearing at the center

Table 1.10 Maceral analysis of some Chinese coal samples (%)

Coal sample	Vitrinite	Semivitrinite	Fusinite and semifusinite	Liptinite	Mineral Group
Benxi coal	85–86	–	11–12	0	2–4
Hegang coal	70–83	–	9–15	1–4	6–11
Beipiao coal	50–63	3–10	17–26	3–6	5–15
Fushun coal	90–93	–	0–1	3–8	0–3
Fengfeng coal	77–85	–	15–23	0–1	–
Jiawang coal	65–81	1–10	7–20	4–9	0–6
Huainan coal	50–60	7–13	9–20	8–20	2–7

of the microscope of the electrical counter. By measuring the number of macerals in the measurement area on the specimen, the volume percentage of a component can be calculated as follows:

$$V = \frac{n}{N} \times 100\,\%,$$

where

V is the volume percentage of the measured component,
n is the total number of points of the measured component,
N is the sum of points of all components on the specimen.

Quantitative analysis is performed based on the different colors and structures of different macerals under a microscope, and the pulverized coal is usually made into a thin optical slice. The microscope magnification is ×400 to ×500. Four hundred to five hundred points are measured on the thin optical slice, the count is divided into five major groups, namely vitrinite, semivitrinite, inertinite, liptinite, and minerals, and then the percentages are calculated. The maceral analytical data for some coal samples are shown in Table 1.10. Vitrinite and semivitrinite together usually account for about 80 % [4].

1.3.6 Chemical Composition of Coal Macerals

The proximate and ultimate analytical results for coal macerals show regular changes for different macerals and coalification degrees. The proximate and ultimate analytical results for several macerals are shown in Table 1.11. With increasing coalification, the volatile matter (V_{daf}) and hydrogen (H_{daf}) contents of vitrinite decrease, whereas the content of carbon (C_{daf}) and carbon to hydrocarbon ratio (C_{daf}/H_{daf}) increase. Various macerals with the same or similar coalifications have significantly different analytical indicators such as V_{daf}, C_{daf}, H_{daf}, and C_{daf}/H_{daf}. In addition, when the C_{daf} of vitrinite is less than 87 %, the aromaticity, f_a is little changed, but when C_{daf} is greater than 87 %, f_a increases with increasing coalification [5].

Table 1.11 Chemical analytical data for coal macerals[a]

Maceral	Degree of coalification	Production area	V_{daf} (%)	C_{daf} (%)	H_{daf} (%)	C_{daf}/H_{daf}	f_a^b
Vitrinite	Sub-bituminous	Fushun west open-pit mine	41.89	79.23	5.42	14.6	0.71
	Gas coal	Leping Zhongjiashan	37.31	84.91	5.88	14.4	0.72
	Gas coal	Hegang Xingshan	36.69	84.36	5.69	14.8	0.72
	Fat coal	Fengfeng third mine	32.69	88.04	5.52	16.0	0.74
	Coking coal	Fengfeng fifth mine	21.91	89.26	4.92	18.1	0.84
	Lean coal	Fengfeng fourth mine	17.88	90.73	4.82	18.8	0.87
	Meager coal	Zibo Longqua	13.49	91.31	4.37	20.9	0.91
Resinite	Sub-bituminous	Fushun west open-pit mine	99.01	80.73	10.1	8.0	0.01
Sporinite	Gas coal	Xuangang	61.80	86.24	7.84	11.0	0.39
Barkinite	Gas coal	Leping Zhongjiashan	49.47	87.27	7.03	12.4	0.56
Fusinite	Gas coal	Leping Zhongjiashan	21.47	88.63	4.43	20.0	0.86
Fusinite	Gas coal	Hegang Xingshan	18.97	88.51	3.88	22.8	0.88

[a]daf: dry ash free basis

[b]Aromaticity: $f_a = \frac{1200\times(100-V_{daf})}{1240\times C_{daf}}$

When the coalification is the same, inertinite has the highest carbon content, followed by exinite, and that of vitrinite is slightly lower than that of exinite. Exinite has the highest hydrogen and volatile matter contents, followed by vitrinite, and inertinite has the lowest contents. The oxygen content is highest for vitrinite, and then inertinite, and lowest for exinite. With increasing coalification, the oxygen and volatile matter contents of all the macerals, and the hydrogen content of exinite, decrease. The hydrogen contents of vitrinite and inertinite initially increase and then decrease; the sizes of the decreases vary for different macerals, but finally tend to be uniform [6].

Vitrinite is a typical organic maceral in coal, so its chemical structure has been widely studied [7]. The characteristics and chemical structures of liptinite and inertinite differ from those of vitrinite (Table 1.12). The liptinite structure is characterized by a relatively high H/C atomic ratio, low aromaticity, low oxygen content, and a high content of aliphatic and cycloaliphatic structures. According to the author's direct measurements, the aromaticities of the organic macerals in the same type of coal (Pingshuo bituminous coal), in descending order, are as follows: inertinite (0.75) > vitrinite (0.67) > exinite (0.39); [8] this is similar to the results reported in the literature. [9] During coalification, the structure and nature of exinite gradually move closer to those of vitrinite. When the carbon content approaches 90 %, the differences disappear. The inertinite group includes fusinite microsomes and macrosomes and other microstructural ingredients. These changes occurred in

Table 1.12 Comparison of compositions and structures of different macerals. Reprinted from Ref. [7], Copyright 2008, with permission from Metallurgical Industry Press

C in coal (%)	Maceral[a]	Elemental composition (%)					*H/C*	f_{ar}^{C} [b]
		C	*H*	*O*	*N*	*S*		
81.5	V	81.5	5.15	11.7	1.25	0.40	0.753	0.83
	E	82.2	7.40	8.50	1.30	0.60	1.073	0.61
	M	83.6	3.95	10.5	1.35	0.60	0.563	0.91
85.0	V	85.0	5.40	8.00	1.20	0.40	0.757	0.85
	E	85.7	6.50	5.80	1.40	0.60	0.950	0.73
	M	87.2	4.15	6.70	1.35	0.60	0.566	0.92
87.0	V	87.0	5.35	5.90	1.25	0.50	0.732	0.86
	E	87.7	5.85	4.40	1.45	0.60	0.793	0.83
	M	89.1	4.20	4.70	1.40	0.60	0.561	0.93
89.0	V	89.0	5.10	4.00	1.30	0.60	0.683	0.88
	E	89.6	5.20	3.30	1.30	0.60	0.691	0.87
	M	90.8	4.10	3.20	1.30	0.60	0.537	0.94
90.0	V	90.0	4.94	3.20	1.35	0.50	0.655	0.90
	E	90.4	4.50	2.80	1.30	0.60	0.646	0.90
	M	91.5	3.65	2.60	1.35	0.60	0.514	0.95

[a]V, vitrinite; E, liptinite; M, microsomes in inertinite
[b]f_{ar}^{C} data are obtained using a classic approach, so they are much higher, and are only for comparison

the early coal-forming period, so the changes during coalification are not obvious. Inertinite has a high carbon content, low hydrogen content, and high aromaticity. Compared with vitrinite and exinite in the same coal sample, the aromatic layer chip size and the degree of parallel orientation of inertinite are close to or even greater than those of anthracite.

1.3.7 Relationship Between Coal Type and Rank

From a chemical point of view, during the coalification process, oxygen and hydrogen in the organic functional groups of the initial materials are removed from the framework structure of aromatic carbons and escape in the form of gas molecules, e.g., H_2O, CO_x and CH_4, causing regular changes in the carbon content during coalification. As a result, the carbon content can serve as an index for determining coal rank. The changes in the macerals of primary coal-forming materials with different chemical compositions are completely different from each other during coalification. The properties of coals from different primary materials are significantly different from each other, even those subjected to the same degree of metamorphism. Several types of coal can therefore be easily observed in the same coal bed. Similarly, because of the different effects of metamorphism, coals formed from the same primary material or with the same macerals may result in different final products. The most important factors in determining coal properties are therefore coal type and coal rank, two closely related parameters. The coal type depends on the content and shape of organic macerals, whereas the coal rank is mainly dependent on the degree of metamorphism. The nature of the coal is determined by two factors: the coal type, which shows the proportions and compositions of various macerals, and the metamorphic degree, which reflects the geochemistry and geological factors. These factors strongly influence the physical and chemical composition of the coal that is formed, which in turn has implications for the reactivities of various coal types.

References

1. van Krevelen DW (1981) Coal. Elsevier scientific publishing company, Amsterdam, p 45
2. Peng SP et al (2014) Coal resource and water resource. In: Xie KC (ed) Strategy consulting report of clean and high-efficiency coal exploitation and utilization in China. Science Press Ltd., Beijing
3. Xie KC, Ling DQ (1990) Coal gasification kinetics and effects of minerals. Shanxi Science Education Press, Shanxi, p 326
4. Zhu PZ, Gao JS (1984) Coal chemistry. Shanghai Science and Technology Press, Shanghai, p 78
5. Zhu PZ, Gao JS (1984) Coal chemistry. Shanghai Science and Technology Press, Shanghai, p 74
6. van Krevelen DW (1981) Coal. Elsevier Scientific Publishing Company, Amsterdam, p 115
7. He XM (2010) Coal chemistry. Metallurgical Industry Press, Beijing, p 216
8. Xie KC, Li WY, Zhu SY (1994) Pyrolysis studies on coal macerals using DTA- GC, 13C-NMR and SEM techniques. Fuel Sci Technol Int 12(9):1159–1168
9. van Krevelen DW (1981) Coal. Elsevier Scientific Publishing Company, Amsterdam, p 327

Chapter 2
Physical Characteristics of Coal

Abstract To understand the behavior of coal, characterizing and understanding its physical properties is of paramount importance. The bulk properties of coal, and in particular the mechanical and thermal properties, have a significant impact on the various processes that are discussed in later chapters. In addition, the microstructure and porosity of coal are intimately involved in the complex heterogeneous reactivity of coal. In this chapter, the physical structures and properties of various coal types are described in terms of the most commonly used physical characterization techniques, including X-ray diffraction, X-ray scattering, surface area, and porosity measurement techniques. Other physical properties of coal such as optical, electrical and magnetic properties can also offer insight into the chemical structure and composition of coal. The physical properties of coal, including the mechanical, electromagnetic, and optical properties, are related to structural models of coal. The bulk thermodynamic properties of coal combustion are also introduced. Finally, the use of computational methods to determine the physical properties of coal is discussed.

2.1 Physical Properties of Coal

2.1.1 Mechanical Properties of Coal

The bulk mechanical properties of coal have major implications for designing coal-based processes. The mechanical properties of coal are related to coal type.

2.1.1.1 Mechanical Strength

The mechanical strength of coal refers to its capacity to resist external forces and is related to physical properties of coal such as shatter indices and grind ability index.

K.-C. Xie, *Structure and Reactivity of Coal*, DOI 10.1007/978-3-662-47337-5_2

Table 2.1 Grading standards for mechanical strength of coal

Grade	Mechanical strength of coal	Proportion of particles >25 mm (%)
Grade I	High-strength coal	>65
Grade II	Medium-strength coal	50–65
Grade III	Low-strength coal	30–50
Grade IV	Ultralow-strength coal	≤30

The shatter indices of coal can be determined using the drop method. The method is to let coal lumps of size 60–100 mm fall freely from a point 2 m above a steel plate, sieve them with a sieve of 25 mm, and repeat the process for coal samples of size greater than 25 mm. After repeating the process three times, the mass of coal samples larger than 25 mm is determined, and the percentage with respect to the mass of the original coal samples is taken as the shatter indices of the coal. The grading standards for determining the mechanical strength of coal using the drop test are shown in Table 2.1.

The grind ability index of coal is determined using the Hardgrave method. The method is to break 1 kg of coal samples of size 6 mm in a Hardgrave grind ability analyzer, grade by grade, until all of them pass through a 1.25 mm screen, weigh the coal samples of size 0.63–1.25 mm, calculate the percentage of their mass to the total mass of the coal samples, and determine the corresponding grind ability index from a standard curve.

The mechanical strength of coal is related to factors such as the degree of coalification, lithotype, mineral content, and weathering. The mechanical strengths of high-rank and low-rank coals are greater than those of medium-rank fat coals and coking coal. In terms of the macroscopic lithotype of coal, the mechanical strength of fusain is lowest, followed by vitrain, and that of durain is the highest. The mechanical strengths of coals with high mineral contents are high, and are reduced by weathering.

2.1.1.2 Density

Density is an important parameter, and it reflects the nature and structure of a material. The density depends on the closeness of the molecular structure and the molecular arrangement and there is also a relationship between density and degree of coalification. The density can also be used for structural analysis of coal, using statistical methods. The coal density is the coal mass per unit volume. Coal volume has different meanings in different situations, because of the inhomogeneity of coal, so coal density has various definitions.

The true relative density (TRD) of coal refers to the coal mass per unit volume, excluding the pores in the coal. It is an important indicator for calculating the average mass of a coal seam and in coal quality research. The TRD can be determined in aqueous media using a pycnomete. When different substances (for example, helium, methanol, water, *n*-hexane, and benzene) are used as the replacement substances for determining coal density, the values obtained vary. Usually, the result obtained using helium as the replacement substance is taken as the TRD (also known as the helium density). The diameters of the smallest pores in coal are about 0.5–1 nm, whereas the diameter of the helium molecule is 0.178 nm; therefore helium can completely penetrate the porous structure of coal. The general ranges of the TRDs of various types of peat, lignite, bituminous coal, and anthracite are about 0.72, 0.8–1.35, 1.25–1.50, and 1.36–1.80 g cm^{-3}, respectively.

The apparent relative density (ARD) of coal is the coal mass per unit volume, including the pores in the coal. This parameter is necessary for calculating coal reserves and in the transportation, crushing, and combustion of coal. The ARD can be determined using the wax-coating method. The coal porosity can be calculated using the TRD and ARD of the coal:

$$\text{Porosity} = \frac{\text{True relative density} - \text{Apparent relative density}}{\text{True relative density}} \times 100\,\%.$$

The bulk density (BD) of coal is the ratio of the total mass of coal grains filling a container using the free-stacking method to the vessel volume. The BD is used when estimating the mass of a coal pile or calculating the coal capacity of a coke oven.

For the same coal sample, the value of the TRD of the coal is highest, followed by that of the ARD, and the value of BD is lowest. The densities of minerals are significantly higher than that of organic matter, so the content and composition of the minerals in coal has a significant influence on the coal density. In the study of coal structure, it is usually necessary to eliminate the impact of minerals. The density must be corrected roughly as follows: for every 1 % increase in coal ash, the coal density will increase by 0.01 %.

The general relationship between the density and the degree of coalification for various macerals is as follow [1]: the TRD of inertinite is the highest, followed by those of vitrinite and exinite; when the carbon content on a dry ash-free basis (C_{daf}) is greater than 90 %, the three become similar and increase sharply, indicating that their structures have undergone profound changes.

The general relationship between vitrinite density and the degree of coalification is as follows [2]: the vitrinite density begins to decrease slowly with increasing coalification, mainly because the reduction in the oxygen content is greater than the increase in the carbon content, and the atomic weight of oxygen is greater than that of carbon. The density of coal containing 85–87 % carbon is at least 1.3 g cm^{-3}. For coal with a carbon content greater than 90 %, the density increases sharply with increasing coalification, mainly because of the increasing amount of more compact aromatic structures.

2.1.1.3 Hardness

Coal hardness reflects the coal's ability to withstand external mechanical actions. The representation and determination of coal hardness differ depending on the applied mechanical force.

The scratch hardness (Mohs hardness) is the relative hardness determined by scratching the coal surface with 10 types of standard mineral. The scratch hardness of coal is usually between 1 and 4. Coal hardness is related to coalification. Lignite, which has a low coalification degree, and coking coal, with medium coalification, have the lowest scratch hardnesses of 2–2.5, whereas anthracite has the highest scratch hardness that is close to 4.

The micro Vickers hardness is referred to as the microhardness (symbol MH or Hm). It is determined by pressing a diamond indenter with a static load into the maceral under a microscope. The greater the indentation is, the lower the microhardness of the coal. The value of the microhardness is expressed by the load capacity per contact area between the indenter and the coal (in kilograms per square millimeter). The microhardness determined by the indentation method is widely used in the study of coal chemistry.

2.1.1.4 Elasticity

Coal elasticity is the deformation produced under an external force, and the degree of recovery after removal of the external force. The elasticity of a material is related to its structure; in particular, there is a close relationship with the binding force between constituent molecules. The determination of coal elasticity is very important in the study of coal structure, e.g., the elastic modulus of coal is an indication of the characteristics of the chemical bonds between the units of coal structure.

The methods for determining material elasticity are divided into static and dynamic methods. The static method determines the relationship between stress and strain; for example, it can determine the bending degrees of coal briquettes produced under different loads. The dynamic method is based on determining the transmission speed of sound in the coal. The coal elastic modulus can be calculated as

$$v = k\sqrt{\frac{E}{\rho}},$$

where

v is the transmission speed of sound in the coal (m s^{-1});

k is a constant;

E is the coal elastic modulus (10^{-5} N cm^{-2});

ρ is the coal density (g cm^{-3}).

Fine cracks or other factors can lower the static elastic modulus, therefore the dynamic elastic modulus is considered to be more reliable. The elastic moduli of low-coalification and bituminous coals are generally isotropic, whereas that of high-coalification coal is anisotropic. Different macerals have different elasticity values. The values increase in the order liptinite, vitrinite, and inertinite. However, with increasing coalification degree, the differences gradually become smaller. In addition, the coal elasticity increases with increasing mineral and moisture content, because minerals have high densities.

2.1.2 Thermal Properties of Coal

2.1.2.1 Thermal Conductivity

The thermal conductivity of coal includes two basic constants: the thermal conductivity coefficient λ (kJ m^{-1} h^{-1} K^{-1}) and the thermal diffusivity coefficient α (m^2 h^{-1}). The thermal conductivity coefficient λ is the transmission rate for a temperature difference of 1 K over a unit distance when heat is transferred from a high-temperature area to a low-temperature area; it is the direct conduction speed of heat in an object, representing the heat-sinking capability of the object; $c\rho$ represents the heat-storage capacity of the object. The thermal diffusivity coefficient α is the ratio of the heat-sinking capability to the heat-storage capacity of the material and it represents the temperature-changing (heating or cooling) capability of an object; λ and α are related as follows:

$$\alpha = \frac{\lambda}{c\rho},$$

where c is the mass heat capacity of the coal (kJ kg^{-1} K^{-1}) and ρ is the coal density (kg m^{-3}).

The λ and α values of medium-coalification bituminous coal can be calculated using the empirical equation

$$\lambda = 0.0003 + \frac{at}{1000} + \frac{bt^2}{1000^2},$$

where a and b are specific constants; the values of a and b for caking coal are equivalent, i.e., 0.0016, and for weakly caking coal, the values of a and b are 0.0013 and 0.0010, respectively; t is the temperature (°C).

The thermal diffusivity coefficient α of medium-coalification bituminous coal is given by the empirical equations

$$\alpha = 4.4 \times 10^{-4}[1 + 0.0003(t - 20)] \quad \text{when } t = 20\text{--}400\,^\circ\text{C},$$
$$\alpha = 5.0 \times 10^{-4}[1 + 0.0033(t - 400)] \quad \text{when } t = 400\text{--}1000\,^\circ\text{C}.$$

The thermal conductivity coefficient of coal is related to its moisture and ash contents, temperature, and type. The thermal conductivity coefficient of water is approximately 25 times that of air. The thermal conductivity coefficient of coal therefore increases with increasing moisture content. The thermal conductivities of minerals are much higher than those of organic materials; therefore if the amount of coal ash increases, the thermal conductivity coefficient increases. There is a positive relationship between the thermal conductivity coefficient and temperature of the coal, i.e., the thermal conductivity coefficient increases with increasing temperature. The thermal diffusivity coefficients of various coals have roughly similar variation rules. These variation rules reflect the characteristics of the internal structure of the coal. In the metamorphic process, the structure of the organic matter in coal gradually becomes more compact and regular; therefore the thermal conductivity gradually increases, approaching that of graphite.

2.1.2.2 Specific Heat

The quantity of heat needed to raise the temperature of a unit mass of coal by 1 K is referred to as the specific heat of the coal, The specific heat of coal at room temperature ranges from 1.00 to 1.266 kJ kg^{-1} K^{-1}. The specific heat of coal changes with changes in coalification, moisture, ash, and temperature. The specific heat of coal at room temperature decreases with increasing coalification (expressed by the carbon content) [3]. The specific heat of coal increases approximately linearly with increasing moisture content, because the specific heat of water is high. The mass heat capacity decreases with increasing coal ash content, because the mass heat capacity of minerals is generally from 0.70 to 0.84 kJ kg^{-1} K^{-1} at room temperature.

The specific heat of coal varies with temperature. In the temperature range 0–350 °C, the mass heat capacity increases and reaches a maximum at about 350 °C. From 350 to 1000 °C, the specific heat decreases. Thermal decomposition of coal occurs above 350 °C, and it finally reaches the specific heat of graphite, 0.71 kJ kg^{-1} K^{-1} [3].

2.1.3 Optical Properties of Coal

The optical properties of coal provide important information on coal structure, such as coalification, anisotropy, sizes and arrangement of aromatic layers. They can also reflect the shape, orientation, and agglomeration of the internal particles of the coal.

This section focuses on the reflectivity and refractive index of coal. Spectroscopic properties such as infrared (IR) properties will be discussed elsewhere.

2.1.3.1 Coal Reflectivity

The capacity of a polished coal surface to reflect vertically incident light is referred to as the reflective capacity of the coal. The visual performance under a microscope is the brightness of the polished surface. The reflective capacities of different coal types are different. The reflectivity, R, of coal is defined as:

$$R = \frac{I_r}{I_i} \times 100\,\%,$$

where I_r is intensity of the reflected light and I_i is the intensity of the incident light.

Reflectivity is an important property of opaque minerals, and is also an important indicator of the coalification degree. The reflectivity is usually determined using a relative method: the reflected light intensity of a standard sheet with a known reflectivity is measured under a certain intensity of incident light (typically monochromatic polarized light) and compared with the intensity of reflected light of the material under investigation. Materials such as optical glasses, quartz, and diamond are commonly used as standards. The equation for calculating the reflectivity is

$$R = \frac{I}{I_0} R_0 \times 100\,\%,$$

where I is the intensity of the reflected light from the material being examined, I_0 is the intensity of reflected light of the standard material, and R_0 is the reflectivity of the standard material.

As the resolution of coal in an oil medium is far better than that in air, the reflectivity is generally measured under an oil-immersion objective (R^{o}). At least 20 points for each optical coal sample are used to measure the maximum reflectivity. Usually, the average value, $\bar{R}^{o}_{max}$, of the maximum reflectivity measured under an oil-immersion objective is used as the analytical indicator. The range of reflectivities of different Chinese coal types is shown in Table 2.2.

The reflectivity of vitrinite in bituminous coal is related to other commonly used coal classification indicators such as volatile matter content determined on an ash-free basis (V_{daf}), C_{daf}, and calorific value. The general relationships between the maximum average reflectivity of vitrinite of Chinese coal under an oil-immersion microscope correlates well with the coalification, which is a good indicator of coal rank [4].

Table 2.2 Reflectivities (%) of different types of Chinese coal

Coal type	Metamorphic stage	Reflectivity	Coal type	Metamorphic stage	Reflectivity
Lignite	0	0.40–0.50	Lean coking coal	VI	1.50–1.69
Long-flame coal	I	0.50–0.65	Lean coal	VII	1.69–1.90
Gas coal	II	0.65–0.80	Meager coal	VIII	1.90–2.50
Gas-fat coal	III	0.80–0.90	Anthracitic	IX	2.50–4.00
Fat coal	IV	0.90–1.20	Anthracitic	X	4.00–6.00
Coking coal	V	1.20–1.50	Anthracitic	XI	>6.00

2.1.3.2 Refractive Index of Coal

The definition of the refractive index is the ratio of the sine of the incident angle to that of the refraction angle when light passes through a material interface, is refracted at the interface, and enters the interior of the substance. The molecular refraction can be obtained from the additive refractive index, an important property in analytical studies of coal structure. The refractive index of coal cannot be directly measured, but the reflectivity and the refractive index of vertical incident light are related as follows:

$$R = \frac{(n - n_0)^2 + n^2K^2}{(n + n_0)^2 + n^2K^2},$$

where

R is the coal reflectivity (%);
n_0 is the standard medium refractive index, $n_0 = 1.514$ for cedar oil;
n is the refractive index of the coal;
K is the light absorption rate of the coal (%).

Based on the reflectivities of incident light measured in air and cedar oil, two equations can be obtained using the above equation. Simultaneously solving the equations gives n and K.

The refractive index increases with increasing coalification. When the carbon content is higher than 85 %, the increase is large [5].

Generally, lignite is optically isotropic. Coal is transformed from bituminous coal to anthracitic coal with increasing coalification. The layered structure of the aromatic nuclei in the molecular structure improves, and the arrangement becomes regular. Anisotropy of the optical properties parallel or perpendicular to the aromatic layers gradually becomes apparent. The reflectivity and refractive index can both reflect such changes, which are determined by the internal structure of the coal.

2.1.4 Electrical and Magnetic Properties of Coal

The electrical properties of coal include the conductivity and dielectric constant. The study of the electrical properties of coal can theoretically provide information such as the semiconducting nature of the coal, the size of aromatic structures in the coal, and the anisotropy. The main magnetic properties of coal are diamagnetism and paramagnetism.

2.1.4.1 Coal Conductivity

The conductivity of coal is the degree of difficulty of passing a current through the coal. The conductivity of a substance is commonly expressed in terms of the electrical resistivity ρ (Ω cm) or electrical conductivity κ (Ω^{-1} cm^{-1}). The reciprocal of the electrical resistivity is the electrical conductivity, i.e.,

$$\kappa = \frac{1}{\rho},$$

In terms of electrical conductivity, coal is generally considered to be close to a semiconductor. The κ value of young lignite is approximately $1 \times 10^{-14}\ \Omega^{-1}\ \text{cm}^{-1}$.

2.1.4.2 Dielectric Constant of Coal

The ratio of the charging capacity of a substance between two parallel plate electrodes to the charging capacity when there is a vacuum is between the parallel plate electrodes is referred to as the dielectric constant, ε, of the substance. The relationship between ε and the refractive index, n, of a non-polar insulator is

$$\varepsilon = n^2.$$

2.1.4.3 Magnetic Properties of Coal

The internal magnetic field intensity, B, of a material placed in a magnetic field of intensity H is referred to as the magnetic induction intensity:

$$B = H + H' = H + 4\pi d\chi H,$$

where H' is the additional magnetic field intensity induced by magnetization of the magnetic medium. When $H' > 0$, a magnetic medium with a magnetic field H with the same direction as that of H' is referred to as a paramagnetic substance; when

$H' < 0$, a magnetic medium with a field H with the opposite direction to that of H' is referred to as a diamagnetic substance.

In chemistry, the specific magnetic susceptibility χ ($cm^3\ g^{-1}$) and molar magnetic susceptibility χ_m ($cm^3\ mol^{-1}$) are generally used to indicate the magnetic properties of a substance. The relationship between the specific magnetic susceptibility and the molar magnetic susceptibility is

$$\chi_m = \chi \cdot M,$$

where M is the molar mass of the substance ($g\ mol^{-1}$).

2.1.5 Surface Properties of Coal

During the process of coal change, chemical reactions occur on the reactant surfaces and coal solid surfaces. Most of the surfaces are related to the pore structure, i.e., the diffusion of products and penetration of reaction media are related to the coal pore structure in most cases. The porosity also affects the physical and chemical properties of coal. The surface and body of coal are full of holes composed of organic matter and minerals, and coal is a porous solid substance with different pore size distributions; micropores are the main type of pore. The correlation between coal porosity and coal rank is of great practical significance. This section describes the pore structures in coal.

2.1.5.1 Classification and Forms of Pores in Coal

There is no single standard for classifying the pores in coal. Dubinin suggests the following classification, which is widely used in the study of coal chemistry, to distinguish different-sized pores of a porous adsorbent: large pores, of diameter >20.0 nm; transitional pores, of diameter 2.0–20.0 nm; and micropores, of diameter <2.0 nm. Some researchers divide pores into macropores and micropores, and the dividing point is an effective pore radius of 10.0 nm.

The International Union of Pure and Applied Chemistry (IUPAC) proposed the following new standard aperture classification in 1978.

Micropores (<2 nm). The micropore size and pore diameter range can be measured and calculated using small-angle X-ray scattering (SAXS), CO_2 adsorption, or the pure helium adsorption.

Mesopores (2–50 nm). These can be observed using scanning electron microscopy (SEM), measured using transmission electron microscopy (TEM), and quantified using N_2 adsorption, small-angle neutron scattering (SANS), or SAXS.

Macropores (>50 nm). Pores larger than 1000 nm can be observed with an optical microscope, and smaller pores can be studied using SEM. The diameters of larger pores can be quantified using image analysis technology or measured using the mercury intrusion method.

Physical properties of micropores. Micropores are almost the same size as the adsorbed molecules. The volume of a micropore is about 0.2–0.6 $cm^3\ g^{-1}$, the pore quantity is 10^{20}, and the total surface area of micropores in coal-based activated carbon is about 500–1000 $m^2\ g^{-1}$.

Physical properties of mesopores. A meniscus can be formed in the pore. A decrease in the vapor pressure causes capillary condensation, and liquefies the adsorbates. The pore volume is small, about 0.015–0.15 $cm^3\ g^{-1}$; the surface area is also small.

Physical properties of macropores. No obvious meniscus can be formed in the pores; no capillary condensation occurs. The causes of these pores have the following categories: holes and fissures. The hole types are air, residual plant material, corrosion, mold, intercrystal, interbioblast, and condensation water loss. Fissures are endophytic and tectonic.

Pores in coal have different forms: some pores form passages, some are blind holes, some are closed holes, and some are open holes. These hole structures can be observed using an electron or optical microscope.

2.1.5.2 Pore Size and Distribution in Coal

Surface area is one of the most important factors determining coal surface properties and micropore structure. The overall porosity can be calculated from the difference between the densities measured using helium and mercury:

$$V_t = \frac{1}{\rho_{Hg}} - \frac{1}{\rho_{He}},$$

where V_t is the total pore volume ($cm^3\ g^{-1}$), ρ_{He} is the true density (measured using mercury, $cm^3\ g^{-1}$), and ρ_{He} is the apparent density (measured using helium, $cm^3\ g^{-1}$).

The pore size distribution of coal also depends on the coalification. Lignite pores are relatively uniform, long flame coal has micropores and few macropores and mesopores, bituminous coal has a large number of micropores, and anthracite is usually dominated by micropores, with only about 10 % of the pore volume contributed by macropores and mesopores.

2.1.5.3 Coal Surface Area

The surface area of coal consists of internal and external surface areas. However, the external surface area accounts for a very small proportion, and the internal surface area is the major component. Very fine capillaries and pores are formed within coal during coal generation. There are large amounts of such capillaries and pores, and they are deeply and widely distributed; they have complex and well-developed internal structures. The coal internal surface area is the entire surface area of the internal pore structure, and is generally expressed by the specific surface area ($m^2\ g^{-1}$). It is closely related to the microscopic structure and chemical reactivity of coal, and is one of the important physical indicators for coal.

There are many methods for determining the coal specific surface area, such as the wetting heat, Brunauer–Emmett–Teller (BET), gas chromatography, and micropore volume methods. The BET method is the classical method. As the coalification changes, the coal internal surface area increases significantly in both the early and late stages (lignite and anthracite), but decreases in the middle stages (medium-coalification coal); this reflects changes in the spatial arrangement of molecules during coalification. The results obtained using different gases and temperatures vary, and can be unreliable.

2.1.5.4 Distribution of Porosity and Pore Size in Coal

Coal contains many pores. The percentage of the total volume of these pores compared with the entire coal volume is referred to as the coal porosity; this can also be expressed as the void volume ($cm^3\ g^{-1}$) in a unit mass of coal. The porosity can be calculated using the substitution method, TRD, or ARD. The principle of the substitution method is as follows: helium can fill all the coal pores, but mercury cannot enter the pores at all, so densities are obtained using helium and mercury as substitutes. The coal porosity is obtained as

$$\text{Porosity} = \frac{d_{\text{He}} - d_{\text{Hg}}}{d_{\text{He}}} \times 100\,\%,$$

where d_{He} and d_{Hg} are the coal densities ($g\ cm^{-3}$) obtained using helium and mercury, respectively.

There is a minimum near $C_{\text{daf}} = 90\ \%$, where the porosity is lowest, less than 3 %. Generally, the porosity of young bituminous coal is greater than 10 %. The amount of pores decreases with increasing coalification, because the coal structure becomes more compact under metamorphism. As coalification increases further, the porosity tends to increase, because of the increase in coal fractures with increasing coalification [6].

The pore diameters in coal are non-uniform in size, and can be roughly divided into three categories: micropores, transitional pores, and macropores. The pore diameter distribution can be determined using the mercury intrusion method or the liquid N_2 isothermal adsorption method. The mercury intrusion method is used to determine the distribution of pores of diameter 10–1000 nm. The liquid N_2 isothermal adsorption method can only be used to measure the volumes of transitional pores, and then translate them into pore diameters. Micropore diameters cannot be directly measured, but can be obtained indirectly by subtraction.

The pore diameter distributions of coals vary with different coalification degrees [7]: when the C_{daf} content is less than 75 %, macropores are predominant in lignite, with hardly any transitional pores; for coals with C_{daf} contents in the range 75–82 %, transitional pores are well developed, and the main contributors to the total pore volume are transitional pores and micropores; for coals with C_{daf} contents in the range 88–91 %, micropores remain predominant, and their volume accounts for more than 70 % of the total volume, with few transitional pores. With increasing coalification, the coal pore diameter decreases, the micropore proportion increases, and the physical structure of the coal gradually becomes more compact.

2.1.5.5 Coal Wettability

When coal is in contact with a liquid, the wetting of the coal differs depending on the properties of the liquid and the coal surface, and the liquid–solid surface properties. At the intersection of a gas, liquid, and solid, the included angle between the liquid surface and the solid–liquid interface is called the contact angle, θ. It is determined from the relative values of the interfacial tensions at the coal, liquid, and solid–liquid interfaces.

$$\cos\theta = \frac{\gamma_{\mathrm{coal}} - \gamma_{\mathrm{s-l}}}{\gamma_{\mathrm{l}}},$$

where

γ_{coal} is the surface tension of the coal (N m^{-1});

γ_{l} is the surface tension of the liquid (N m^{-1});

γ_{s-l} is the interfacial tension of the coal and the liquid (N m^{-1}).

2.1.5.6 Heat of Coal Wetting

The heat released when coal is wetted is referred to as the heat of coal wetting. The heat of coal wetting is the heat released when 1 g of coal is wetted; the unit is joules per gram. The heat of coal wetting is usually directly determined using a calorimeter. The heat of coal wetting is the result of interactions between the liquid and the coal surface, mainly van der Waals forces. The value of the heat of wetting is related to the liquid type and the specific surface area of the coal. The measured value of the heat of wetting can therefore be used to determine the total surface area of the pores in the coal. The coal surface areas obtained using the heat of wetting and BET methods show that 0.39–0.42 J is approximately equivalent to 1 m^2, and the inner surface area of the coal is calculated to be approximately 10–200 m^2 g^{-1}.

2.1.6 Coal Models

Among the models describing the physical structure of coal, the Hirsch model and the two-phase model are the most typical.

2.1.6.1 Hirsch Model

Hirsch [8] classified coal with different degrees of coalification into three physical structures, based on the physical model obtained using X-ray diffraction (XRD).

Open structure. This is observed for bituminous coal with a low degree of coalification, and features small aromatic lamellae and a large proportion of irregular amorphous structures. The aromatic lamellae are bound by cross bonds and more or less arbitrarily oriented in all directions, forming a porous three-dimensional (3D) structure.

Liquid structure. This is observed for bituminous coal with an intermediate degree of coalification, and is characterized by the orientation of some aromatic layer and formation of microcrystals comprising two or more lamellae. There are few cross bonds among the lamellae, resulting in high mobility. This type of coal has low porosity, low mechanical strength, and tends to be plastic during pyrolysis.

Anthracite structure. This is observed for anthracitic coal and is characterized by an increased number of aromatic lamellae and increased orientation. Condensation reactions result in a large number of micropores, so the porosity is higher than those of the former two structures.

The Hirsch model visually reflects the physical structural features of coal and explains many phenomena. However, the term "aromatic lamella" is not defined precisely, and the model fails to show the heterogeneity of the molecular constitution of coal.

2.1.6.2 Two-Phase Model

From the chemical point of view, coal has a 3D mesh structure composed of celluloses, lignins, lipids, proteins, tannins, and resins. In terms of the two components: one is a 3D carbon structure that consists of polycyclic aromatic hydrocarbons and hydrogenated aromatic hydrocarbons connected by aliphatic and ether chains; the other component is low-molecular-weight compounds in the network voids. Based on this idea, Haenel [9] found that ^1H nuclear magnetic resonance (NMR) spectroscopy shows that there are two coal proton relaxation times, one fast and one slow, and suggested a two-phase model, which is also known as the host–guest or macromolecular–molecular model, as shown in Fig. 2.1. In the figure, the macromolecular network is the stationary phase, and the micromolecular network is the mobile phase. This model suggests that the coal structure is an immobile, three-dimensionally crosslinked macromolecular phase, with embedded small

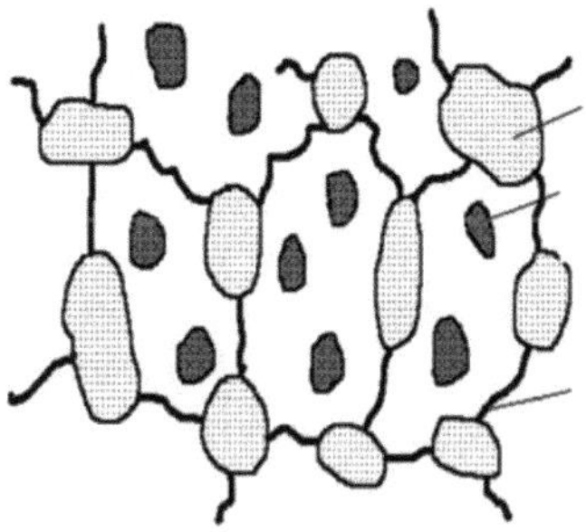

Fig. 2.1 Two-phase model. Reprinted from Ref. [9], Copyright 1992, with permission from Elsevier

mobile-phase molecules of various structures. The polyaromatic ring is the host in the coal, and is the same for similar coal types. However, the mobile-phase small molecules, which act as guests, vary depending on the coal type. The host and guest can be separated by extraction with different solvents. This model shows that crosslinking of the molecules in coal is essential for covalent bonds and that physical associations are essential for intermolecular forces, and explains the bonding properties of coal during solvent expansion processes. However, the idea of a mobile phase consisting of small molecules with low molecular weights is still controversial.

2.2 Physical Methods for Coal Structure Characterization

Physical research methods can directly provide molecular structural information on coal and are usually used in combination with other research methods. Recently, progress in coal structural studies has mostly been achieved using physical methods, as a result of significant progress in instrumental and analytical techniques.

2.2.1 Infrared Spectroscopy

IR absorption occurs when the energy (hv) of IR light is absorbed by molecules. The energy of IR light is not high enough to cause an electronic transition, but matches the energy level required to induce bond vibrations. The sample is irradiated with IR light with continuous frequencies, and characteristic absorptions are observed at specific frequencies. Infrared absorption follows the Beer law.

The absorption frequency and intensity in an IR spectrum depend on the types of bond vibration and the adjacent substituents. Knowledge of the characteristic absorptions of various functional groups and the peak displacement patterns provides a basic understanding of molecular structure, and, in some circumstance, enable quantitative analysis of the compound or specific functional groups.

The IR absorption frequencies of various common functional groups are in the range 4000–400 cm^{-1}; for convenience, this range is usually divided into four sections, as shown in Table 2.3.

The absorption peak at 3450 cm^{-1} is assigned to hydroxyl groups. The hydroxyl groups in coal are usually cross-linked via hydrogen bonds. IR spectroscopy has been widely used in the analysis of coals and coal derivatives, and the absorption peak shifts in the range 3300 cm^{-1} (–OH groups) to 3450 cm^{-1} can offer useful information about the functional groups present. The absorption intensity weakens with increasing coalification, which indicates that the number of hydroxyl groups decreases.

The absorption peak at 3030 cm^{-1} is attributed to C–H bonds in aromatic rings. The peaks at 870, 820, and 750 cm^{-1} are also associated with aromatic C–H vibrations. For coal samples with a low degree of coalification, the absorption

Table 2.3 Assignment of infrared absorption peaks of coal

Wavenumber (cm^{-1})	Wavelength (μm)	Peak assignment
>5000	<2.0	Multiple frequency absorptions and weak main frequencies
3300	3.0	–OH (or –NH) bonded, phenols
3030	3.30	Aromatic CH
2950 (shoulder)	3.38	$-CH_3$
2920 2860	3.42 3.50	Naphthene or alkane $-CH_3$
2780–2350	3.6–4.25	Carboxylic groups
1900 1780 1700	5.25 5.6 5.9	Aromatics, 1,2- and 1,2,4-substituted Carboxylic group >C=O
610	6.2	>C=O, HO– in carboxylic acid (bonded),–O-substituted aromatic C=C
1590–1470	6.3–6.8	Aromatics
1460	6.58	$-CH_2$ and $-CH_3$, or inorganic carboxylate
1375	7.72	$-CH_3$
1330–1110	7.5–9.0	C–O in phenols, alcohols, ethers, and esters, ash
1040– 910 860 833 (weak) 815 750 700 (weak)	9.6– 11.0 11.6 12.6 12.3 13.3 14 0.3	CH in substituted aromatics, ash

intensity at 3030 cm^{-1} is weak, but increases significantly with increasing coalification.

The absorption peaks at 2925, 1450, and 1380 cm^{-1} are contributed by C–H bonds in alkanes and naphthene. The intensities of all these peaks initially increase with coalification but then decrease significantly as the degree of coalification approaches 81.5 %. The absorption ratio A_{3030}/A_{2925} is consistent with the H_a/H_{al} ratio, and reflects the degree of coalification [10]. The peak at 1380 cm^{-1} is the characteristic IR absorption of methyl groups, and can be used for quantitative calculation of the methyl group content.

The strong absorption at 1600 cm^{-1} is still not definitively assigned. The absorption gradually weakens with increasing degree of coalification.

The absorptions between 1000 and 1300 cm^{-1} are attributed to ether bonds.

The absorption between 900 and 650 cm^{-1}, normally consists of three wide absorption peaks, are attributed to aromatic rings.

IR spectroscopy also confirms that aliphatic C=C and C≡C bonds are not present in coal molecules; the amounts of carboxylic acid and methoxy groups in bitumite (C > 80 %) are low; there is a significant amount of carboxylic acid groups in lignite, but these groups are not detected in coal above bitumite.

IR spectroscopy is sometimes used for quantitative analysis. The absorptions at 1380 and 900–650 cm^{-1} are usually used for calculating the content of hydrogen in methyl groups and aromatic rings. The hydrogen in hydroxyl groups is measured by chemical titration, and the numbers of hydrogen atoms in methylene and methenyl groups are calculated using the equation:

$$H_{CH_2,CH} = H_{total} - (H_{CH_3} + H_{OH} + H_{ar}).$$

2.2.1.1 Infrared Spectroscopic Studies of Functional Groups in Coal

Eight coal samples with various coalification degrees, ranging from peat to anthracite, were studied using Fourier-transform (FT) IR spectroscopy (Bio-Rad FTS165). The proximate and ultimate analyses of the samples are shown in Table 2.4. The amount of sample to be used in the analysis was determined using a series of pilot IR experiments. The amount of sample was considered to be suitable when a linear correlation between the absorption intensity and sample amount at 3450 cm^{-1} was achieved. The results show that for coal samples in the range 0.8–1.5 mg cm^{-2}, the IR absorption follows the Beer law. The coal test samples were prepared by mixing KBr (1:180) and pressing. The coal samples and KBr were dried in a vacuum oven at 110 °C overnight before use.

The standard curves obtained are used for quantitative analysis of functional groups.

The raw IR spectrum is processed further to remove distortions. Spectrum 1 in Fig. 2.2 is the original spectrum; the baseline is clearly tilted. After calibration using the standard normal variate and Savitsky–Golay methods (spectrum 2), the baseline

Table 2.4 Coal proximate and ultimate analyses (%)

Coal sample	Proximate analysis			Ultimate analysis				
	M_{ad}	A_d	V_d	C_d	H_d	O_d	N_d	S_{td}
Huachuan peat	8.02	49.38	40.11	26.40	2.97	19.15	1.86	0.24
Pinzhuang lignite	13.49	5.72	43.68	66.44	4.57	21.90	1.08	0.29
Fuxin long flame coal	2.36	14.24	32.27	68.08	4.27	11.61	0.91	0.89
Xingzhi fat coal	0.81	14.52	29.52	70.69	4.45	7.33	1.13	1.88
Zaozhuang coking coal	1.25	7.96	33.60	76.25	4.67	9.55	1.05	0.52
French lean coal	0.79	6.82	11.05	75.85	4.85	7.56	1.45	1.01
Fengfeng meager coal	0.87	17.54	12.22	72.72	3.36	5.04	1.05	0.29
Jincheng anthracite	0.72	14.52	7.93	77.95	2.93	0.57	0.94	0.36

ad air dried basis; *d* dry basis; *td* total dry basis

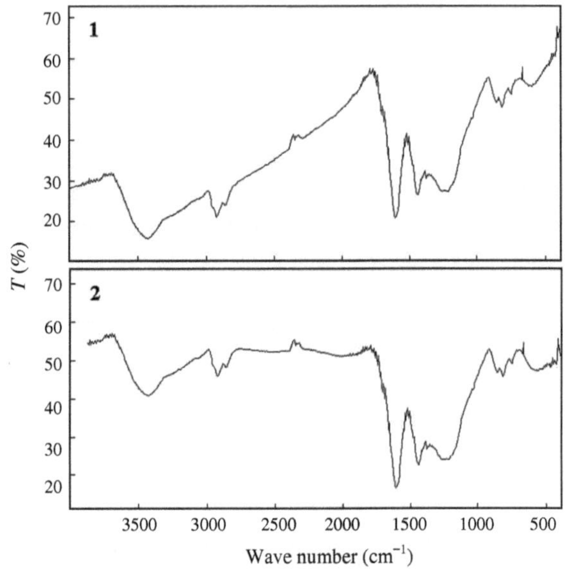

Fig. 2.2 IR spectra of coal samples: *1* without calibration; *2* calibrated

is smoother and most of the distortion is removed. The IR spectroscopic information for the top seven coal samples are shown in Table 2.5 (except Jincheng anthracitic).

There is a broad and strong absorption at 3410 cm^{-1} in most of the coal spectra, which is assigned to hydrogen-bonded hydroxyl group vibrations. In the spectrum of peat coal, a strong absorption at 3285 cm^{-1}, assigned to the wagging vibrations of hydroxyl groups, is also present. This suggests that there are large numbers of condensed hydroxyl groups in peat coal, forming a coal–OH–OH–coal internal network structure. No absorption above 3600 cm^{-1} was observed in the peat coal spectrum, suggesting that free hydroxyl groups have been totally removed. With increasing coal rank from lignite to jet coal, the hydroxyl group absorption

Table 2.5 Relationship between coal type and the wave number and content of OH group

Coal	Free OH group (cm^{-1})	Content	Hydrogen bond (cm^{-1})	Content	Poly OH1 (cm^{-1})	Content	Poly OH2 (cm^{-1})	Content
Peat	3634	3.17	3518	3.23	3409	12.90	3292	7.20
Lignite	3622	2.35	3540	5.19	3425	14.81	3228	21.34
Long flame coal	3622	0.80	3524	9.75	3421	16.89	3230	5.28
Fat coal	3620	1.02	3544	0.84	3446	10.22	3244	5.68
Coking coal	3524	0.33	3412	5.87	3301	0.39	3213	1.73
Meager coal	3585	0.00	3544	0.27	3438	13.34	3227	1.73
Lean coal	3590	1.83	3443	33.28	3352	0.16	3247	4.43
Anthracite[a]	–	–	–	–	–	–	–	–

[a]Below detection limit

gradually splits into two absorption bands, at 3408 and 3221 cm^{-1}, and then gradually weakens and disappears completely in anthracite.

Hydrogen bonds are common linkages in coal structures, and play an important role in stabilizing and deconstructing the coal molecular network. As seen from Table 2.5, the amounts of intermolecular hydrogen bonds are highest in meager coal and lowest in Lean coal. The total content of polymerized hydroxyl groups (poly OH1 and poly OH2), which form the coal–OH–OH–coal internal network structure, decreases with increasing coal rank.

Ratio of Aromatic to Aliphatic Hydrogens

The ratio of H_{ar} to H_{al}, where H_{ar} refers to hydrogen atoms attached to aromatic backbones, and H_{al} refers to hydrogen atoms in alkane chains, is a key parameter in the coal average molecular structure. The ratio usually changes with increasing coal rank. The contents of H_{ar} and H_{al} reflect the coal activity during pyrolysis and the molecular-weight distribution of the pyrolysis products.

The absorption at 3065 cm^{-1} is assigned to aromatic C–H stretching vibrations and indicates the content of condensed aromatic structures. A strong absorption at 3065 cm^{-1} usually indicates low pyrolysis activity of the coal. In this section, the absorption intensities between 3030 and 3060 cm^{-1} are used to calculate the H_{ar} content.

The sharp absorptions at 2920 and 2850 are attributed to alkane-chain C–H symmetric stretching and asymmetric stretching vibrations. In this section, the intensity of the peak at 2920 cm^{-1} is used to calculate the H_{al} content and the absorption at 2850 cm^{-1} is used for CH_2 and CH. The correlation between H_{ar}/H_{al} and coal rank is illustrated in Fig. 2.3.

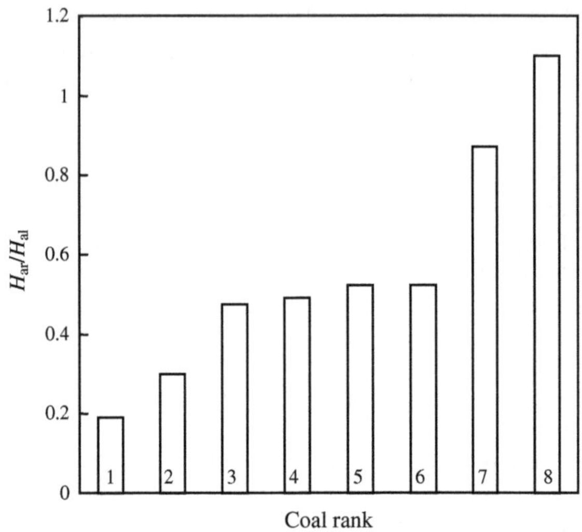

Fig. 2.3 Correlation between H_{ar}/H_{al} and coal rank: *1* peat; *2* lignite; *3* long flame coal; *4* fat coal; *5* coking coal; *6* meager coal; *7* lean coal; and *8* anthracite

Assignment of the absorption at 1618 cm^{-1} remains controversial. Brown [11] and Solomon [12] believed that the 1618 cm^{-1} absorption arose from the rocking vibrations of hydroxyl-substituted aromatic backbones. However, Friedel [13] proposed it was more reasonable to assign the absorption to conjugated carbonyl groups. This author agrees with Brown and Solomon's conclusion, because a typical absorption at 1580 cm^{-1}, an important indicator of conjugated C=C structures, was not observed in the coal IR spectra. Eolfson [14] thought that an increase in the number of aromatic ring substituents weakened the intensities of absorption bands. The strong absorption band at 1450 cm^{-1} shifts to lower frequency and gradually disappears with increasing ring condensation degree.

The oxygen-containing functional groups in coal include carboxyl, carbonyl, hydroxyl, and ether groups. The ether group absorption is normally at 1225 cm^{-1}, caused by an asymmetric stretching vibration. The C=O absorption in peat is at 1716 cm^{-1}, assigned to C=O in aliphatic chains, which confirms that peat coal is of low coal rank, without condensed aromatic structures. Siskin et al. [15] confirmed that the rupture of ether bonds usually occurred on bonds linked to a single benzene ring. Figure 2.4 shows the relationship between the number of ether bonds and the coal rank. Figure 2.4 shows that the ether bond abundance is very low in all coals except peat and lignite. The low abundance of ether bonds does not indicate low pyrolysis reactivity.

Length of CH_2 Chain

The bimodal absorption at 725 cm^{-1} is usually assigned to CH_2 rocking vibrations, and its intensity is proportional to the amount of consecutive CH_2 groups. The CH_2 absorption shifts from 725 to 750 cm^{-1}, indicating increasing polarity of the coal

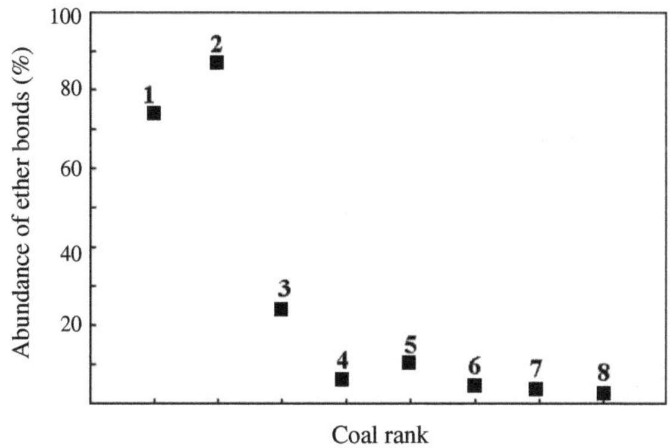

Fig. 2.4 Relationship between number of ether bonds and coal rank: *1* peat; *2* lignite; *3* long-flame coal; *4* fat coal; *5* coking coal; *6* lean coal; *7* meager coal; and *8* anthracite

structure. In high-rank coal, the CH_2 absorption peak is sharp, the coal structure is unitary and CH_2–CH_2 bond cleavage depends on the bond energy rather than on the effects of adjacent functional groups. As Fig. 2.5 shows, the number of CH_2 groups decreases with increasing coal rank. The CH_2 absorption wavenumber increases as the coal rank increases, and reaches 779 cm^{-1} in lignite. In the lignite structure, large amounts of unsaturated electron-donating groups surround CH_2 groups, and these electron-donating groups significantly affect the cleavage of CH_2–CH_2 bonds. Heredy and Neuworth [16] showed that CH_2 attached to a single benzene ring was least active, CH_2 linked to a phenanthrene ring was moderately active, and a single CH_2 was more active than consecutive CH_2 groups in an aliphatic chain.

2.2.2 Nuclear Magnetic Resonance Spectroscopy

Since its invention in 1946, NMR has become a powerful tool for the determination of organic structures. The NMR spectrum provides a great deal of information about the structure of a compound, and some structural information can only be

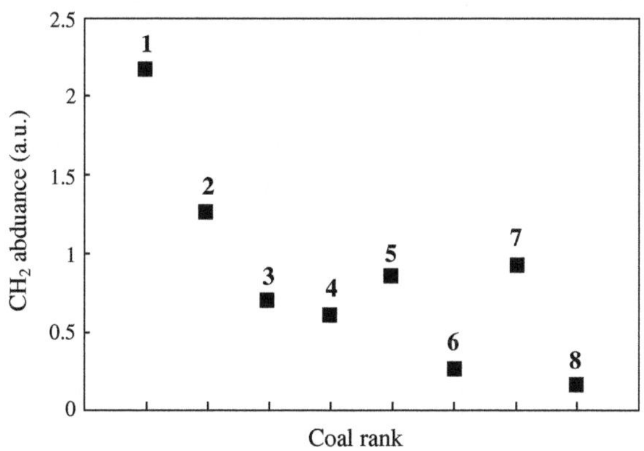

Fig. 2.5 Correlation between abundance of CH_2 and coal rank: *1* peat; *2* lignite; *3* long flame coal; *4* fat coal, *5* coking coal; *6* lean coal; *7* meager coal; and *8* anthracite

obtained using NMR spectroscopy. NMR spectroscopy is widely used in research on coal structure.

NMR spectra provide information on chemical shifts, peak areas, and coupling and/or splitting. Peak coupling and splitting provide detailed information on the fine chemical structure. When NMR is used in the analysis of coal samples, it is usually a challenge to analyze the fine chemical structure, because the absorption signals overlap. In this section, therefore, only a brief introduction to chemical shifts and peak areas is given.

In NMR spectroscopy, nuclei in the same chemical environment generate only one absorption signal, and nuclei located in different chemical environments absorb magnetic waves at various frequencies, because of the magnetic shielding effects of adjacent atoms. A typical ^{1}H NMR spectrum is shown in Fig. 2.6.

Many factors, including the adjacent functional groups, conjugated bonds, and hydrogen bonding, change the nuclear chemical shift, as a result of changes in the chemical environment surrounding the target nucleus.

The area under a peak is represented by the integration curve and proportional to the number of hydrogens contributing to that peak. This provides useful information for assigning the absorption peaks and classifying the environment of the nucleus.

The major elements (C and H) in coal samples have nuclei that are active in NMR spectroscopy, with significant differences among the signal intensities. The ^{1}H nucleus has a spin quantum number of 1/2, and a natural abundance of 99.985 %. It is therefore easy to measure the ^{1}H NMR signal. The ^{12}C nucleus is not active in NMR and does not generate absorption signals, because its nucleus spin quantum number is 0. The ^{13}C nucleus has a spin quantum number of 1/2, and a natural abundance of 1.107 %, making it more difficult to measure ^{13}C NMR signals.

^{1}H NMR was first used in coal structural analysis by Newman [17] in 1955, and has been widely used since then. ^{1}H NMR provides detailed information on the types and distribution of hydrogen atoms in coal and coal derivatives; for example, the chemical shifts of aromatic hydrogen atoms are 6–10 ppm; the chemical shifts

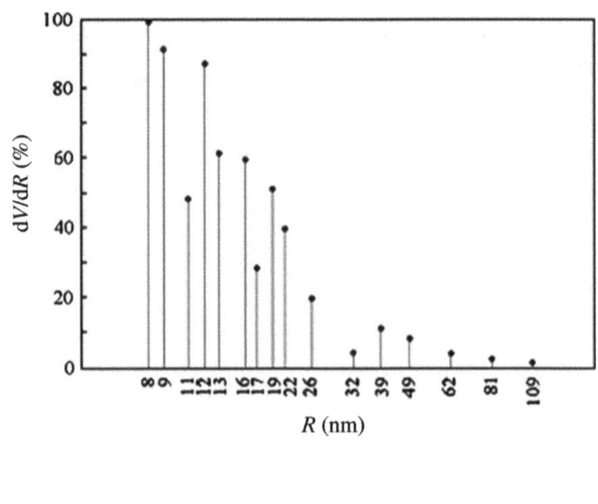

Fig. 2.6 Pore size distribution of Jincheng anthracite

of hydrogen atoms at the α positions of alkyl-group side-chains bonded to aromatic carbons are 2–4 ppm, and those of hydrogen atoms in alkyl groups attached at the β or further positions to aromatic carbons are around 0.2–2 ppm. Brown et al. [18] calculated three structural parameters using the following equations:

$$f_a = \frac{\frac{C}{H} - \frac{H_\alpha^*}{X} - \frac{H_0^*}{Y}}{\frac{C}{H}},$$

$$\sigma = \frac{\frac{H_\alpha^*}{X} + \frac{O}{H}}{\frac{H_\alpha^*}{X} + \frac{O}{H} + H_{ar}^*},$$

$$\frac{H_{aru}}{C_{ar}} = \frac{\frac{H_\alpha^*}{X} + H_{ar}^* + \frac{O}{H}}{\frac{C}{H} - \frac{H_\alpha^*}{X} - \frac{H_0^*}{Y}},$$

where

f_a is the percentage of aromatic carbons, i.e., the ratio of aromatic carbons to total carbons;

σ is the percentage of substituted aromatic carbons, i.e., the ratio of substituted aromatic carbons to total substitutable aromatic carbons;

H_{aru}/C_{ar} is the condensation degree of aromatics, i.e., the hydrogen/carbon ratio in unsubstituted aromatics;

H_{ar}^* is defined as H_{ar}/H, i.e., the ratio of aromatic hydrogens to total hydrogens;

H_α^* is defined as H_α/H, i.e., the ratio of H_α (hydrogen atoms at the α positions in alkyl side chains bonded to aromatic carbons) to total hydrogen atoms;

H_0^* is defined as H_0/H, i.e., the ratio of H_0 (hydrogen atoms at the β or further positions in alkyl side chains bonded to aromatic carbons) to total hydrogen atoms;

X, Y are the ratios of hydrogen/carbon at the α or β positions of alkyl side chains bonded to aromatic carbons, usually assumed to be 2.

In Table 2.6 [19], the calculated key structural parameters of pyridine extracts from various coals are listed. With increasing coalification, H_{ar} and H_α gradually increase, and H_0 gradually decreases. This suggests that the extent of the aromatic structure increases and the alkyl side chains become shorter as the degree of coalification increases.

Coal extract samples were used in the ^{1}H NMR test, because dissolved samples are typically required. The chemical shifts of highly active hydrogen atoms, usually attached to the carbons in ether and ester groups and involved in crosslinking, are listed in Table 2.7.

^{13}C NMR spectroscopy provides direct information about the carbon backbone. Both liquid and solid samples can be used in ^{13}C NMR spectroscopy, so interference originating from the solvent extraction process can be eliminated. The chemical shift of ^{13}C may reach 200 ppm, which is a much broader range than that

Table 2.6 Hydrogen distribution and structural parameters of pyridine extracts

C_{daf} (%)	Extraction yield (%)	H distribution			Structural parameters		
		H^*_{ar}	H^*_{α}	H^*_0	f_a	σ	H_{aru}/C_{ar}
61.5	13.8	0.07	0.12	0.75	0.41	0.74	0.93
70.3	16.6	0.18	0.20	0.56	0.61	0.55	0.69
75.5	15.8	0.21	0.20	0.53	0.62	0.52	0.72
76.3	6.7	0.20	0.30	0.44	0.64	0.59	0.76
76.7	16.7	0.10	0.21	0.64	0.53	0.67	0.60
80.7	12.8	0.27	0.22	0.45	0.70	0.45	0.65
82.6	21.4	0.35	0.26	0.36	0.73	0.37	0.68
84.0	18.5	0.30	0.25	0.43	0.69	0.41	0.67
85.1	20.9	0.27	0.29	0.39	0.72	0.47	0.59
86.1	19.3	0.32	0.28	0.37	0.73	0.37	0.57
90.0	2.8	0.55	0.31	0.13	0.85	0.27	0.63
90.4	2.5	0.50	0.30	0.19	0.83	0.26	0.57

Table 2.7 Chemical shifts for typical active protons (ppm)

Compound	δ	Compound	δ
ROH	0.5–5.5	Ar–SH	3.0–4.0
ArOH	10.5–16	RSO_3H	11.0–12.0
Ar–OH	4–8	RNH_2, R_2NH	0.4–3.5
C=C–OH (bonded)	15–19	$ArNH_2$, Ar_2NH, ArNHR	2.9–4.8
RCOOH	10–13	$CONH_2$, $ArCONH_2$ R	5.0–6.5
=N–OH	7.4–10.2	RCONHR, ArCONHR	6.0–8.2
R–SH	0.9–2.5	RCONHAr, ArCONHAr	7.8–9.4

of the ^1H nucleus. Overlap of absorption peaks is rarely observed in ^{13}C NMR spectra, which is helpful in analyzing individual carbons. However, ^{13}C NMR suffers from a low signal-to-noise ratio and low sensitivity (only 1/5800 that of ^1H NMR), but this can be greatly improved using Fourier-transform technology.

2.2.3 X-ray Diffraction

In a crystal structure, atoms are arranged in a periodic repeated pattern. The crystal lattice spacing is of the same order of magnitude (around 10 nm), so diffraction occurs when X-rays pass through the crystal. The diffraction pattern is recorded photographically for analysis of the crystal structural parameters. The fundamental theory of XRD is expressed by Bragg's law:

$$2d \sin \theta = n\lambda,$$

where

d is he spacing between layers of atoms (nm);

θ is the angle between the incident rays and the crystal surface;

n is an integer (1, 2, 3, and so on), because the reflected waves from different layers are perfectly in phase with each other;

λ is the X-ray wavelength (nm).

Graphite has a clear crystal structure and exhibits nine XRD peaks which are related to different crystal facets. Although coal is not crystalline, it still has diffraction peaks, which show the arrangement of the carbon atoms. Lignite and bituminous coal only have two diffraction peaks, and anthracite has four peaks.

Macerals from the same coal sample have significantly different molecular structures. This difference is observed in the XRD patterns. The inertinite group shows four diffraction peaks, which are similar to those from anthracite. Vitrinite only shows three diffraction peaks, and the peaks from exinite are indistinct.

As the degree of coalification increases, diffraction patterns become increasingly distinctive and similar to that of graphite, and the aromatic structure of the coal becomes more microcrystalline. The aromatic microcrystalline structure is composed of various layers of polyaromatic rings, and is described by three key parameters, L_a, L_c, and d, where L_a is the horizontal dimension of the aromatic microcrystal, L_c is the vertical dimension, and d is the average distance between the polyaromatic layers. The parameters L_a, L_c, and d can be calculated from the XRD pattern. With increasing degree of coalification, L_a and L_c increase, and d gradually decreases. The distance between polyaromatic layers increases with increasing coalification [20]; The parameters for the aromatic microcrystals of various coal macerals also differ significantly. Table 2.8 [21] lists the L_a, L_c, and d values for the macerals from Yanzhou coal (the maceral density gradually increases from group E to I_2).

Table 2.8 Percentages of aromatic carbon and microcrystal parameters for Yanzhou macerals

Maceral	f_a^{Ca}	d (nm)	L_c (nm)	L_a (nm)
E	0.56	0.492	0.638	1.824
V_1	0.65	0.372	0.658	1.768
V_2	0.69	0.360	0.790	1.490
V_3	0.70	0.360	0.802	1.976
I_1	0.73	0.357	0.946	1.868
I_2	0.77	0.355	1.298	1.998

[a]Calculated from ^{13}C NMR spectra

2.2.4 Small-Angle X-ray Scattering

SAXS is a scattering technique based on the deflection of collimated X-ray radiation from the straight trajectory after it interacts with structures that are much larger than the X-ray wavelength. Reich et al. [22] was the first to use SAXS in the study of coal structures. SAXS gives information about the sizes, shapes, and orientations of structures in a sample.

SAXS has significant advantages in porosity studies. No sample pretreatment is required and closed pores are measurable using SAXS. SAXS can be used under various conditions, with no pressure and temperature restrictions. It is applicable to solution and in situ reaction systems. In SANS, another small-angle scattering technique, neutrons are used instead of X-rays. SANS is very similar to SAXS in many respects and provides reliable porosity data. The limitation of small-angle scattering techniques is that they cannot be used to analyze the ultramicropore structures in coal. Bale et al. [23] first used SAXS techniques for fractal analysis. Johnston et al. [24] used SAXS to study drying and thermal processes of Victoria lignite. In the drying process, the volume fractal dimension and the area fractal dimension are significant before and after dehydration.

2.2.5 Electron Microscopy

Electron microscopy is used to investigate the micro-structures of a wide range of solid surfaces. The pore structures or spatial structures of large molecules in coal samples have also been extensively studied using electron microscopy. In work performed at Tokyo University, Japan three organic maceral groups from Pingshuo gas coal were investigated by high resolution transmission microscopy and was confirmed to be an effective method for studying coal internal structures and coalification processes. Other researchers have performed fractal analysis of coking coal using electron microscopy [25]. The 3D profiles of solid surfaces can be observed using the secondary electron imaging technology in scanning electron microscopy (SEM). If the incident intensity of the electron is I_p, the intensity of the secondary electron, I_s, is calculated using the following equation:

$$I_S = C\frac{I_P}{\cos\theta},$$

where θ is the angle between the electron beam and the solid surface, and C is a constant. When the electron bean is scanning through a concave–convex surface, θ is changing, so the amount of secondary electrons at each imaging point changes.

When a surface with fractal features is analyzed using electron microscopy, the SEM images reveal gray levels and gradients. The use of graphics-processing software and computational tools enables the fractal dimension to be estimated using the discrete fractal Brownian increment random field mathematical model.

2.2.6 Surface Methods

2.2.6.1 Gas Adsorption

Gas adsorption is a classic method for investigating surface structures. Normally, low-temperature (77 K) N_2 adsorption or room-temperature (298 K) CO_2 adsorption is used, with the BET adsorption model:

$$\frac{X}{V(1-X)} = \frac{1}{V_m C} + \frac{(C-1)}{V_m C} X,$$

where

X is the relative pressure, i.e., the ratio of the equilibrium and the saturation pressures of the adsorbates at the adsorption temperature, P/P_s;
V is the adsorbed gas quantity at a relative pressure (m^3);
V_m is the saturated monolayer adsorbed gas quantity (m^3);
C is the BET constant related to adsorption heat.

The BET equation can be plotted as a straight line with $X/V(1-X)$ on the y-axis and X on the x-axis. The value of the slope, a, and the y-intercept, b, of the line are used to calculate the monolayer adsorbed gas quantity V_m:

$$V_m = \frac{1}{a+b},$$

If the adsorbed gas is N_2, which has a molecular cross-sectional area of 1.62×10^{-19} m^2, the surface area in the BET model, S_{BET}, can be calculated using the equation:

$$S_{BET} = 4.353 \times \frac{V_m}{m},$$

where m is the mass of the sample (g). Anderson [26] used an empirical constant, K, in the classic BET equation for coal sample analysis to improve the accuracy (K is usually 0.8 for coal analysis):

$$\frac{X}{V(1-KX)} = \frac{1}{V_m CK} + \frac{(C-1)}{V_m C} X,$$

N_2 adsorption requires a relatively long equilibrium period and is inappropriate for micropore analysis. CO_2 adsorption at room temperature is also widely used in coal surface analysis. Gan et al. [7] studied coal surfaces using N_2 and CO_2 adsorption methods, and found that the surface areas obtained with N_2 were 1–88.4 $m^2\ g^{-1}$, which were notably lower than those of 96–426 $m^2\ g^{-1}$ obtained using CO_2 adsorption. These results suggest that only the macropores and mesopores in coal adsorbed N_2 molecules.

The D–P equation is also commonly used to calculate surface areas:

$$\ln V = \ln V_0 - \left(\frac{BT^2}{\beta^2}\right) \ln^2\left(\frac{P_s}{P}\right),$$

where

- V is the adsorbed gas volume (m^3) under pressure P;
- V_0 is the volume of micropores (m^3);
- T is the adsorption temperature (K);
- P_s is the saturation pressure (Pa) of adsorbates at the adsorption temperature;
- B is a constant related to the average micropore size;
- β is a coefficient.

Walker et al. [27] used CO_2 adsorption at 298 K to study several coals and calculated the surface areas using both the BET and D–P equations. Surface area tests with 28 types of Chinese coal showed an empirical relationship between the results obtained using the BET and D–P equations [28]:

$$S_{\text{D–P}} = -14.06 + 2.11 S_{\text{BET}},$$

Although CO_2 adsorption is widely used in coal surface area analysis, it has been shown to be inconsistent in various independent studies. Dietz et al. [29] and Ghetti et al. [30] suggested that the surface hydroxyl groups and ash content of coal affected the CO_2 adsorption. However, Chinese researchers repeated the CO_2 adsorption tests with Chinese coal samples and found no significant variations in the results, confirming the feasibility of using CO_2 adsorption for coal surface area analysis. Gas adsorption was the first method used for surface fractal analysis.

Normally, N_2 and CO_2 are used as the adsorbed gas in fractal dimension analysis. Unlike the case of surface area analysis, discussed previously, Fairbridge et al. [31] found that using different adsorbed gases did not affect the results of the fractal dimension analysis. In a study of 28 types of Chinese coal, fractal dimension analysis was performed based on the data collected for surface area analysis, using mathematic tools, and the results agree with the fundamental theory of coal structure [32].

2.2.6.2 Mercury Porosimetry

Macropores cannot be analyzed using gas adsorption methods. Mercury porosimetry was developed for the study of macropores, based on the external pressure needed to force the liquid into a pore against the opposing force of the liquid's surface tension. The force balance equation is

$$p_{Hg} = \frac{-2\gamma \cos\theta}{r},$$

where

P_{Hg} is the forcing pressure (N m^{-2});
γ is the surface tension of the liquid (0.486 N m^{-1} is usually used);
θ is the contact angle of mercury (140° is usually used);
r is the pore radius (m).

The pressure required to intrude mercury into the coal pores is inversely proportional to the pore size.

Mercury porosimetry analysis is simple and quick, but it cannot be used for micropore analysis. So, to obtain the overall distribution of all the pores, a combination of gas adsorption and mercury porosimetry is required.

Mercury porosimetry is also commonly used for surface fractal analysis. Friesen et al. [33] showed that the volume of mercury intruded into the pores is related to the fractal dimension.

2.2.6.3 Case Study of Coal Surface Characterization

An investigation of five coals of various coalification degrees is presented: Fuxin long flame coal, Pingshuo gas coal, Xinzhi fat coal, Fengfeng meager coal, and Jincheng anthracitic coal. The proximate and ultimate analyses of the selected coal samples are displayed in Table 2.4. The five coal samples were ground and sieved; particles of diameter around 100 μm were chosen for gas adsorption tests.

Surface analysis was conducted using an automatic adsorption apparatus on 1 g samples in an adsorption tube cooled with liquid N_2. The instrumental physical constant was determined before the analysis. The sample underwent degasing pretreatment under vacuum (<6.7 Pa) at 100–300 °C. The pressure of the high-purity N_2 gas for the adsorption tests was adjusted to 0.2–0.3 MPa.

The key surface structural parameters for five coals are displayed in Table 2.10. Figures 2.6, 2.7, 2.8, 2.9 and 2.10 show the pore diameter distributions of the five coals. Figure 2.11 shows the changes in the raw coal surface area with degree of coalification.

As can be seen from Table 2.9, micropores are the predominant pore type. For all five coals, the most probable diameters are lower than 1 nm and the mean pore diameters are below 3 nm. The coal surface areas are large, but the pore volumes

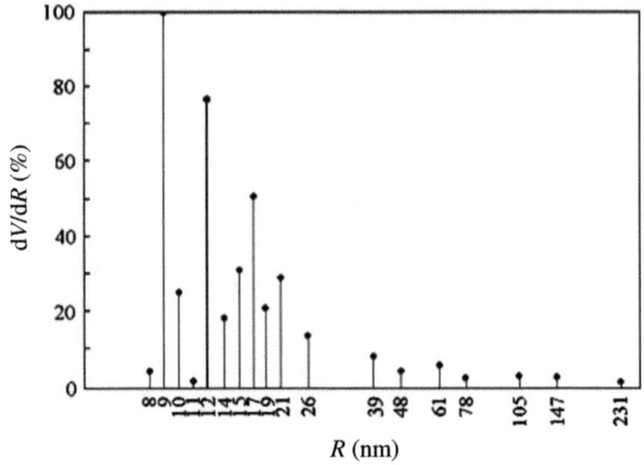

Fig. 2.7 Pore size distribution of Fengfeng meager coal

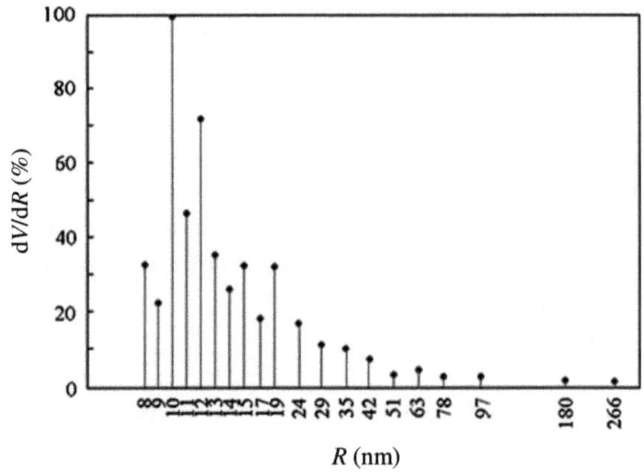

Fig. 2.8 Pore size distribution of Xinzhi fat coal

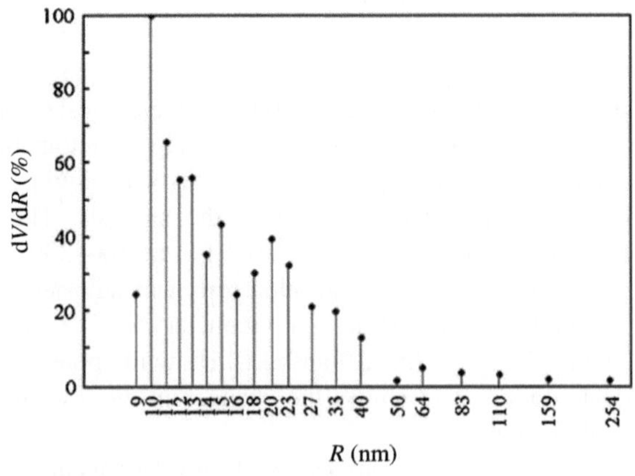

Fig. 2.9 Pore size distribution of Pingshuo gas coal

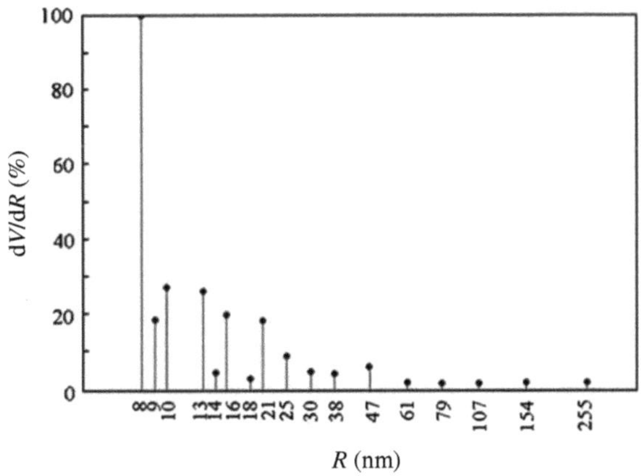

Fig. 2.10 Pore size distribution of Fuxin long flame coal

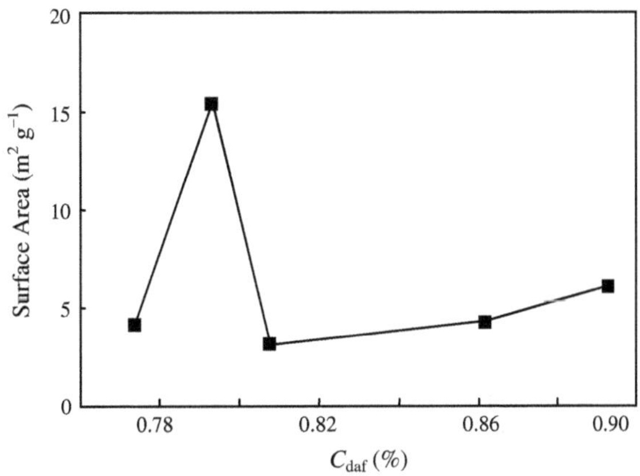

Fig. 2.11 Relationship between surface area and degree of coalification

Table 2.9 Key surface structural parameters of five raw coals

Coal sample	Surface area ($m^2\ g^{-1}$)	Pore volume ($cm^3\ g^{-1}$)	Most probable diameter (nm)	Mean pore diameter (nm)
Fuxin long flame coal	4.23	0.01	0.83	2.57
Xinzhi fat coal	3.12	0.01	0.87	2.73
Pingshuo gas coal	15.5	0.04	0.93	1.32
Fengfeng meager coal	4.30	0.01	0.94	2.76
Jincheng anthracite	6.03	0.01	0.84	2.19

are small, confirming that micropores are the main contributors to the surface area. The data also show that the order of the maceral surface areas is inertinite > exinite > vitrinite. The surface area of exinite is almost three times than those of exinite and vitrinite. Pingshuo gas coal has an inertinite content as high as 25 %, which explains why the surface area and pore volume of Pingshuo gas coal are around four times those of the other four coals. In the other four coals, the changes in surface area are related to the degree of coalification: with increasing degree of coalification, the coal surface area initially decreases and then gradually increases [34]. The surface area is minimum at a carbon content of 85 %.

Xinzhi fat coal, with a carbon content of 83 %, has the lowest surface area (3.122 $m^2\ g^{-1}$), because it has the largest pore volume and most probable pore diameter. Fuxin long flame coal has a large surface area, in line with its low most probable pore diameter. Jinchen anthracite coal has the highest degree of coalification and an even pore size distribution of below 2.5 nm, which suggests that the coal has a high aromatic content with a regular arrangement of large molecules.

2.3 Statistical Constitution Analysis

Natural products can be classified into three categories: micromolecules with uniform molecular structures; large molecules with uniform molecular structures; and mixtures of micromolecules and polymers with various molecular structures. Coal belongs to the last category and contains large molecules with various molecular structures. Waterman [35] first suggested that the characteristics of a material can be described by certain critical structural parameters, which can be obtained using statistical methods. This method was introduced for structural studies of coal by van Krevelen et al. [36], who developed the statistical constitution analysis method, which is one of most powerful methods in coal studies.

2.3.1 Coal Structural Parameters

The complexity of coal structures makes determining the exact molecular structure challenging. Currently, structural parameters that describe an averaged coal structural unit are typically applied. In the following sections, the definitions and formulas for calculating the structural parameters are briefly introduced.

For saturated aliphatic alkane groups:

$$H = 2C + 2,$$

where, H and C represent the number of H and C atoms, respectively.

For the alkanes without conjugated double bonds and acetylene linkage (carbon triple bonds) in the molecules:

$$H = 2C + 2 - 2R - C_a,$$

$$2\left(\frac{R-1}{C}\right) = 2 - f_a - \frac{H}{C},$$

where, R is the number of rings in the structural unit. C_a and C are the number of aromatic and total carbon atoms in the structural unit, respectively. f_a is the aromaticity, defined as C_a/C.

In the above formula, $2(R - 1)/C$ is defined as the ring condensation index. If heteroatoms are included in the molecules, the above formula is still applicable after adjustment of the value of H/C.

For coal or other polymeric compounds composed of a variety of structural units, an averaged structural unit is adopted. By considering the bridge bonds among structural units together with the additional rings within these linkages, the structural parameters of the average structural unit are calculated as:

$$H_u = 2C_u + 2 - 2R_u - C_{au} - 2b_u,$$

$$H_u = \frac{p-1}{p} + r_u,$$

$$2\left(\frac{R_u + r_u - \frac{1}{p}}{C_u}\right) = 2 - f_{au} - \frac{H_u}{C_u},$$

where: H_u, C_u are the number of hydrogen and carbon atoms in the average structural unit, respectively. R_u is the number of ring in the average structural unit; C_{au} is the number of aromatic carbon atoms in the average structural unit; f_{au} is the aromaticity of the average structural unit, defined as C_{au}/C_u; b_u is the number of bridge bonds in an average structural unit, known as polymerisation intensity; r_u is the average number of additional rings that arise from bridging bonds among structural units; p is the number of the structural units included in a polymeric molecule, known as degree of polymerisation.

In coal, the parameter $f_{au} = C_{au}/C_u$ is similar to $f_a = C_a/C$, and H_u/C_u is similar to H/C. For the following calculations in this chapter only f_a and H/C are used. Thus the new parameters b_u, r_u and p are determined for the average structural unit. To further illustrate this relationship, examples with average structural units and associated structural parameters (b_u, r_u and p) are listed in Table 2.10.

The total number of rings in an average structural unit are the sum of rings in individual units and those additional rings from bridging bonds.

$$R'_u = R_u + r_u.$$

For a coal structural unit, $1/p$ is negligibly small so that the formula may be further simplified as:

Table 2.10 Examples of b_u, r_u and p in various polymeric structures

Polymeric structural models	p	r_u	b_u
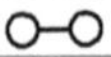	2	0	0.5
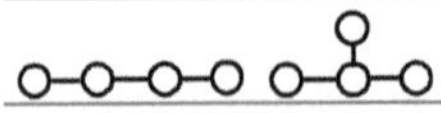	4	0	0.75
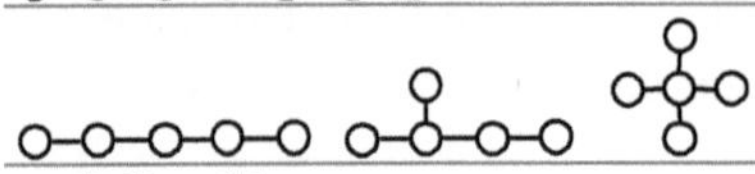	5	0	0.8
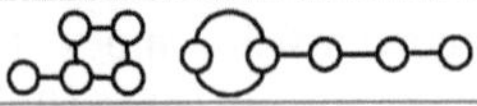	5	1/5	1
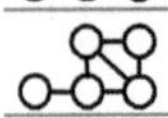	5	2/5	1.2
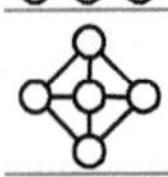	5	4/5	1.6
20 structural units arranged in a linear or branching structure	20	0	0.95
An infinite number of structure units, arranged in a linear or branching structure	∞	0	1
A structure containing a large number of structure units with a ring in the molecule	p	$1/p$	1
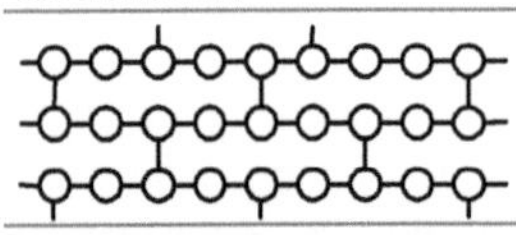	∞	0.25	1.25
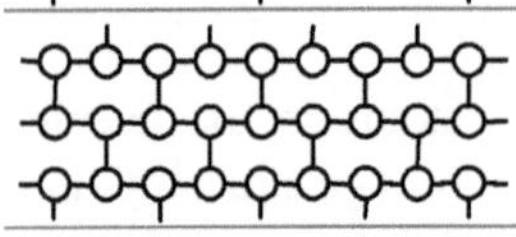	∞	0.5	1.5
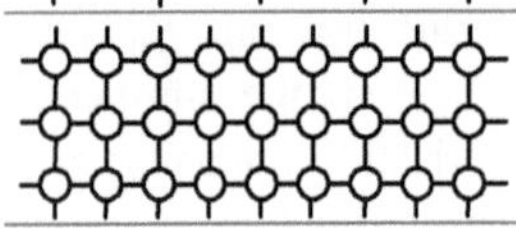	∞	1	2

$$2\left(\frac{R}{C}\right)_u = 2\left(\frac{R'_u}{C_u}\right) = 2 - f_a - \frac{H}{C}.$$

In the above equation, $2(R/C)_u$ is defined as the ring index of the average structural unit and $(R/C)_u$ means the average ring number for each carbon atom.

In Table 2.11, the key structural parameters and their internal relations in the coal matrix are described.

2.3.2 *Principle of Statistical Constitution Analysis*

Statistical constitution analysis of coal is based on the additive molar function developed by van Krevelen [37]:

Table 2.11 Key structural parameters and their internal relations in coal matrix

Category	Structure parameters	Symbol and definition	Extremum	Note
Aromaticity	Fraction of aromatic carbons	$f_a = \frac{C_a}{C}$	0-non aromatic hydrocarbons 1-aromatics	The ratio of aromatic carbon, and hydrogen, and rings to the overall respective values in the structure unit
	Fraction of aromatic hydrogens	$f_{Ha} = \frac{H_a}{H}$		
	Fraction of aromatic rings	$f_{Ra} = \frac{R_a}{R}$		
Degree of ring condensation	Degree of ring condensation	$2\left(\frac{R-1}{C}\right)$	0-Benzene 1-Graphite	Reflects the degree of ring condensation
	Ring index for structural unit	$2\left(\frac{R}{C}\right)_u$	0-Alphatic alkane 1-Graphite	The average ring number of each carbon atom in a structure unit
	Aromatic ring compactness	$4\left(\frac{R_a+\frac{1}{2}}{C}\right)-1$	0-cata type condensed aromatics 1-peri type condensed aromatics	Reflects of how many aromatic rings are formed for a certain number of aromatic carbons
Molecular level	Size of aromatic unit	C_{au}	–	Aromatic carbons in structure unit
	Intensity of polymerization	b	0-monomer ≤1-liner polymer >1-cross-linked polymer	The number of bridge bonds in the average structure unit across the coal matrix
	Degree of polymerization	p	–	The number of structural units in the coal matrix

$$\mathrm{MF} = C\varphi_C + HC\varphi_H + O\varphi_O + \cdots + \sum X_i\varphi_{X_i},$$

where

MF	is the additive molar function;
C, H, and O	are the numbers of C, H, O atoms;
X_i	is the contribution factor i in an average structural unit;
φ_C φ_H, and φ_O	are the contributions from C, H, and O atoms to the additive molar function;
φ_{X_i}	is the contributions from contribution factor i.

The structural parameters of an average structural unit can be developed using the MF function, even without a coal molecular weight. For example, coal true density, d, is used as the additive molar function to calculate the ring number (R) and fraction of aromatic carbons (f_a). van Krevelen obtained the following equations, with the adoption of some reasonable assumptions:

$$\frac{M_C}{d} = 9.9 + 3.1\frac{H}{C} + 3.75\frac{O}{C} - \left(9.1 - 3.65\frac{H}{C}\right) \cdot \frac{R}{C},$$

$$M_C = \frac{M}{C} = \frac{1200}{C_{daf}},$$

where M_C is the single-carbon molecular-weight, describing the molecular weight reflected by every single carbon atom; d is the true density (g cm^{-3}). van Krevelen et al. [38] verified the reliability of the above equations using 18 polymers with known structural parameters. The calculated densities were very close to their actual densities. In practical applications, the atom ratio, carbon content, and true density are determined by laboratory tests.

The correlations between f_a and the degree of coalification obtained using four methods are summarized. The results from statistical constitution analysis are slightly higher than those from other methods, but follow the same trend. The f_a increases with increasing degree of coalification. When the carbon content is higher than 95 %, f_a is almost equal to 1, which suggests that only anthracite is highly aromatized [39].

Different macerals of the same coal have different structural parameters. The fractions of aromatic carbons and the ring condensation index increase with increasing degree of coalification for almost all macerals. Exinite has a much lower aromaticity than vitrinite, and its fraction of aromatic carbon changes significantly in response to changes in the coalification degree. Fusinite is close to the highest stage of coalification [40]. Coal with a high degree of reduction usually has a high content of hydrogen atoms, making the H/C ratio an important indicator of the degree of reduction in this type of coal.

References

1. van Krevelen DW et al (1954) Fuel 1954(33):79
2. Zhu PZ, Gao JS (1984) Coal chemistry. Shanghai Science and Technology Press, Shanghai, p 74
3. Yu JS (2000) Coal chemistry. Metallurgical Industry Press, Beijing, p 108
4. He XM (2010) Coal chemistry. Metallurgical Industry Press, Beijing, p 79
5. van Krevelen DW (1981) Coal. Elsevier Scientific Publishing Company, Amsterdam, p 351
6. Wilfrid F (1954) Coal. Edward Arnold LTD, London, p 678
7. Gan H et al (1972) Fuel 51:272
8. Hirsch PB (1954) Proc R Soc Ser A 226:143
9. Haenel MW (1992) Fuel 71(11):1211
10. Brown JK (1959) Fuel 38(1):55–63
11. Brown JK (1955) J Chem Soc 744–752
12. Solomon PR (1979) ACS Div Fuel Chem Prepr 24:154–159

13. Friedel RA (1966) In: Kendall DN (ed) Applied infrared spectroscopy. Wiley, New York, p 55
14. Eolfson RM (1957) Cat J Chem 35:926
15. Siskin M et al (1983) Fuel 62:1321
16. Heredy LA, Neuworth MB (1962) Fuel 41(3):221–231
17. Newman PC (1955) Nature 175:645
18. Brown JK et al (1960) Fuel 39:79
19. He XM (2010) Coal chemistry. Metallurgical Industry Press, Beijing, p 153
20. van Krevelen DW (1981) Coal. Elsevier Scientific Publishing Company, Amsterdam, p 338
21. Chen P (2001) Nature, classification and utilization of Chinese coal. Chemical Industry Press, Beijing
22. Reich MH et al (1993) J Colloid Interf Sci 155:146
23. Bale HD et al (1984) Phys Rev Lett 53:569
24. Johnston PR et al (1993) J Colloid Interf Sci 155:146
25. Ren YZ (1995) Previews of heat and mass transfer 16(3):366
26. Anderson RB (1946) J Am Chem Soc 68:686
27. Walker PL et al (1970) Fuel 49:91
28. Qiu JS et al (1991) J Fuel Chem Technol 19(3):253
29. Dietz VR et al (1964) Carbon 1:245
30. Ghetti P et al (1985) Fuel 64:950
31. Fairbridge C et al (1986) Fuel 65(12):1759
32. Xu LJ et al (1995) Coal Convers 15(1):33
33. Friesen WI et al (1993) J Colloid Interf Sci 160:226
34. Nandi SP et al (1971) Fuel 50:345
35. Waterman HI (1935) J Inst Petrol Technol 21:661
36. van Krevelen DW et al (1959) Brennstoff Chem 40:155
37. van Krevelen DW (1981) Coal. Elsevier Scientific Publishing Company, Amsterdam, p 309
38. van Krevelen DW et al (1952) Brennstoff Chem 33:260
39. Yu JS (2000) Coal chemistry. Metallurgical Industry Press, Beijing, p 161
40. van Krevelen DW (1981) Coal. Elsevier Scientific Publishing Company, Amsterdam, p 342

Chapter 3
Chemical Characteristics of Coal

Abstract The physical properties of coal are closely linked to the chemical constituents and make up. These features are both strongly influenced by the geochemical history of the coal, and are important factors for determining how coal behaves under different types of reaction conditions. This chapter focuses on the study of coal from the perspective of understanding the chemical components that make up coal. The carbon-based materials and other elements that constitute various coal types are described. The operation of physical and chemical techniques used to determine chemical structure are also introduced, including infrared and nuclear magnetic resonance spectroscopies, and elemental analysis and solvent extraction methods. The analysis of the results of methods and their interpretation with respect to the coal composition are discussed. This background is supplemented with real examples of the application of these methods to characterize various coal types. This chapter is intended to give the reader an overview of the origins of the compositional and chemical analysis techniques presented in later chapters.

3.1 Chemical Properties

3.1.1 Moisture in Coal

Water is a key part of many coal types with various degrees and the moisture content of coal can be divided into three categories: surface moisture, inherent moisture, and constitutional water.

3.1.1.1 Surface Moisture

The surface moisture (also known as free moisture) of coal is the moisture attached to the particle surfaces and large capillary pores (of diameter greater than 10^{-5} cm)

K.-C. Xie, *Structure and Reactivity of Coal*, DOI 10.1007/978-3-662-47337-5_3

in the mining, transport, storage, and washing of coal; it is represented by the symbol M_f (%). Surface moisture combines mechanically with coal, depending on the external conditions, but not on the coal quality itself. The vapor pressure is equivalent to the vapor pressure of normal water, which is easily evaporated. When the coal is placed in the air at room temperature, the surface moisture evaporates until it reaches a balance with the relative humidity of the air. In this case, the moisture loss is surface moisture. Coal containing surface moisture is called as-received coal, and the coal after surface moisture loss is called air-dried coal.

3.1.1.2 Inherent Moisture

The inherent moisture (also known as the moisture in air-dried coal) of coal is the moisture attached to or condensed on the capillaries or pores (of diameter less than 10^{-5} cm) on the internal surfaces of coal particles; it is expressed as M_{inh} or M_{ad} (%). The inherent moisture combines physicochemically with coal, and is related to the essential coal characteristics (Table 3.1). The greater the internal surface area is and the more smaller capillary pores there are, the higher the inherent moisture will be. The vapor pressure of inherent water is less than that of normal water; it is difficult to evaporate and can only can be evaporated by heating to 105–110 °C. Coal that has lost its inherent moisture is called dry coal. The water lost when a coal sample is dried in air at 105–110 °C is the inherent moisture. The inherent moisture of coal also depends on the external conditions. The inherent moisture reaches a maximum at a certain humidity and temperature. At this point, the inherent moisture is called the moisture-holding capacity (MHC).

The sum of the surface and inherent moistures of coal is called the total moisture, M_t, of the coal.

The constitutional water in coal is the moisture remaining after the total moisture determination, which is chemically combined with minerals; it is also called the water of crystallization or combined water. It combines chemically with inorganic matter. The constitutional water content is small, and is lost at high temperatures, for example, gypsum ($CaSO_4 \cdot 2H_2O$) decomposes and loses water of crystallization at

Table 3.1 Inherent moisture contents (%) of different types of coal

Coal type		M_{inh}
Peat		12–45
Lignite		5–25.4
Bituminous coal	Long flame coal	0.9–8.7
	Gas coal	0.6–4.9
	Fat coal	0.5–3.2
	Coking coal	0.4–2.6
	Lean coal	0.3–1.6
	Meager coal	*a.c.* 0.6
Anthracite		0.1–4.0

163 °C, but kaolin ($Al_2O_3 \cdot 2SiO_2 \cdot 2H_2O$) loses constitutional water at 450–600 °C. The constitutional water is not generally considered in the proximate analysis of coal.

3.1.2 Mineral Matter and Ash in Coal

Mineral matter (MM, %) is a generic term for all the inorganic substances, excluding moisture, in coal. The main components are generally clay, kaolinite, pyrite, and calcite. Coal ash refers to the residue of minerals in coal after a series of decompositions and chemical combinations at certain temperatures, during which all the combustible substances in coal are completely combusted, and is represented by the symbol A (%). Ash consists of the various products of coal from predetermined operations, and is composed of oxides and other salts. It is not inherent in coal and cannot be regarded as the mineral content; it is more precisely called the ash yield. However, it is clear that coal ash is closely related to the minerals in coal.

Traditional methods for ash determination are divided into slow ashing methods and rapid ashing methods. The specific steps and procedures involved in the two methods are slightly different, but the main steps are to weigh a certain amount of air-dried coal sample, place it in a muffle furnace, heat to 815 ± 10 °C, and burn until a constant weight is reached. The percentage of residue mass, based on the coal sample mass, is taken as the ash yield A_{ad} (%).

When coal is used as a fuel or the raw material for processing and transformation, in almost every application, it is the organic matter in the coal that is used. Coal minerals or ash are therefore considered to be hazardous wastes, and attempts are made to reduce their quantities or remove them. The methods for removing the mineral in coal are divided into two major categories: physical washing methods and chemical purification methods. Physical washing methods involve separation of the coal from the gangue, based on the differences in density or surface properties, and include hydraulic jigging, dense medium separation, froth flotation, and magnetic separation. Chemical purification methods remove the minerals mainly by using the differences between the chemical natures of the organic matter and minerals in coal; methods include hydrofluoric acid and hydrochloric acid treatment, and alkaline solvent extraction methods. However, it has been found that the minerals in coal have some beneficial effects in certain processes, and even coal ash has become more and more widely used.

3.1.2.1 Sources of Minerals in Coal

There are three general sources of minerals in coal: primary minerals, secondary minerals, and foreign minerals.

Primary minerals refer to the minerals present in the coal-forming plants, mainly alkali metal and alkaline-earth metal salts. Primary minerals form part of the molecular structure of coal and are closely combined with the organic matter. They are finely dispersed in the coal, and are difficult to wash and remove using mechanical methods. Such mineral contents are small, usually only 1–2 %. Secondary minerals are the minerals mixed into the coal seam from the external environment during the coal-forming process; they are present in coal in various forms, for example, kaolin, calcite, pyrite, orpiment, feldspar, and mica. Foreign minerals refer to the gangues mixed with coal in the top and bottom plates, and the tonstein layer in the coal-mining process; the main constituents are SiO_2, Al_2O_3, $CaCO_3$, $CaSO_4$, and FeS_2.

3.1.2.2 Calculation and Determination of Coal Mineral Contents

The mineral content of coal is different from the ash content, but there is a relationship between the two. The mineral content of coal can be calculated from the ash content using the following empirical equations:

$$\begin{aligned} \mathrm{MM} &= 1.08A + 0.55S_{\mathrm{t}}, \\ \mathrm{MM} &= 1.10A + 0.5S_{\mathrm{p}}, \\ \mathrm{MM} &= 1.13\mathrm{A} + 0.47S_{\mathrm{p}} + 0.5\mathrm{Cl}; \end{aligned}$$

where

MM is the mineral content (%);
A is the ash yield (%);
S_{t} is the total sulfur content (%);
S_{p} is sulfur content from iron sulfide (%);
Cl is the chlorine content (%).

The mineral content can also be directly determined. The coal sample is treated with hydrochloric acid and hydrofluoric acid, to partly remove minerals. Under these conditions, the organic matter in the coal is not affected. The coal mass loss after the acid treatment is calculated, and the partly demineralized coal is burned to determine the undissolved part of the minerals.

There is also a plasma low-temperature ashing method for determining the mineral content. The principle of the method is that an active gas plasma is formed by passing oxygen gas through a radio frequency and discharging; the plasma then flows through the coal sample at about 150 °C. At this time, the organic matter in coal is lost by oxidation, but the minerals are basically unchanged, except for the loss of water of crystallization. This method can also be used to correct various results for coal, and convert the data to a dry mineral-free basis (dmmf).

3.1.2.3 Composition of Ash in Coal

During high-temperature coal combustion, most minerals undergo a variety of chemical reactions, and are converted to a mixture of ash and untransformed minerals.

Around 30–36 % of the chlorine in coal exists in the form of organic chlorine, and is easily decomposed in high-temperature ashing to generate HCl or Cl_2. The uncombined sulfur in the coal will be lost in the form of SO_2, and the alkali metal oxides and mercury in the coal will partly volatilize above 700 °C.

In accordance with the chemical reactions of minerals that occur in the high-temperature combustion of coal, the ash of coal consists mainly of metallic and non-metallic oxides and salts. In industrial production, coal ash refers to the large quantity of coal slag obtained when coal is used as a boiler fuel and gasification feedstock. The chemical compositions of coal ash and ash from coal are similar, and the main ingredients are SiO_2, Al_2O_3, CaO, and MgO. The sum of the ingredients accounts for more than 95 % of the coal ash, along with small amounts of K_2O, Na_2O, SO_3, P_2O_5, and some microelement compounds. The main components of Chinese coal ash are listed in Table 3.2 [1].

3.1.3 Volatile Matter and Fixed Carbon in Coal

3.1.3.1 Volatile Matter

The yield of volatile organic substances after heating coal isolated from the air under specified conditions is referred to as the volatile matter (*V*). The volatiles generated from coal under such conditions include the gaseous products from pyrolysis of organic matter in coal, water vapor generated from the moisture in coal [2], and CO_2 from the decomposition of carbonate minerals. The volatile matter is therefore part of the coal volatiles, but is not equivalent to the volatiles. In addition,

Table 3.2 Contents (%) of main components of Chinese coal ash

Coal ash composition	Lignite		Hard coal	
	Min.	Max.	Min.	Max.
SiO_2	10	60	15	>80
Al_2O_3	5	35	5	50
Fe_2O_3	4	25	1	65
CaO	5	40	0.5	35
MgO	0.1	3	<0.1	5
TiO_2	0.2	4	0.1	6
SO_3	0.6	35	<0.1	15
P_2O_5	0.04	2.5	0.01	5
KNaO	0.09	10	<0.1	10

volatile matter is not an inherent substance in coal, but a product of coal pyrolysis under specific heating conditions. It is therefore more precise to call the volatile matter of coal the volatile matter yield.

A certain amount of air-dried coal sample is weighed and then heated in isolation from the air at 900 ± 10 °C for 7 min. The moisture content (M_{ad}) of the coal sample is subtracted from the percentage of reduced mass, based on the coal sample mass, to obtain the volatile matter yield (V_{ad}). In the determination of volatile matter, the solids left in the crucible are referred to as the char residue.

In the determination of the volatile yield, not only the organic matter in the coal decomposes, but also the minerals can be decomposed. Under normal circumstances, there are few products from mineral decomposition, and they have little effect; however, for coal with a high carbonate content, the CO_2 generated by decomposition must be corrected for. Hydrochloric acid can be used to treat the coal sample to release the CO_2 before determining the volatile matter; for example, when the CO_2 content in the form of carbonate in the coal is greater than or equivalent to 2 %, the CO_2 content is deduced as follows.

$$V_{ad} = V'_{ad} - (CO_2)_{ad}, \text{ when the content of } CO_2 \text{ is } 2\text{–}12\,\%$$

$$V_{ad} = V'_{ad} - [(CO_2)_{ad} - (CO_2)'_{ad}], \text{ when the content of } CO_2 \text{ is greater than } 12\,\%,$$

where

V'_{ad} is the volatile matter yield of an uncorrected air-dried coal sample (%);
$(CO_2)_{ad}$ is the CO_2 content of carbonate in the coal sample (%);
$(CO_2)'_{ad}$ is the percentage CO_2 content of carbonate in the char residue, converted to that in the total coal sample (%).

3.1.3.2 Fixed Carbon

The residue remaining after subtracting the ash from the char residue for determination of the volatile matter in a coal sample is referred to as the fixed carbon (FC, %).

$$FC_{ad} = 100 - (M_{ad} + A_{ad} + V_{ad}),$$

where

FC_{ad} is the fixed carbon content of the air-dried coal sample (%);
M_{ad} is the moisture content of the air-dried coal sample (%);
A_{ad} is the ash yield of the air-dried coal sample (%);
V_{ad} is the volatile matter yield of the air-dried coal sample (%).

The fixed carbon is actually the solid pyrolysis products from the organic matter in coal under a certain heating system, and is part of the char residue. In terms of element composition, the fixed carbon contains not only carbon, but also hydrogen, oxygen, nitrogen, and other elements. The fixed carbon content and the carbon content of the organic matter in coal are therefore two different concepts. Generally, the fixed carbon content of coal is less than the carbon content of the organic matter in the coal; the two values are only similar for high-coalification coals.

3.1.3.3 Volatile Matter/Fixed Carbon and Coal Rank

The volatile matter yield is closely related to the coal rank. China and many other countries in the world take the volatile matter yield as the first coal classification index for determining the coal rank. The suitability of various coal processes can be preliminarily evaluated based on the volatile matter yield and char residue characteristics. Use of the volatile matter yield in conjunction with other indicators can predict and estimate the yields of the main products during coal carbonization, and can also give the calorific value during coal combustion.

The ratio of fixed carbon to the volatile matter of the coal is referred to as the fuel ratio, expressed as FC_{daf}/V_{daf}. The fuel ratios of various coals are roughly as follows: lignite, 0.6–1.5; long-flame coal, 1.0–1.7; gas coal, 1.0–2.3; coking coal, 2.0–4.6; lean coal, 4.0–6.2; meager coal, 4–9; and anthracitic coal, 9–29. The fuel ratio can be used to evaluate the combustion characteristics of coal.

3.1.4 Elemental Composition

3.1.4.1 Elemental Composition of Coal

The organic matter in coal consists mainly of carbon, hydrogen, oxygen, nitrogen, sulfur, and some other elements. Carbon, hydrogen, and oxygen account for more than 95 % of the organic matter in coal. The elemental composition of the organic matter is related to the origin and type of coal, coal petrographic composition, and degree of coalification (Table 3.3; [3]). In coal quality analysis and research, it is therefore important to understand the elemental composition of the organic matter in coal through ultimate analysis. Elemental analysis cannot identify the compounds in coal organic matter, nor can it fully determine the nature of the coal. However, a combination of elemental analytical data and data obtained using other techniques can help us to understand some of the properties of coal. Elemental analysis can be used to calculate the calorific value, theoretical combustion temperature, and composition of the combustion products of coal, and to estimate the yields of chemical products from coking; it can also to be used as an auxiliary indicator for coal classification.

Table 3.3 Main elemental compositions of various coals (%)

Coal type	C_{daf}	H_{daf}	O_{daf}
Peat	55.0–62.0	5.3–6.5	27.0–34.0
Young lignite	60.0–70.0	5.5–6.6	20.0–23.0
Old lignite	70.0–76.5	4.5–6.0	15.0–30.0
Long flame coal	77.0–81.0	4.5–6.0	10.0–15.0
Gas coal	79.0–85.0	5.4–6.8	8.0–12.0
Fat coal	82.0–89.0	4.8–6.0	4.0–9.0
Coking coal	86.5–91.0	4.5–5.5	3.5–6.5
Lean coal	88.0–92.5	4.3–5.0	3.0–5.0
Meager coal	88.0–92.7	4.0–4.7	2.0–5.0
Young anthracitic	89.0–93.0	3.2–4.0	2.0–4.0
Typical anthracitic	93.0–95.0	2.0–3.2	2.0–3.0
Old anthracitic	95.0–98.0	0.8–2.0	1.0–2.0

3.1.4.2 Carbon

Carbon is the main element in coal organic matter. The skeletons of the polycyclic aromatic hydrocarbons in coal structural units consist of carbon. Carbon is the main material involved in coke formation. In coal combustion, it is the main source of the calorific value.

The carbon content increases regularly with increasing coalification (Table 3.3). In the same coal, the carbon contents of various macerals vary. Generally, the C_{daf} (daf: dry ash-free basis) of inertinite is the highest, followed by that of vitrinite, and that of exinite is the lowest. There is a negative correlation between the carbon content and volatile matter, so the carbon content can be used as a classification index for the characterization of coalification. In some cases, the carbon content is more accurate than the volatile yield for determining coalification.

3.1.4.3 Hydrogen

Hydrogen is the second most important element in coal. Generally, hydrogen accounts for less than 7 % of organic matter in humic coal by mass, however, because it has the lowest atomic weight, the atomic percentage is similar to that of carbon. Hydrogen is an important element in the macromolecular skeleton and side chains of coal. Hydrogen is more reactive than carbon, and its heat of combustion per unit mass is greater than that of carbon.

Hydrogen content of coal is also closely related to coalification. The hydrogen content gradually decreases with increasing coalification (Table 3.3). After the mesometamorphic bituminous coal stage, this trend becomes more obvious. In the gas coal and gas fat coal phases, the hydrogen content can be as high as 6.5 %; in the high metamorphic bituminous coal phase, the hydrogen content can decrease to

1 % or less. The hydrogen contents of various macerals vary considerably. For coals of the same coalification degree, the H_{daf} of exinite is the highest, followed by that of vitrinite, and that of inertinite is the lowest.

From mesometamorphic bituminous coal to anthracitic coal, there is a good correlation between the hydrogen content and the carbon content. The empirical equations can be obtained from linear regression:

$$H_{\mathrm{daf}} = 26.10 - 0.241C_{\mathrm{daf}} \text{ for mesometamorphic bituminous coal}$$

$$H_{\mathrm{daf}} = 44.73 - 0.448C_{\mathrm{daf}} \text{ for anthracite}$$

3.1.4.4 Oxygen

Oxygen is mainly present in the form of carboxyl (–COOH), hydroxyl (–OH), carbonyl (>C=O), methoxy (–OCH_3), and ether (–C–O–C–) groups. However, some oxygen is incorporated into carbon skeletons, in the form of heterocyclic rings. The total amount and form of oxygen present in coal directly affects the behavior of the coal. The oxygen content decreases significantly with increasing coalification (Table 3.3). The oxygen content on an ash-free basis, O_{daf}, is 27.0–34.0 % in peat, 2.0–15.0 % in the bituminous coal stage, and 1.0–3.0 % in the anthracitic coal stage. In the study of coalification evolution, the O/C and H/C atomic ratios are usually used to describe variations in the coal elemental composition, and the decarboxylation, dehydration, and demethylation reactions of coal [4].

Oxygen is highly reactive and plays an important role in the processing and use of coal. During liquefaction of low-coalification coal, a large amount of hydrogen is consumed because of the high oxygen content, leading to generation of water. In the coking process, where oxidation increases the oxygen content of the coal, the cohesiveness of coal is reduced, or even disappears. The oxygen is not involved in coal combustion, but constrains the combustible elements such as carbon and hydrogen. Coal with high oxygen content is the preferred raw material for producing aromatic carboxylic acids and humic substances.

The oxygen content of various macerals are related to coalification. For bituminous coal of medium-metamorphic grade, the O_{daf} of vitrinite is highest, followed by that of inertinite, and that of exinite is lowest. For high-metamorphic bituminous coal and anthracitic coal, the O_{daf} of vitrinite is still highest, but that of exinite is slightly higher than that of inertinite.

Similar to hydrogen, there is a correlation between the oxygen and carbon content of coal (but for anthracitic coal, the negative correlation between oxygen and carbon is not obvious):

$$O_{\mathrm{daf}} = 85.0 - 0.9C_{\mathrm{daf}} \text{ for bituminous coal}$$

$$O_{\mathrm{daf}} = 80.38 - 0.84C_{\mathrm{daf}} \text{ for lignite and long flame coal}$$

3.1.4.5 Nitrogen

The nitrogen content of coal is often low, usually about 0.5–3.0 %. The organic nitrides in coal are present as relatively stable heterocyclic rings and complex non-cyclic structures [5]. The main source of nitrogen in coal is protein, and other constituents of the precursor plants and animals. Alkaloids and chlorophyll in plants, and the ring structures in other tissues, contain nitrogen. Such compounds are fairly stable and do not change during coalification, but are converted to nitrides and retained in the coal. Nitrogen present as proteins is found only in peat and lignite, with only very small amounts found in bituminous coal. The nitrogen content of coal tends to decrease with increasing coalification; a regular trend becomes obvious after the high-metamorphic bituminous coal stage. The correlations between macerals and nitrogen contents are irregular. However, according to a study by the author [6], nitrogen is present in the form of pyrroles and pyridines in vitrinite, amino groups and pyridines in exinite, and amino groups and pyrroles in inertinite, respectively.

During coal transformation, the nitrogen in the coal can generate amines, nitrogen-containing heterocycles, nitrogen-containing polycyclic compounds, and cyanides. In the combustion and gasification of coal, nitrogen is converted into the environmental pollutant NO_x. In coal liquefaction, hydrogen is used to minimize the nitrogen content in the product. In coal coking, some of the nitrogen escapes as nitrogen gas, ammonia, HCN, and other organic nitrides, and the rest of the nitrogen enters the coal tar or remains in the coke. In coking chemical products, the ammonia yield is related to the nitrogen content and its form in the coal. Nitrogen-containing compounds in coal tar include pyridines and quinolines, whereas in coke they have complex structures in coke.

For most Chinese coals, there is a relationship between the nitrogen and hydrogen content:

$$N_{\text{daf}} = 0.3H_{\text{daf}}.$$

The difference between the calculated and measured nitrogen content is generally less than ±0.3 %.

3.1.4.6 Sulfur

Sulfur exists in coal as organic sulfur and inorganic sulfur; the main forms are shown in Table 3.4. Organic sulfur refers to sulfur combined with the organic structures of the coal, and the structure and composition are complex. Organic sulfur is mainly derived from proteins in coal plants and microorganisms. The total sulfur content of plants is generally less than 0.5 %. The sulfur in most coals with sulfur contents of less than 0.5 % is generally organic sulfur. Organic sulfur and the organic matter in coal are paragenetic, evenly distributed, and therefore difficult to separate [7]. According to our studies, organic sulfur in the three basic types of

Table 3.4 Occurrence and classification of sulfur in coal

Type		Name		Chemical formula	Distribution conditions
Inorganic (S_I)	Non-combustible	Sulfate (S_S)	Gypsum ferrous sulfate	$CaSO_4 \cdot 2H_2O$ $FeSO_4 \cdot 7H_2O$	Unevenly distributed in coal
	Combustible	Elemental (S_E)			
		Sulfide (S_P)	Pyrite Marcasite Magnetite Galena	FeS_2, tetragonal FeS_2, orthorhombic Fe_7S_8 PbS	
Organic (S_O)		Mercaptan		R–SH	Evenly distributed in coal
		Thioether	Thioether Disulfide Bithioether	R_1–S–R_2 R_1–S–S–R_2 R_1–S–CH_2–S–R_2	
		Heterocyclic sulfide	Thiophene Quinone sulfide		
		Others	Thioketone		

organic macerals is present mainly as thiophenes, mercaptans, and thioethers. Inorganic sulfur in coal comes from various sulfur-containing compounds in minerals, mainly sulfides, small amounts of sulfates, and occasionally elemental sulfur. Sulfide is mainly present as pyrite, which is widespread in coal. High-sulfur coal contains a large proportion of sulfide. Sulfate is mainly present as gypsum, and a small amount of ferrous sulfate. The sulfate content of Chinese coal is generally less than 0.1 %.

The sulfur in coal can be divided into combustible and non-combustible sulfur, or into volatile and fixed sulfur, based on volatility in the retorting process. The forms of sulfur in coal are listed in Table 3.4. The sum of the various forms of sulfur in the coal is called the total sulfur, and varies from 0.1 to 10 %. The sulfur content is related to the depositional environment in which the coal was formed. In China, the sulfur content of coal is lower in northern regions, and gradually increases towards the south.

Sulfur in coal is harmful during coking, gasification, combustion, storage, and transport. The sulfur content is therefore an important indicator of coal quality. In coal coking, about 60 % of the sulfur enters the coke, and the presence of sulfur causes temper embrittlement of cast iron. In coal gasification, the hydrogen sulfide and sulfur dioxide produced not only corrodes equipment, but also poisons catalysts, affecting the operation and product quality. In coal combustion, sulfur is converted to sulfur dioxide and emitted into the atmosphere, corroding metal equipment and facilities, polluting the environment. Coals with high pyrite sulfur contents oxidize easily and undergo spontaneous combustion, increasing the amount of coal ash and lowering the calorific value. High-sulfur coal accounts for a certain proportion of global reserves, therefore it is important to develop efficient and economical desulfurization methods, and methods for recycling and using sulfur.

3.1.5 *Calorific Value*

The calorific value of coal is the heat released when a unit mass of coal is completely combusted, and it is represented by the symbol Q. The calorific value of coal is an important indicator for coal quality and in thermal calculations. In the combustion or conversion of coal, the calorific value is commonly used to calculate the thermal equilibrium, coal consumption, and thermal efficiency. For coal used for power generation, the calorific value is the main basis for determining the price.

The calorific value of coal is generally determined using oxygen bomb calorimetry. In the measurement, a certain amount of coal is loaded into a stainless-steel pressure-resistant bomb-type container, and the container is pressurized to 2.6–2.8 MPa with an oxygen tank, to provide sufficient oxygen for the complete combustion of the coal sample in the container. A metal wire is used to heat the cylinder with a current, to ignite the coal sample; the test sample is completely combusted in the excess oxygen under pressure, producing CO_2 and water. The heat generated by combustion is absorbed by a certain amount of water that surrounds the oxygen bomb. After conducting a series of temperature corrections in accordance with the rise in water temperature, the heat generated by the combustion of a unit mass of coal can be calculated, and is denoted by $Q_{b,v,ad}$.

3.1.5.1 Gross Calorific Value and Net Calorific Value of Coal

There are different conditions for the combustion reaction (mainly differences in the constant pressure and constant volume), so the status of the products of combustion are designated differently (mainly liquid and vapor). There are several different definitions of the calorific value of coal, depending on different purposes.

The constant-volume gross calorific value of coal is the heat released by complete combustion of a unit mass of coal, measured under constant volume in the cylinder, and assuming that all the vaporous water produced in the combustion is condensed to liquid water at the same temperature.

If the coal sample is combusted in an open system, the nitrogen and sulfur in the coal will escape in the form of free nitrogen and SO_2. The combustion of coal in the cylinder is performed at high temperature and under high pressure, so the nitrogen in the test sample and the air in the cylinder will generate nitrogen oxides, which dissolve in water to produce dilute nitric acid. Similarly, the sulfur in the coal will generate dilute sulfuric acid. Both the generation of dilute nitric and sulfuric acids and dissolution in the water previously added to the cylinder are exothermic reactions. Subtracting the formation and dissolution heats of nitric acid and sulfuric acid from the calorific value of the cylinder gives the constant-volume gross calorific value of the coal. The calculation equation is

$$Q_{gr,v,ad} = Q_{b,v,ad} - (94.1S_{b,ad} + \alpha Q_{b,v,ad}),$$

where

$Q_{grv,ad}$ is the constant-volume gross calorific value of the coal on an air-dried basis (J g^{-1});

$Q_{b,ad}$ is the bomb calorific value of the coal on an air-dried basis (J g^{-1});

$S_{b,ad}$ is the sulfur content of the coal on an air-dried basis, measured from the lotion by washing the bomb cylinder (%);

94.1 is the correction factor for every 1 % of sulfur in the coal (J);

α is the correction factor for formation heat of nitric acid: 0.0010 for anthracitic coal, and 0.0015 for other coals.

The constant-volume net calorific value of coal refers to the heat released by complete combustion of a unit mass of coal measured under constant volume in a cylinder, assuming that all the water produced in the combustion exists as vaporous water at the same temperature. The definition of the constant-volume low calorific value assumes that the water is discharged with the combustion exhaust gas in a vaporous state in the conventional combustion of coal. Its value can be obtained by subtracting the vaporization heat of water from the gross calorific value. In industry, the net calorific value on an as-received basis is mainly used:

$$Q_{net,v,ar} = (Q_{gr,v,ad} - 206H_{ad}) - \frac{100 - M_{t,ar}}{100 - M_{ad}} - 23M_{t,ar},$$

where

$Q_{net,v,ar}$ is the constant-volume low calorific value of the coal on an as-received basis (J g^{-1});

$Q_{gr,v,ad}$ is the constant-volume high calorific value of the coal on an air-dried basis (J g^{-1});

H_{ad} is the hydrogen content of the coal on an air-dried basis (%);

$M_{t,ar}$ is the total moisture content of the coal on an as-received basis (%);

M_{ad} is the moisture content of the coal on an air-dried basis (%).

The constant-pressure net calorific value of coal refers to the heat released by complete combustion of a unit mass of coal measured at constant pressure, and assuming that all the water produced in the combustion exists as vaporous water at the same temperature. The definition of the constant-pressure net calorific value assumes that the coal is under a constant pressure but not a constant volume in an actual combustion. Its value is related to the expansion work of the gas produced, which can be obtained by the conversion of gross calorific value:

$$Q_{\mathrm{net,p,ar}} = \left(Q_{\mathrm{gr,v,ad}} - 212H_{\mathrm{ad}} - 0.80O_{\mathrm{ad}}\right)\frac{100 - M_{\mathrm{t,ar}}}{100 - M_{\mathrm{ad}}} - 24.5M_{\mathrm{t,ar}},$$

where

$Q_{\mathrm{net,p,ar}}$ is the constant-pressure net low calorific value of the coal on an as-received basis (J g^{-1});

$Q_{\mathrm{gr,v,ad}}$ is the constant-volume high gross calorific value of coal on an air dried basis (J g^{-1});

H_{ad} is the hydrogen content of the coal on an air-dried basis (%);

O_{ad} is the oxygen content of the coal on an air-dried basis (%);

$M_{\mathrm{t,ar}}$ is the total moisture content of the coal on an as-received basis (%);

M_{ad} is the moisture content of the coal on an air-dried basis (%).

3.1.5.2 Estimation of Calorific Value

The oxygen bomb method used in the direct determination of the calorific value of coal is accurate and objective. However, like other thermodynamic determination procedures, it is cumbersome and the conditions are harsh. Much research has therefore been done on using proximate or ultimate analytical data for approximate calculations of calorific values, and various empirical equations for calculating the calorific value of coal have been proposed. The errors between such calculated results and the measured values are small. Example of these empirical equations are given below [8].

The empirical equation for calculating the calorific value of bituminous coal using proximate analytical data is

$$Q_{\mathrm{net,v,ad}} = \left[100K - (K+6)\times(M_{\mathrm{ad}} + A_{\mathrm{ad}}) - 3V_{\mathrm{ad}} - 40M_{\mathrm{ad}}\right]\times 4.1868,$$

where K is a constant between 72.5 and 85.5, which can be obtained from the V_{daf} and tables of char residue characteristics. In addition, when V_{daf} is less than 35 % and M_{ad} is greater than 3 %, $40M_{\mathrm{ad}}$ needs to be subtracted.

The empirical equation for calculating $Q_{\mathrm{gr,ad}}$ of anthracitic coal using proximate analytical data is

$$\mathrm{Q}_{\mathrm{gr,ad}} = K_0 - 80M_{\mathrm{ad}} - 90A_{\mathrm{ad}},$$

where K_0 is determined from V_{daf}.

The empirical equation for calculating the calorific values of lignite, bituminous coal, and anthracitic coal using ultimate analytical data is

$$Q_{\mathrm{net,v,ar}} = [80(\text{or } 78.1)C_{\mathrm{daf}} + 310(\text{or } 300)H_{\mathrm{daf}} + 15S_{\mathrm{daf}} - 25O_{\mathrm{daf}} - 5(A_{\mathrm{d}} - 10)] \times 4.1868,$$

For coals with C_{daf} greater than 95 % and/or H_{daf} less than 1.5 %, 78.1 is used for the C_{daf} coefficient; 80 is used for other coals. For coals with C_{daf} less than 77 %, 300 is used as the H_{daf} coefficient; 310 is used for other coals. The correction factor for ash in the last term in the equation is only used for coals with A_d greater than 10 %; in all other cases, a correction for ash is not necessary.

3.1.5.3 Coal Calorific Value and Coal Rank

The calorific value of coal is closely related to the coal quality, and changes regularly with increasing coalification (Table 3.5). The calorific value increases with increasing coalification from lignite to coking coal, reaches a maximum in the coking stage, then gradually decreases with increasing coalification to highly metamorphic anthracitic coal. However, the changes are small. This trend is closely related to the elemental composition of the coal. From the lignite to coking coal stages, the carbon content continuously increases, and the oxygen content decreases substantially. However, the reduction in the hydrogen content is small, so the calorific value of the coal increases. From coking coal to highly metamorphic anthracite, the increase in the carbon content and decrease in the oxygen content are small, but the hydrogen content decreases significantly. The calorific value of hydrogen is 3.7 times that of carbon, so the calorific value of coal decreases slowly with increasing coalification.

The relationship between the calorific value of coal and the amount of volatile matter is parabolic. When V_{daf} is in the range 20–30 %, equivalent to the coking coal stage, the calorific value is maximum. When V_{daf} is less than 20 %, the calorific value decreases slightly with decreasing V_{daf}; when V_{daf} is greater than 30 %, the calorific value decreases significantly with increasing V_{daf}.

In humic coal, the calorific value of exinite is the highest, followed by vitrinite, and then inertinite. The calorific value of coal also decreases with increasing mineral and water contents, and degree of weathering. Generally, for every 1 %

Table 3.5 Calorific values of various coals (MJ kg^{-1})

Coal type	$Q_{gr,v,daf}$
Lignite	25.12–30.56
Long-flame coal	30.14–33.49
Gas coal	32.24–35.59
Fat coal	34.33–36.84
Coking coal	35.17–37.05
Lean coal	34.96–36.63
Meager coal	34.75–36.43
Anthracitic coal	32.24–36.22

increase in coal ash, the calorific value is reduced by about 370 J g^{-1}; for every 1 % increase in moisture in the coal, the calorific value is also reduced by about 370 J g^{-1}.

3.1.6 Basis of Coal Analysis Indicators

There are wide variations in the moisture and ash contents of coal, and if the basis adopted for analysis is not specified, the analytical results will be significantly different for the same type of coal. To enable comparison of analytical data from different sources, the basis of the actual analysis of the coal sample or of the theoretical conversion of the coal sample must be given when reporting the analytical results. So far, a standard set of terms and symbols for coal quality analysis has been compiled (Tables 3.6).

The definitions of various bases and the proximate and ultimate analytical compositions of coal obtained on various bases are described below.

As-Received Basis: When the coal is in the received state, this is referred to as the as-received basis, it is presented by subscript ar. On this basis,

$$V_{\mathrm{ar}} + \mathrm{FC}_{\mathrm{ar}} + A_{\mathrm{ar}} + M_{\mathrm{ar}} = 100,$$
$$C_{\mathrm{ar}} + H_{\mathrm{ar}} + O_{\mathrm{ar}} + N_{\mathrm{ar}} + S_{\mathrm{ar}} + A_{\mathrm{ar}} + M_{\mathrm{ar}} = 100.$$

Air-Dried Basis: When the coal is in the air-dried state, this is referred to as the air-dried basis, it is presented by subscript of ad. On this basis,

$$V_{\mathrm{ad}} + \mathrm{FC}_{\mathrm{ad}} + A_{\mathrm{ad}} + M_{\mathrm{ad}} = 100,$$
$$C_{\mathrm{ad}} + H_{\mathrm{ad}} + O_{\mathrm{ad}} + N_{\mathrm{ad}} + S_{\mathrm{ad}} + A_{\mathrm{ad}} + M_{\mathrm{ad}} = 100.$$

Table 3.6 Names and symbols of common indicators

Name	Symbol/unit		Meaning of subscript	Symbol
Ash	*A*	%	External or free	f
Apparent relative density	ARD	–	Inherent	inh
Yield of benzene extract	EB	%	Organic	o
Fixed carbon	FC	%	Pyrite	p
Moisture	*W*	%	Sulfate	s
Moisture-holding capacity	*MHC*	%	Constant volume, gross	gr,v
Mineral	MM	%	Constant pressure, net	net,p
Calorific value	*Q*	J g^{-1}	Constant volume, net	net,v
Yield of tar	Tar	%	Total	t
True relative density	TRD	–	Bomb Cylinder	b
Volatile matter	*V*	%		

Dry Basis: When the coal is in a completely water-free state, this is referred to as the dry basis, it is presented by subscript of d. On this basis,

$$V_d + FC_d + A_d = 100,$$
$$C_d + H_d + O_d + N_d + S_d + A_d = 100.$$

Dry Ash-Free Basis: When the coal is in a water-free and ash-free state, this is referred to as the dry ash-free basis, it is presented by subscript of daf. On this basis,

$$V_{daf} + FC_{daf} = 100,$$
$$C_{daf} + H_{daf} + O_{daf} + N_{daf} + S_{daf} = 100.$$

Dry Mineral-Matter Free Basis: When the coal is in a water-free and mineral matter-free state, this is referred to as the dry mineral-matter free basis. On this basis,

$$V_{dmmf} + FC_{dmmf} = 100,$$
$$C_{dmmf} + H_{dmmf} + O_{dmmf} + N_{dmmf} + S_{dmmf} = 100.$$

The relationships among the above bases are shown in Fig. 3.1. There is a conversion relationship between the analytical data obtained using different bases. From the analytical datum X_1 obtained on a known basis (known basis), the analytical datum X_2 can be obtained on another basis (required basis) through a certain conversion.

The chemical characteristics of coal are determined by the chemical properties of various components of coal, and the different chemical reactions that occur among them under certain conditions. These characteristics determine the properties of coal, and also the basis of the conversion and chemical-processing of coal.

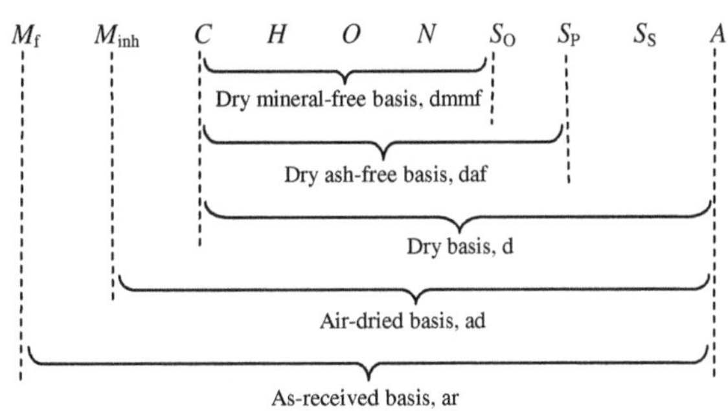

Fig. 3.1 Relationships among common bases

3.1.7 Functional Groups in Coal

In addition to alkyl side chains, the periphery of the coal structural unit contains other functional groups, primarily oxygen-containing, and small amounts of nitrogen-containing and sulfur-containing functional groups. As the oxygen content and its form significantly influence the nature of the coal, particularly for coal with a low degree of coalification, analysis of functional groups usually focuses on oxygen-containing functional groups.

3.1.7.1 Oxygen-Containing Functional Groups

The main oxygen-containing functional groups in coal are carboxyl, hydroxyl, carbonyl, and methoxy groups, and ether bonds.

Carboxyl groups are present in peat, lignite, and weathered coal, but bituminous coal contains almost no carboxyl groups (when the carbon content is greater than 78 %, there are no carboxyl groups). Hydroxyl groups are found in peat and lignite, and are the main oxygen-containing functional group in bituminous coal; it is generally believed that the hydroxyl groups exist primarily in the organic matter of coal and the vast majority of coal contains only phenolic hydroxyl groups [9], with few alcoholic hydroxyl groups. Carbonyl groups are present during the whole coalification process, from peat to anthracitic coal (in coal with a high degree of coalification, most carbonyl groups are present as benzoquinonyl groups); they are acid free, present in lower amounts than carboxyl groups, but widely distributed. Methoxy groups are only present in peat and soft lignite, and disappear faster than carboxyl groups with increasing coalification. Ether bonds are relatively stable chemically and thermally. They are also known as non-reactive oxygen.

The distribution of oxygen-containing functional groups in coal is shown in Fig. 3.2. The figure shows that the methoxy groups disappear first, and are almost

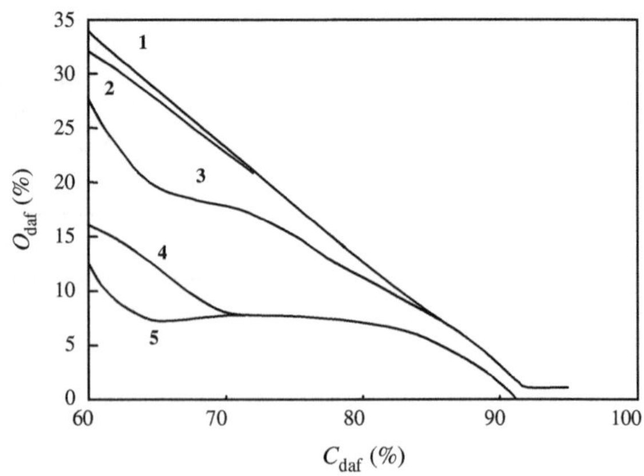

Fig. 3.2 Correlation between distribution of oxygen-containing functional groups and degree of coalification: *1* $-OCH_3$; *2* $-COOH$; *3* $>C=O$; *4* $-O-$; and *5* $-OH$

absent from old lignite; then the carboxyl groups disappear. The presence of carboxyl groups is one of the main characteristics of lignite, and they are not present in typical bituminous coal. Hydroxyl and carbonyl groups are present throughout the bituminous coal phase. Despite the depletion in numbers during coalification, they are even found in anthracite. Furthermore, some carbonyl groups, together with benzoquinonyl groups, have redox properties. The non-reactive oxygen is contained primarily in ether bonds and furan heterocyclic rings, which are also present in the entire coal-forming process. The number of oxygen-containing functional groups in coal declines significantly with increasing coalification; the amount of hydroxyl groups declines most, followed by carbonyl and carboxyl groups. During the bituminous phase, the oxygen-containing functional groups exist primarily in the form of non-reactive oxygen (ether bonds and heterocyclic oxygen). When the carbon content reaches 92 %, all the oxygen-containing groups are present as non-reactive oxygen [10].

Most functional groups in coal can be determined by chemical methods. However, with the development of modern instrumental techniques, functional groups are usually determined using instrumental analysis methods, which are more convenient and accurate. In this section, typical reactions of functional groups are introduced.

Carboxyl Group (–COOH)

Thc carboxyl group is acidic, and is usually determined by the reaction of the coal sample with calcium acetate. The acetic acid formed is then titrated with a standard aqueous alkaline solution:

$$2RCOOH + Ca(CH_3COO)_2 \xrightarrow{1-2d} (ROO)_2Ca\downarrow + 2CH_3COOH$$

The unit of the content of carboxyl groups is mmol g^{-1}; this also applies to the other functional groups described in the following sections.

Hydroxyl Group (–OH)

Hydroxyl groups are determined by the reaction of the coal sample with $Ba(OH)_2$ solution, which reacts with both carboxyl groups and phenolic hydroxyl groups. The amount of phenolic hydroxyl groups is calculated by subtracting the amount of carboxyl groups from the total number of acidic groups.

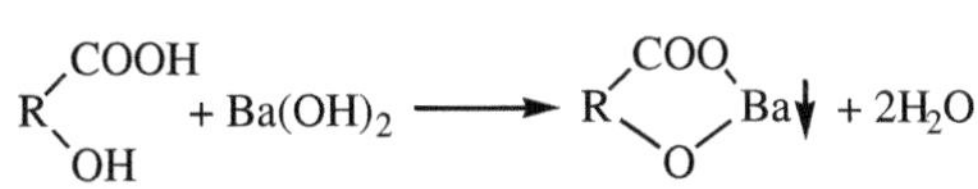

The amount of alcoholic hydroxyl groups is obtained by subtracting the amounts of phenolic hydroxyl groups from the total hydroxyl groups, which is achieved by titration of the coal with acetic anhydride. Another method is the reaction of the coal sample with KOH–C_2H_5OH solution. The hydroxyl and carboxyl groups react with KOH to form potassium salts, which are insoluble in alcohol. The reacted coal is then placed in water and partly neutralized with CO_2; only potassium hydroxyl groups react with CO_2, to form K_2CO_3. The K_2CO_3 is further quantitatively titrated with a standard acidic solution.

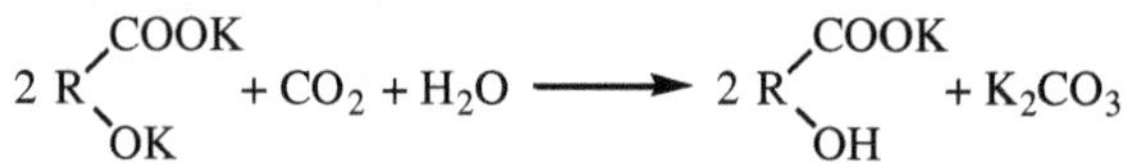

Carbonyl Group (>C=O)

Carbonyl groups are determined by the reaction of the coal sample with phenylhydrazine solution:

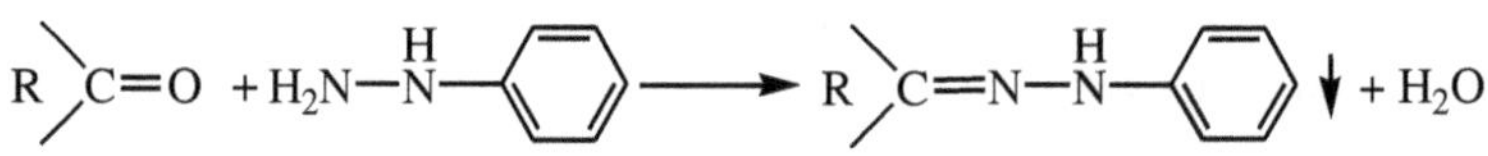

The excess phenylhydrazine is oxidized to produce nitrogen gas. The amount of phenylhydrazine that reacts with carbonyl groups is calculated from the volume of nitrogen gas and the total amount of phenylhydrazine. It is also possible to determine the nitrogen content of the coal before and after the reaction. The amount of carbonyl groups is then calculated from the increase in the amount of nitrogen.

Methoxy Group (–OCH_3)

Methoxy groups react with HI to form CH_3I, which is analyzed using iodimetry:

$$ROCH_3 + HI \rightarrow ROH + CH_3I$$
$$CH_3I + 3Br_2 + H_2O \rightarrow HIO_3 + 5HBr + CH_3Br$$
$$HIO_3 + 5HI \rightarrow 3I_2 + 3H_2O$$

Ether Links (–O–)

Ether links are relatively stable thermally and chemically. An optimum analytical method has not yet been developed. One method is hydrolysis with HI.

$$\mathrm{R{-}O{-}R' + HI \rightarrow ROH + R'I}$$
$$\mathrm{R'I + NaOH \rightarrow R'OH + NaI}$$

The number of ether groups is then calculated from the increase in hydroxyl groups in the coal or the amount of iodide combined with the coal. This method is not sufficiently accurate, because hydrolysis is incomplete for all types of ether group.

3.1.7.2 Alkyl Side Chains

Coal spectroscopic data show that the basic structural unit of coal is based upon alkyl side chains, which can be oxidized to carboxylic acids under mild oxidation conditions (150 °C, oxygen). The average length of the alkyl side chains of different coal types can be calculated using elemental analysis and infrared (IR) spectroscopy to show that the lengths of the alkyl side chains decrease rapidly with increasing degree of coalification.

The proportion of alkyl carbon to total carbon also decreases accordingly: When C% is equal to 80 %, methyl carbon accounts for 4–5 % of total carbon and about 75 % of alkyl carbon; when C% is equal to 90 %, methyl carbon accounts for 3 % of total carbon and more than 80 % of alkyl carbon. In addition to methyl, there are groups such as ethyl and propyl, and the higher the carbon atom number, the lower the proportion.

3.1.7.3 Sulfur-Containing and Nitrogen-Containing Functional Groups in Coal

The properties of sulfur are similar to those of oxygen, so the types of sulfur containing functional groups in coal are similar to those of oxygen-containing functional groups, including thiols (R–SH), thioethers, disulfides (R–S–S–R′), thioquinones, and other heterocyclic sulfur compounds. The sulfur content of coal is lower than the oxygen content, and sulfur is more difficult to assay, so the distribution of organic sulfur in coal has not yet been fully clarified. The distribution of organic sulfur types is shown in Table 3.7. In general, the organic sulfur in lignite is present primarily as mercaptans and aliphatic thioethers, and that in bituminous coal is in the form of thiophene rings (primarily dibenzothiophenes).

The proportion of nitrogen in coal is 1–2 %, of which 50–75 % exists in the form of six-membered heterocyclic pyridine or quinoline rings, with the balance provided by amino, imino, and nitrile groups, and five-membered heterocyclic pyrroles and carbazoles. The nitrogen-containing structures are relatively stable, so quantitative determination is difficult. This issue will be discussed in subsequent chapters.

Table 3.7 Distribution of organic sulfur species (%)

Coal type	Organic sulfur	Sulfur-containing structures in coal				
		Aliphatic thiol R–SH	Aromatic thiol Ar–SH	Aliphatic thioether R–S–R′	Aromatic thioether Ar–S–Ar′	Thiophene 
Illinois bituminous	3.20	7	15	18	2	58
Kentucky sub-bituminous	1.43	18	6	17	4	55
Pittsburgh bituminous	1.85	10	3	7	2	78
Xitian lignite	1.48	30	30	25.5	–	14.5
Texas young lignite	0.80	6.5	21	17	24	31.5

3.2 Chemical Methods for Coal Structure Determination

Methods used to analyze coal structure are also related to understand the chemical properties of coal. It is difficult to understand coal structure purely in chemical terms, because coal contains a variety structures and varies with coal type, degree of metamorphism, and microscopic composition [11]. Initially, studies of the chemical structure of coal aimed to produce monomers by pyrolysis, hydrolysis, or other chemical depolymerization processes. Some other attempts to obtain other simple structures called tectonic units were also made using copolymerization; however, these approaches failed. Later, researchers tried to fully characterize the carbon skeleton structures in coal and identify the types of connections between the carbon skeleton and other elements to determine the amount of testable functional groups, and study how the functional group content is related to in coal type and degree of coalification. These research methods were based on chemical reactions such as alkylation, acetylation, and selective oxidation of coal, and physical methods such as gas chromatography-mass spectrometry (GC-MS) and nuclear magnetic resonance (NMR) spectroscopy. Studies changed from those of planar combinations of aromatic rings and bridging bonds to three-dimensional (3D) analysis of coal structure. Rough, qualitative, empirical, and semi-quantitative content studies have since developed into precise morphological studies. Over the past 20 years, new techniques and methods such as scanning transmission X-ray microscopy [12], particle-induced X-ray emission analysis/particle-induced γ-ray emission analysis [13], diffuse reflectance Fourier-transform IR [14], laser ionization mass analysis [15], nuclear magnetic resonance imaging [16], and quantitative structure–activity relationship analysis have been introduced. With the constant development of research methods, our understanding of coal structure is being deepened continuously. This chapter starts from functional groups.

3.2.1 Solvent Extraction

In the early stages of coal research, soluble molecules in coal were separated through solvent extraction and thoroughly analyzed, providing abundant information on coal structure. Extraction is still widely used in the development of high-value coal-derived products.

Solvent extraction processes are classified into several categories, according to the solvent selected, and the extraction temperature and pressure.

3.2.1.1 General Extraction

General extraction is usually carried out below 100 °C, using low-boiling-point solvents such as benzene, chloroform, and ethanol. Only small amounts (≤1–2 %) of materials are extracted using this method, with resins and waxes being the major components of the extract. The extract obtained does not reflect the coal composition.

When bitumite is directly extracted with chloroform, less than 1 % of the material is extracted. If the coal is first pyrolyzed at 400–410 °C (high volatiles) or 430–450 °C (medium volatiles), and then extracted with chloroform, the extraction yield increases to 4–7 %. At room pressure, benzene extraction of bitumite provides an extraction yield of less than 1 %. If the extraction temperature and pressure are increased, e.g., to 285 °C and 5.56×10^6 Pa, the extraction yield increases to 5–10 %. The compositions of the extracts obtained using chloroform and benzene solvents are very similar.

3.2.1.2 Specific Extraction

Specific extraction includes extractions using specific nucleophilic solvents (electron donors) such as pyridine, phenolic compounds, and aniline at below 200 °C. The yield in specific extraction reaches 20–40 %, or even over 50 %. Specific extraction is an important method and is widely used in coal studies, because it gives high yields of extracts, with no changes in the molecular structures.

Table 3.8 shows the analytical results for extracts obtained by pyridine extraction of coals with various degrees of coalification. The extraction yield depends on the degree of coalification [17]. Normally, high extraction yields from low-coalification coals such as bitumite can be achieved, with the extract elemental composition close to that of the raw coal. Pyridine is a strong electron donor and prone to form hydrogen bonds; this increases the solubility and therefore the extraction yield.

Table 3.8 Ultimate analysis of extracts from pyridine extraction and extraction yields

Coal sample	Ultimate analysis (%, daf)					Yield (%)
	C	*H*	*N*	*S*	*O* (difference)	
Raw 1	76.3	5.2	1.3	0.7	16.5	6.7
Extractive 1	77.5	6.4	1.7	0.4	14.0	
Raw 2	82.6	5.9	1.7	1.6	8.1	21.4
Extractive 2	83.9	6.2	1.8	0.4	7.7	
Raw 3	85.1	5.3	1.5	1.0	7.1	20.9
Extractive 3	83.0	5.8	2.3	0.4	8.5	
Raw 4	86.1	5.5	1.8	1.0	5.6	19.3
Extractive 4	87.1	6.0	1.8	0.5	4.6	
Raw 5	90.0	4.4	0.7	3.8	1.1	2.8
Extractive 5	88.4	5.2	1.8	0.4	4.2	
Raw 6	90.4	4.6	1.4	0.5	3.1	2.5
Extractive 6	90.5	5.3	1.7	0.4	2.1	

Aniline is also a good electron donor and gives a relatively high extraction yield from coals with low degrees of coalification.

3.2.1.3 Supercritical Extraction

A supercritical fluid is a substance at a temperature and pressure above its critical point, where distinct liquid and gas phases do not exist. It can effuse through solids, like a gas can, and dissolve materials, like a liquid can. In addition, close to the critical point, small changes in the pressure or temperature result in large changes in the density, allowing many of properties of a supercritical fluid to be fine-tuned. Supercritical extraction is usually performed at around 400 °C, and extraction yields over 30 % are obtained. At this temperature, a distillation process and a liquid–liquid extraction process are also included, to avoid unnecessary pyrolysis reactions.

The results of a typical supercritical extraction are listed in Table 3.9 [18].

3.2.1.4 Pyrolysis and Hydrogenation Extractions

Pyrolytic extraction is usually carried out above 300 °C using polyaromatic compounds, phenanthrene, anthracene, quinoline, and coal tar fractions as the solvents. Because partial pyrolysis occurs during extraction, the extraction yield is higher than 60 % in most cases, and even as high as 90 % in some cases.

Hydrogenation extraction is carried out under a hydrogen atmosphere at temperatures above 300 °C, using tetrahydronaphthalene or 9,10-dihydrophenanthrene as the solvent. Pyrolysis and hydrogenation reactions both occur during the extraction, so the hydrogen extraction yields are normally very high.

Table 3.9 Analysis of feed coal, extract, and residue of typical supercritical extraction

Index	Feed coal	Extract	Residue
C_{daf} %	82.7	84	84.6
H_{daf} %	5	6.9	4.4
O_{daf} %	9	6.8	7.8
N_{daf} %	1.85	1.25	0.9
S_{ar} %	1.55	0.95	1.45
H/C	0.72	0.98	0.63
OH_{daf} %	5.2	4.4	4.8
A_d %	4.1	0.005	5
V_{daf} %	37.4	–	25
Molecular weight	–	490	–

In both methods, the coal molecules are changed and are significantly different from those in the original coal. So, these two methods have rarely been used in recent coal structural studies, but are useful in studies of coal liquefaction and other coal conversion processes.

3.2.1.5 Coal Extraction Yield and Solvent Properties

Solution Parameters of Coal and Solvent

According to chemical thermodynamics, dissolution only occurs when the mixing free energy of the solvent and the solute, ΔG, is less than 0. $\Delta G = \Delta H - T\Delta S$, therefore the enthalpy ΔH should be as small as possible. For the dissolution process, ΔH can be calculated as

$$\Delta H = \frac{n_1 V_{m,1} \cdot n_2 V_{m.2}}{n_1 V_{m,1} + n_2 V_{m.2}} \left(\delta_1^{\frac{1}{2}} - \delta_2^{\frac{1}{2}}\right)^2,$$

$$\delta_1 = \frac{\Delta E_1}{V_{m,2}} \quad \delta_2 = \frac{\Delta E_2}{V_{m,2}},$$

where

n is the molecular fraction, 1 represents the solvent, and 2 represents the solute (1 and 2 represent the same below);
V_m is the solvent molar volume ($m^3\ mol^{-1}$);
δ is the solubility index of the solvent ($J\ m^{-3}$);
ΔE_1 is the cohesive energy of the solvent ($J\ mol^{-1}$).

From the above equation, in order to make ΔH as small as possible, δ_1 and δ_2 should be as close as possible.

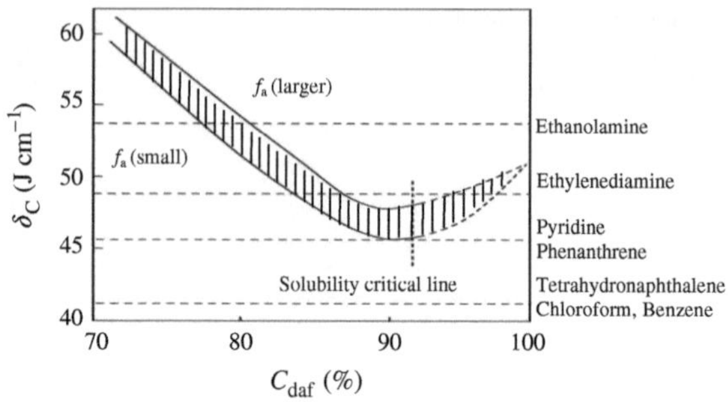

Fig. 3.3 Relationship between coal solubility index and degree of coalification

The coal solubility index, δ_C, is related to the degree of coalification. In Fig. 3.3, δ_C, restricted by aromaticity, is the area surrounded by the two curves. When the coal C_{daf} is at 70 %, δ_C is around 63 J cm^{-3}. As the coalification degree increases, δ_C gradually decreases. The δ_C reaches its minimum value of 46 J cm^{-3} when C_{daf} is 90 %.

The δ_C of coal with a low coalification degree is around 54 J cm^{-3}, which is very close to the δ values of ethanolamine and ethylenediamine. Ethanolamine (53.8 J cm^{-3}) and ethylenediamine (48.8 J cm^{-3}) are therefore good solvents for the extraction of coal with a low degree of coalification, giving a relatively high extraction yield. The δ of pyridine is 45.6 J cm^{-3}, very close to that of coal with a medium degree of coalification. The yield from pyridine extraction of this type of coal is therefore higher than those obtained using other solvents. The δ values of benzene and chloroform (about 41.3 J cm^{-3})are much lower than that of coal, resulting in very low extraction yields.

Electron Donor–Acceptor Abilities of Solvent

The coal extraction yield of a solvent is related to its electron-donating (DN) or electron-accepting (AN) abilities. In Table 3.10, the relationship between the extraction yield and the solvent DN and AN capabilities is shown (high volatile bitumite is used). It can be seen that the extraction yield follows the same trend as the DN − AN values.

The empirical "rule of similarity" applies to the dissolution of a substance. The similarity can be a similarity in structure, molecular weight, molecular geometry, or physical properties. A typical example for elucidating this rule is that non-polar solutes are soluble in non-polar solvents and polar solutes are soluble in polar solvents. Coal contains many aromatic molecules and heteroatoms, so that aromatic hydrocarbons, especially heterocyclic hydrocarbons and phenols, are excellent solvents for coal extraction and give high extraction yields. Conversely, the extraction yields with non-aromatic hydrocarbons are low.

In addition, solvents with large electro negativities are prone to form hydrogen bonds with the coal surface, which improves the extraction yield. Amines, pyridine,

Table 3.10 Relationships among extraction yields of high-volatility bitumite and solvent electron-donating (DN) and electron-accepting (AN) capabilities

Solvent	Extraction yield W_1 (daf, %)	DN	AN	DN − AN
n-Hexane	0.0	0	0	0
Water	0.0	33.0	54.8	−21.8
Nitromethane	0.0	2.7	20.5	−17.8
Isopropanol	0.0	20.0	33.5	−13.5
Acetic acid	0.9	–	52.9	–
Methanol	0.1	19.0	41.3	−22.3
Benzene	0.1	0.1	8.2	8.1
Ethanol	0.2	20.5	37.1	16.6
Chloroform	0.35	–	23.1	–
Acetone	1.7	17.0	12.5	+4.5
Tetrahydrofuran	0.8	20.0	8.0	+12.0
Ethyl ether	11.4	19.2	3.9	+15.3
Pyridine	12.5	33.1	14.2	+18.9
Dimethylformamide	15.2	26.6	16.0	+10.6
Ethylenediamine	22.4	55.0	20.9	+34.1
Methylpyrrolidone	35.0	27.3	13.3	+14.0

and phenolic compounds give much higher extraction yields than heteroatom-free aromatic solvents.

3.2.2 *Computational Chemistry*

Computer-aided molecular design (CAMD) is now widely used in the design of new drugs, enzymes, inhibitors, and protein. CAMD not only provides a 3D view of the molecule, but also determines the stable configuration by optimizing the molecule based on molecular dynamics theory. Important breakthroughs have been achieved in the quantitative and visual analyses of coal matrices using CAMD.

3.2.2.1 Molecular Dynamics for Molecular Structural Model of Coal

Based on Pauling's assumption regarding atomic radii and bond angles, Spiro [19] built 3D space-filling models by extending the structural models of coal proposed by Given [20], Wiser [21], Solomon [22], and Heredy [23]. He found that all the other models had spatial barrier to some extent, except the Solomon model. In Wiser and Heredy models, large numbers of methylene linkages had to be added to sustain the 3D configurations. In Given's model, the quaternary carbon between two rings had to be reconfigured to remove the strong internal stress.

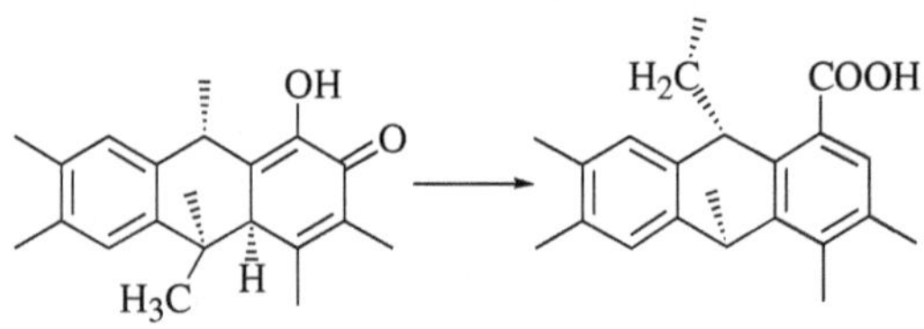

The author checked the space-filling models from the top and side perspectives, and found that all the models contained structural units that were prone to form planar configurations, together with a significant increase in aliphatic, cyclic, or aromatic fragments.

3.2.2.2 Identification of Coal Configuration with Minimum Energy

Carlson and Granoff [24] introduced the use of CAMD in coal research; they simulated the coal structure and calculated energies. They built four types of 3D space-filling models of the coal molecular structures proposed by Given, Solomon, and Shinn [25], using BioGraf software. Simulation of the molecular motion at various temperatures (the Dreiding force field approximation was adopted) identified the molecular configurations with the minimum system energies in those four models.

3.2.2.3 Energy of Molecular Structural Model of Coal

Carlson [26] calculated the van der Waals forces and hydrogen-bonding energies of Wiser, Solomon and Shinn models. The van der Waals force, E_{vdw}, can be calculated from the Lennard–Jones potential energy function:

$$E_{vdw}(R) = D_0\left[\left(\frac{R_0}{R}\right)^{12} - 2\left(\frac{R_0}{R}\right)^{6}\right] \cdot S,$$

where

D_0 is the bond strength, which changes according to the type of atom (kJ mol^{-1});

R_0 is the equilibrium bond length, which changes according to the atom pair (nm);

S the switch function, gradually shuts off the van der Waals force when the distance between two atoms is over 0.8 nm.

The hydrogen-bonding energy, E_{hb}, is calculated using the Lennard-Jones energy function:

$$E_{hb}(R) = D_0 \left[5\left(\frac{R_0}{R}\right)^{12} - 6\left(\frac{R_0}{R}\right)^{10} \right] \cos^2(\theta_{AHD}) \cdot S,$$

where

D_0 is the equilibrium bond energy; 39.71 kJ mol^{-1} is used for the hydrogen bonds formed between H and O, N, and S atoms.

R_0 is the equilibrium distance at which the hydrogen bond is initialized; normally an average number of 0.275 nm is used.

θ_{AHD} is the angle between the A–H and H–D bonds in the hydrogen bond.

The energy differences between the van der Waals forces and hydrogen bonds before and after the molecular dynamics calculation of a molecular configuration are defined as the stabilization energies.

The changes in total energy, covalent bond energy, and non-covalent bond energy in the most complicated Shinn model were investigated. The decrease in total energy was mainly contributed by the stabilization energy from interactions among the molecules (hydrogen-bonding and van der Waals forces). The covalent bond energy (including bond stretching, bending, and twisting) was almost unchanged.

There are two types of hydrogen bonds: inter-crystal hydrogen bonds (the hydrogen-bond acceptor and donor are located in different crystal cluster) and intra-crystal hydrogen bonds (the hydrogen-bond acceptor and donor are located in the same crystal cluster). Only inter-crystal hydrogen bonds stabilize the configuration. The role of van der Waals forces is more important than that of hydrogen bonds in the molecular dynamics calculation.

3.2.2.4 Coal Density

Carlson calculated the true densities and porosities of three minimum-energy molecular configurations. The calculated average true density was 1.27 ± 0.04 g cm^{-3}, which matches the experimental values for bitumite (1.28–1.33 g cm^{-3}) and for vitrinite from bitumite (1.25–1.38 g cm^{-3}). The porosities of the three configurations are all less than 1 %. The experimental porosity of No. 8 Pittsburgh bitumite coal obtained by White [27] is also less than 1 %. It should be pointed out that the Carlson method is not applicable for pores of diameter above 1.5 nm, so it is not applicable for calculating the pore sizes and porosities of coals.

3.2.2.5 Computer-Aided Structural Analysis of Coal Macromolecules

Current coal structural analyses based on various structural models provides chemical and structural parameters. However, these results only reflect the average structure of an unevenly distributed complex polymeric system, and inevitably neglect most of the interactions among structural fractions. Some structural parameters such as the density and pore distribution are characteristic spatial parameters of a material and cannot be precisely described by analysis of an average structure. Chemical information on structural fractions can be obtained experimentally using techniques such as NMR, Fourier-transform IR, and pyrolysis/gas chromatography/mass spectrometry (Py/GC/MS); the difficulty in structural elucidation of coal macromolecules is how to combine the information on all the structural fractions. In previous studies, structural models were mostly built based on individual circumstances, and empirical approximations were usually used to simplify the model, leading to various structural models based on one set of analytical data.

In recent years, significant progress has been achieved in computer-aided elucidation of micromolecular structures. Faulon [28] first introduced this technique in coal matrix (macromolecule) structural studies, and developed a computation program for the joint analysis of multiple substructural models.

3.2.2.6 Application of Quantum Chemistry in Investigation of Coal Structure

Quantum chemical calculations are powerful tools in modern laboratory studies. The essence of a quantum chemical calculation is to find solutions to the Schrödinger equation; various approximations and principles are adopted, and different quantum chemical methods have been developed such as ab initio, density functional theory, and semi-empirical methods (complete neglect of differential overlap, intermediate neglect of differential overlap).

Applying quantum chemistry to study coal is challenging for two reasons. First, coal molecules are not well defined so that an average structural unit must be assumed. Deviations originating from the adoption of an average structural unit have not been considered and are sometimes not unmeasurable. Second, a coal matrix is an unevenly distributed polymeric macromolecule, which is incompatible with the common requirement for quantum calculations that a research object should have finite or infinite numbers of molecules in a periodic repeating pattern. Despite these problems, progress in quantum chemistry and modern computing methods, have started to make the application of quantum chemical calculations to coal studies possible. In this section, quantum chemical descriptions of coal structural parameters are briefly introduced. Bond lengths in coal structural units have been studied in detail by quantum chemical calculations. We present investigations of a coal structural unit and extracted molecules using various calculation methods, including quantum chemical and molecular dynamics calculations.

Different bond lengths were obtained in the same molecule. Molecular dynamics or semi-empirical quantum calculation methods gave delocalized π bond lengths that deviated significantly from the actual values. A quantum calculation using ab initio approximations clearly showed that the length of delocalized π bonds are averaged and p–π conjugation also changed the bond length.

Bond orders and overlapping populations have also been investigated using quantum chemical calculations. The bond order of a bond in a structural unit or among various structural units determines the thermal reactivity of coal and the pyrolysis product distribution. Quantum chemical calculations of the bond order and overlapping populations predict the thermal activities of various types of bonds and are more reliable for systems with intermolecular hydrogen bonds or condensed aromatic subunits.

The chemically reactive sites in a coal matrix can be evaluated using quantum chemical calculations. Quantum chemical calculations can be used to obtain the charge distributions, net charge distributions, and population distributions of molecules, together with the distributions of the highest occupied and the lowest unoccupied molecular orbitals. These calculated parameters are normally used as indicators of the active sites for hydrogenation and oxidation, or to speculate on the reaction mechanism.

The optimum molecular configuration generated by a molecular dynamics calculation can be further corrected using a quantum chemical calculation. A molecular dynamics calculation based on classical mechanics, using a force field approximation, is unable to fully consider electronic interactions, and cannot handle transition and excited states. In the molecular dynamics calculation of a complex chemical interaction, empirical constants normally have to be used, resulting in a large system error. The results of quantum chemical calculations can be used for calibration to compensate for these assumptions and to reduce the system error. Theoretically, bulk physical properties of a material, such as the density, refractive index, dielectric constant, IR and NMR spectra, XRD patterns, and other intrinsic properties, can be determined based on quantum chemical calculations. By comparing calculated and experimental values and continually revising the average molecular configuration model, the final model should be able to reflect the coal chemical reactivity precisely.

3.3 Case Study of Coal Chemical Structural Analysis

3.3.1 Macromolecular Network Structure of Pingshuo Coal

A practical example of coal chemical and structural analysis is presented, in which various methods/techniques were used; a combination of physics, chemistry, statistical structural analysis, and stoichiometric calculations was involved. The major objective was to clarify the structure and composition of Pingshuo bituminous coal (produced in Pingshuo, located in Shanxi Province, China), which has large reserves, well-defined coal characteristics, and moderate coal rank. The structures of the macerals, macromolecular network and micromolecules of Pingshuo bituminous coal were investigated.

3.3.1.1 Preparation of Maceral Sample

Pingshuo bituminous coal was selected for investigation of the separation of three organic macerals: vitrinite, inertinite, and exinite. A typical separation procedure was as follows.

(1) Manual picking: performed based on the macroscopic features of the coal. Vitrinite was selected from vitrain, inertine was selected from fusain, and exinite was selected from the whole coal.
(2) Refining: the manually chosen samples were further separated with screening to prepare samples with mean diameters of 0.1753 mm. Microscopic observations showed the predominant maceral groups in the samples with this diameter distribution.
(3) Acid washing: the washing solution consisted of 40 % distilled water, 45 % HCl (30 %), and 15 % HF (40 %). The ratio of the washing solution to the sample was 10:1. The mixture was boiled for 6 h, followed by washing with deionized water until the solution was neutral, and then the sample was dried at 50 °C for 2.5 h.
(4) Gravity separation: the density gradients of the respective macerals were identified from the curve of the density distribution, using $ZnCl_2$ as the specific gravity liquid for the separation; the procedure is shown in Fig. 3.4.

The purities of the three macerals were above 95 %, which completely satisfies the demands of structural analysis. The results of elemental, proximate, and petrographic analyses of the three macerals are listed in Table 3.11.

3.3.1.2 Preparation of Macromolecular Network Sample

The three macerals obtained by separation were ground to 200 mesh, mixed with pyridine in a maceral:pyridine ratio of 1:6, and extracted in an extraction tank in a thermostated water bath, with continuous stirring, for 24 h. The extraction residues were subjected to the same procedure three or four times until the color of the extraction solvent was light yellow. The pyridine was removed using a vacuum rotary evaporator, and the sample was dried in a vacuum oven at 40 °C for 5–6 d, to a constant weight. The dry residue can be regarded as the macromolecular network of the maceral. The elemental analyses of the macromolecular networks of the three macerals are listed in Table 3.12.

The corresponding pyridine extraction yields for the three macerals were 18.20, 6.37 and 25.48 % for vitrinite, inertinite and exinite, respectively. The solubilities of

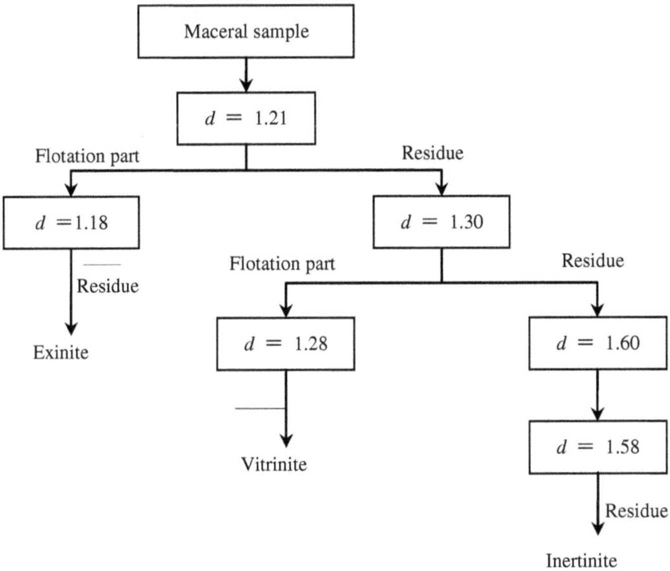

Fig. 3.4 Procedure for maceral separation: d is density (g cm^{-3})

Table 3.11 Ultimate, proximate, and petrographic analyses of Pingshuo Bituminous coal (%)[a]

Sample	Coal	Vitrinite sample	Inertinite sample	Exinite sample
Ultimate analysis				
C_{daf}	79.84	78.46	84.73	73.56
H_{daf}	5.27	5.32	3.96	7.54
O_{daf}	13.03	13.43	9.89	17.22
N_{daf}	1.43	1.22	0.66	0.85
S_{daf}	0.43	0.57	0.76	0.83
Proximate analysis				
A_d	25.10	1.18	1.96	3.67
V_{daf}	41.30	43.30	24.05	52.76
FC_{daf}	45.95	56.03	74.45	45.51
Petrographic analysis				
Vitrinite	69.8	95.2	2.3	2.3
Inertinite	23.1	3.0	96.7	1.4
Exinite	7.1	1.8	1.0	96.3

[a]The densities of vitrinite, inertinite, and exinite are taken as 1.28, 1.58, and 1.18 g cm^{-3}, respectively

Table 3.12 Elemental analyses of pyridine-extracted residues of macerals (daf, %) and *H*/*C*

Pyridine-extracted residues	*C*	*H*	*N*	*O*	*S*	*H*/*C*
Vitrinite	79.13	5.35	2.08	12.84	0.60	0.81
Inertinite	84.14	3.66	0.93	13.19	0.81	0.54
Exinite	76.54	8.24	1.81	13.09	0.32	1.31

the different coal groups are different; exinite is more soluble than inertinite because of its high content of aliphatic hydrocarbons.

3.3.1.3 Macromolecular Networks Structure Characterization of Maceral

Solid ^{13}C NMR Spectra of Macerals

The solid-state ^{13}C NMR spectra obtained for various macerals are shown in Fig. 3.5. The carbon distribution in each maceral (listed in Table 3.13) is determined from assignment of chemical shifts. For example, the chemical shift of aliphatic carbon in the range of 0–80 ppm, and for exinite a strong signal at 30 ppm, indicates the proportion of $(CH_2)_n$ chains in aliphatic carbon or the ratio of $-CH_2$

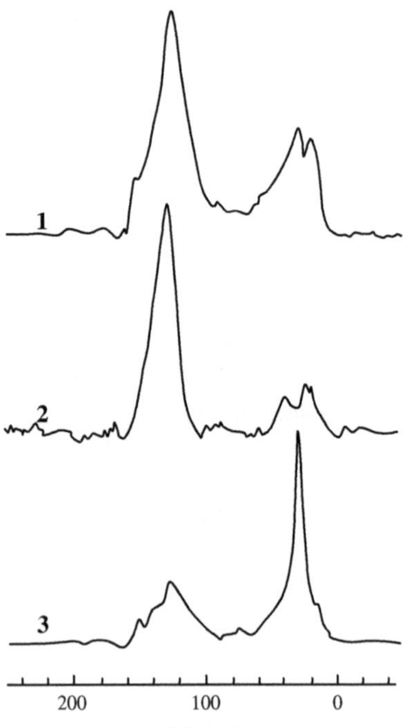

Fig. 3.5 Solid-state ^{13}C NMR spectra of macerals: *1* Vitrinite; *2* Inertinite; *3* Exinite. Reprinted from Ref. [29], Copyright 1994, with permission from Taylor & Francis

Table 3.13 Relative distribution of carbon and aromaticity in maceral sample

Maceral	Carbonyl	C_a	C_{al}	Aromaticity (f_a)
Vitrinite	0.037	0.630	0.333	0.67
Inertinite	0.032	0.715	0.253	0.75
Exinite	0.031	0.354	0.612	0.39

and –CH groups in naphthenic structure. For vitrinite, there are two similar peaks at 20 and 30 ppm, which suggest the presence of similar amounts of long- and short-alkyl chains. For inertinite, there are two aliphatic carbon peaks that are similar to vitrinite, but much lower in intensity than the other two macerals. It should be noted that the relatively higher adsorption peak at 18 ppm can be ascribed to the aliphatic carbon of inertinite bearing shorter alkyl side chains. Though the peak position of aromatic carbons from three maceral groups are very similar, the adsorption peak at 130 ppm for inertinite is much uniform and sharper, which indicates the existence of large amount of carbon and bicyclic aromatic ring. The adsorption at 130 ppm of vitrinite and inertinite are more or less the same, except for the significant broadening and shielding adsorption at left side indicating the abundance of alkyl replaced structure in vitrinite (vs. inertinite). It is complicated to assign the adsorption peaks of exinite, for example, there is a shoulder besides the main adsorption peak at 130 ppm. The shoulder peak at 150 ppm indicates that exinite contains more alkyl substituted aromatic structure.

XRD Analysis of Macromolecular Networks

We investigated the microcrystaline structure of the macerals of Pingshuo bituminous coal by XRD measurements as shown in Fig. 3.6. L_a and L_c were then

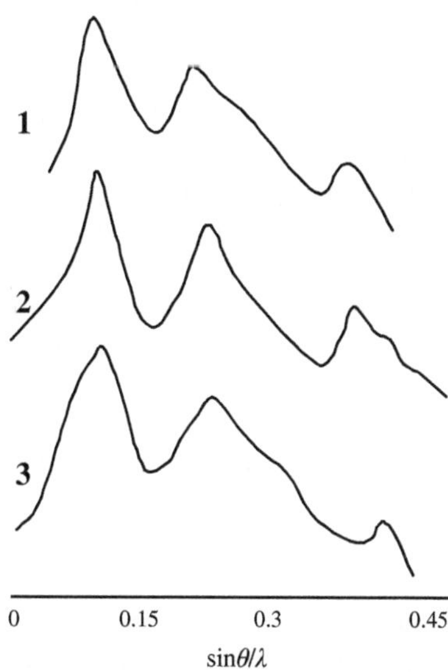

Fig. 3.6 XRD patterns of macerals: *1* Vitrnite; *2* Inertinite; *3* Exinite. Reprinted from Ref. [6], Copyright 1993, with permission from Taylor & Francis

Table 3.14 Crystalline parameters for the macromolecular network of Pingshuo Bituminous coal

Sample		L_a (nm)	L_c (nm)	d_m (nm)	N[a]
Maceral	Vitrinite	1.704	1.25	0.357	4.5
	Inertinite	2.300	1.076	0.370	3.9
	Exinite	1.278	1.150	0.347	4.3
Residue after pyridine extraction	Vitrinite	1.820	1.125	0.367	4.1
	Inertinite	2.191	1.061	0.369	3.9
	Exinite	1.356	1.186	0.462	3.7

[a]The amount of aromatic layer cluster in the microcrystal can be calculated by (L_c/d_m + 1)

calculated from the width of diffraction peaks and d_m was determined from peak position. The crystalline parameters for the residues after pyridine extraction of three macerals group were calculated based on the same methods. The crystalline parameters (listed in Table 3.14) of the residues after pyridine extraction are similar to those of their original macerals. This observation indicates that there is no obvious change in the framework structure of macerals during the extraction process and only the micromolecular phase trapped in the framework has been extracted.

Analysis of Chemical Structure Using XPS

XPS is a versatile tool that can directly identify the chemical states of elements. Pingshuo bituminous coal was examined with respect to the binding energies of C, O, and N, and the percentage peak areas based on C (1s) (Table 3.15).

The chemical states of carbon can be divided into four groups: (1) aromatic and aliphatic hydrocarbons, with peaks at 285.18 eV, and peak areas of 50.38, 41.00, and 5.84 % for vitrinite, inertinite, and exinite, respectively; (2) ether or connected to hydroxyl, with a peak at 286.6 eV, and peak areas of 17 % for all three macerals; (3) carbonyl, with a peak at 288 eV; and (4) carboxyl, with a peak at 289.2 eV. For (3) and (4), inertinite has the highest peak areas, 9.99 and 13.23 %, respectively, in contrast to 3 % for vitrinite and exinite.

The combinations of O and C in exinite are basically COO– and C = O, and are mainly quinone and ether in vitrinite and inertinite. The combinations of N and C are mainly amines and pyrroles in inertinite, pyrroles and pyridines in vitrinite, and amines and pyridines in exinite, respectively.

Analysis of Chemical Structure Using FTIR

IR spectroscopy is a versatile technique with significant advantages in the analysis of functional groups in coal structures. The FTIR spectra of vitrinite, inertinite, and exinite, and the corresponding pyridine extraction residues are shown in Figs. 3.7 and 3.8. The spectra were recorded with a PE-1700 IR spectrometer in the scanning

Table 3.15 Chemical states of C, O, and N in macerals

Chemical state	Vitrinite		Inertinite		Exinite	
	E_b (eV)	Peak area (%)	E_b (eV)	Peak area (%)	E_b (eV)	Peak area (%)
PhC_2H_5, $PhOC_2H_5$		–	249.15	2.77	–	–
$(CH_3)_3C$, CH_3CH_2	–	–	292.05	3.66	–	–
Quinone, COO^-, $PhCH_3$	290.72	1.05	290.65	2.77	290.71	1.04
C=O, CH_3COO^-	289.28	2.3	289.25	13.23	289.29	2.59
C=O, $CH_3PhOOCH_3$	288.24	3.1	288.05	9.99	288.00	3.92
Ether, C–O, OH	286.61	17.42	286.65	17.12	286.40	17.71
CH_3COOH, C–O	285.74	5.48	–	–	–	–
Ph=N, Ph–N	285.18	50.38	285.15	41.00	285.54	5.84
CH_2CH_2	284.23	17.42	–	–	284.94	49.29
Graphite C, Ph–N	–	–	–	–	284.03	17.45
C-N, CH_2–CH_2	283.01	2.1	283.85	5.89	283.01	2.15
Pyrrole	401.04	63.72	401.04	35.46	–	–
Amine	–	–	400.62	65.54	400.48	62.72
Pyridine	399.79	36.28	–	–	399.01	37.28
COO^-, C=O	–	–	536.44	52.23	–	–
COO^-, C=O	534.63	24.28	534.64	18.40	534.22	23.43
-O-	532.99	61.10	533.27	18.17	533.03	53.08
C-O, Ph_3PO	531.39	14.63	531.96	11.20	531.43	23.49

E_b is the binding energy

range 450–4000 cm^{-1}, using pellets prepared by mixing KBr powder (300 mg) with the sample (1 mg) and pressing under a pressure of 10 kg cm^{-2}.

The main differences among the IR absorptions of the three macerals of Pingshuo bituminous coal reflect the aliphatic hydrogen and hydrogenated aromatic hydrogen contents. For exinite, there is a strong absorption at 2845 cm^{-1}, assigned to symmetric $-CH_2$ stretching, a strong absorption at 2926 cm^{-1}, assigned to asymmetric $-CH_2$ stretching, and a strong absorption at 1450 cm^{-1}, assigned to $-CH_2$ and $-CH_3$ bending. These observations indicate a higher amount of hydrogen in exinite. Moreover, the strong absorption from carbonyl and carboxyl groups near 1791 cm^{-1} implies the presence of such oxygen-containing structures in exinite. For vitrinite, the strong absorption band at 3400 cm^{-1} indicates that the oxygen is present in hydroxyl groups. For inertinite, the strong absorption band at 1600 cm^{-1} shows an abundance of oxygen-containing functional groups on aromatic rings; this absorption is weak for exinite.

The influence of pyridine extraction on the structures was determined by comparing the FTIR spectra of the original samples and the residues after pyridine extraction. The results are as follows.

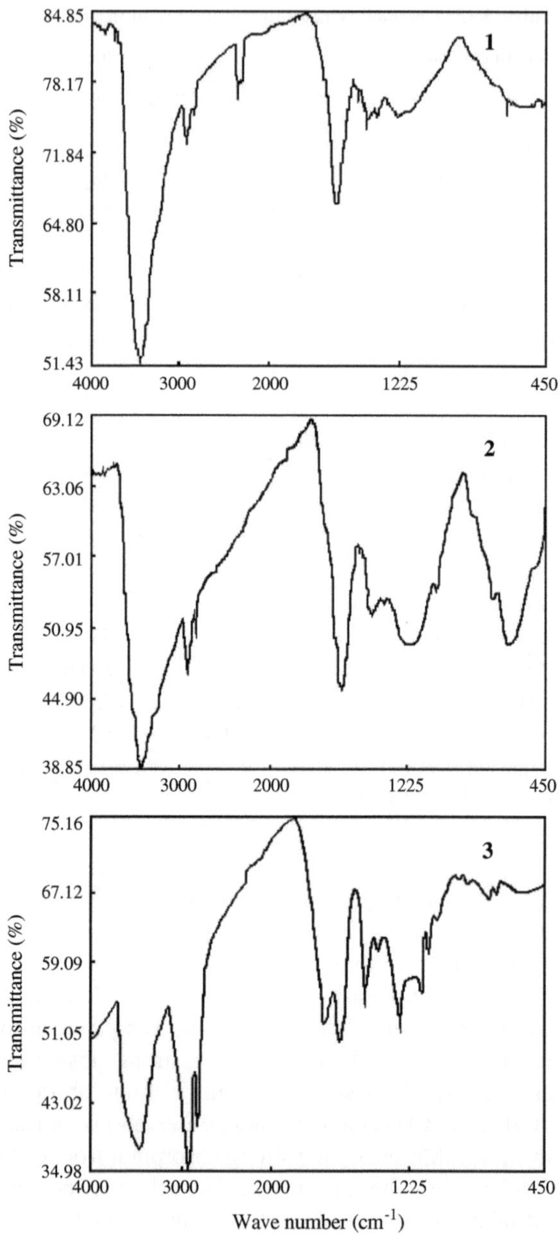

Fig. 3.7 FTIR spectra of macerals of Pingshuo Bituminous coal: *1* Vitrinite; *2* Inertinite; *3* Exinite. Reprinted from Ref. [6], Copyright 1993, with permission from Taylor & Francis

For vitrinite, a stronger absorption band at 3436 cm^{-1} appeared in original sample (compared with the residue), assigned to –OH stretching; which shows the extraction of hydroxyl-containing hydrocarbons. Moreover, the similarity in the range 2919–2854 cm^{-1}, assigned to aliphatic stretching C–H, of the sample and

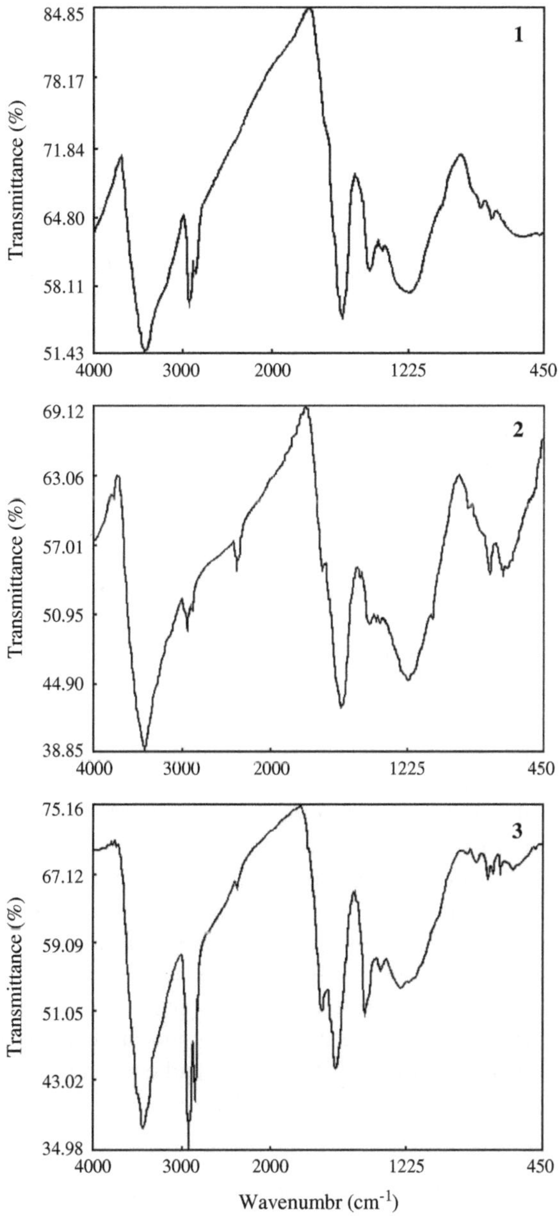

Fig. 3.8 FTIR spectra of the residue after pyridine extraction of maceral: *1* Vitrnite; *2* Inertinite; *3* Exinite

residue suggests that pyridine extraction of $–CH_2$ is difficult. However, the weaker peak at 1376 cm^{-1} for the residue also indicates that groups such as $–CH_3$ and $–CH_2$ have been extracted, because their trapped locations are important in solvent extraction. Similarly, the absence of a peak at 1717 cm^{-1} and weakening of the peak at 1250 cm^{-1} for the residue verify that micromolecules such as carbonyl, phenol, and ether have been extracted by pyridine. It should be mentioned that the stronger peaks at 1600, 1500, and 1450 cm^{-1} for the residue compared with the original sample also indicate that it is much easier for aliphatic structures to be extracted.

In the case of inertinite, the weakening of the peaks at 3430, 2950, and 2920 cm^{-1} and the absence of a peak at 1370 cm^{-1} for the residue show that the functional groups –CH, aliphatic $–CH_3$, $–CH_2$, and –CH were extracted by the solvent.

There is no significant change in the FTIR spectrum of exinite after extraction, because of the high content (above 60 %) of aliphatic hydrocarbons. The weakening of the peaks at 2850 and 1376 cm^{-1} for the residue can be ascribed to the dissolution of low-molecular-weight aliphatic structures. The variations, and corresponding explanations, in the peaks at 1600, 1500, and 1450 cm^{-1} are the same as in the case of vitrinite. Furthermore, the dissolution of carbonyl compound in the solvent is verified by the weakening of the C=O stretching peak at 1720 cm^{-1} and the absence of ether bond and carbonyl stretching peaks at 1120 cm^{-1}.

3.3.1.4 Macromolecular Networks Structural Features of Macerals

Structure Parameter and Three Dimension Model of the Three Macerals

The use of a combination of modern instrumental analysis, mathematical calculations, and modeling tools is an effective strategy for coal structural studies; for example, the experimental data were used to calculate the parameters of the average molecular structure, producing the expected structural parameters of the macromolecular maceral networks (Table 3.16).

The aromatic carbon ratio of vitrinite is 0.67, indicating a graphite-like 3D network of aromatic microcrystals, known as crystal nuclei, which contain four to five aromatic layers (with an average of nine to 11 aromatic rings per aromatic layer), with 0.357 nm between the layers, a perpendicular height of 1.250 nm, long aliphatic $(CH_2)_n$ side chains, and short side chains consisting of CH_3 and CH_2. More specifically, each aromatic nucleus consists of 110 aromatic carbons, 26 aromatic hydrogens, 55 aliphatic carbons, and 108 aliphatic hydrogens, giving an average molecular weight of 2530.

Table 3.16 Structural parameters of maceral microcrystals

Maceral	C_a^a	C_{al}	C	H_a^a	H_{al}	H	R	M^b
Vitrinite	110.8	54.6	165.4	25.8	108.3	134.6	43.7	2529.7
Inertinite	201.9	67.3	269.2	34.8	116.2	151.0	93.8	3812.5
Exinite	62.3	97.5	159.9	19.3	177.3	196.6	31.4	2607.6

[a] $C_a = (100L_a)2/2.62$. $H_a = \sqrt{6}\,C_a^{0.5}$

[b] M is the average molecular weight, calculates as $12C/C_{daf}$ %

The aromatic carbon ratio of inertinite is 0.75, with a more distinct 3D network; each crystal nucleus contains four aromatic layers (with an average of 23 aromatic rings per aromatic layer), with 0.370 nm between the layers, a perpendicular height of 1.076 nm, and only short CH_3 and CH_2 side chains. Each aromatic nucleus consists of 301 aromatic carbons, 35 aromatic hydrogens, 67 aliphatic carbons, and 116 aliphatic hydrogens, resulting in an average molecular weight of 3812.

The aromatic carbon ratio of exinite is only 0.39, and each crystal nucleus consists of three to four aromatic layers (with an average of 62 aromatic rings per aromatic layer), with 0.347 nm between layers, a perpendicular height of 1.150 nm, abundant aliphatic $(CH_2)_n$ structures, and short side chains consisting of CH_3 and CH_2; long-chain alkyl structures predominate. Each aromatic nucleus consists of 62 aromatic carbons, 19 aromatic hydrogens, 98 aliphatic carbons, and 177 aliphatic hydrogens, giving to an average molecular weight of 2607.

Three-dimensional macromolecular network models of the three maceral groups can be constructed, as shown in Fig. 3.9; each layer was optimized using molecular dynamics to give a minimum-energy model. High-resolution transmission electron microscopy (HRTEM) investigations of the maceral macromolecular structures (JEM-ARM1250 microscope, at Tokyo University) showed that the distance between aromatic layers within the microcrystals of inertinite (Pingshuo bituminous coal) is 0.34 nm, which is close to the calculated value of 0.36 nm. This confirms that the HRTEM results and the 3D model of macromolecular network based on the average molecular structure are consistent.

Comparison of the Macromolecular Structure Features of the Three Macerals

Generally, all the macerals of Pingshuo bituminous coal contain 3D networks of microcrystalline structures consisting of three to four aromatic layers, with diameters in the order inertinite > vitrinite > exinite. The microcrystalline structure

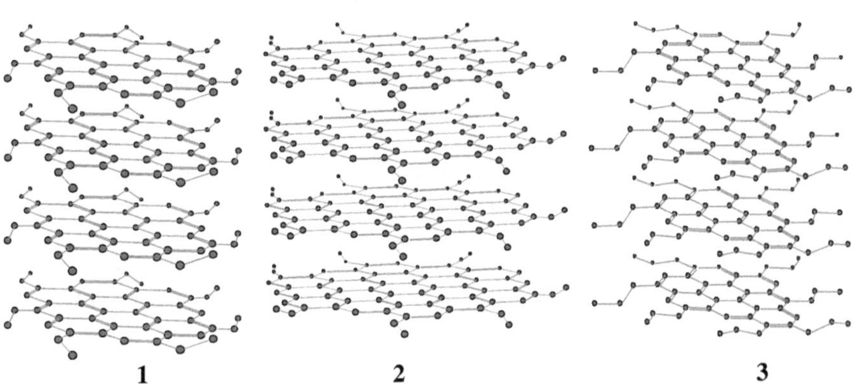

Fig. 3.9 Structural models of macromolecular networks *1* Vitrinite; *2* Inertinite; *3* Exinite

contains C=C, $-CH_2-$, –CH–, –O–, and –S– bridges, and intertwining and interlinking of these bridges further contribute to the macromolecular structure of the coal. There was no significant change in these macromolecular networks after pyridine extraction, which means that the pyridine extraction residue has the same macromolecular network as the original sample, and can therefore be regarded as the macromolecular networks of the macerals.

It should be noted that the aliphatic carbon content of exinite is the highest, with long-chain aliphatic alkyl carbon groups predominating, some of which are connected to the aromatic layer as substituents. The aliphatic alkyl content of inertinite is the lowest, and consists of short-chain alkyl groups. The aliphatic carbon content of vitrinite is moderate (compared with the other two), and consists of long-chain and short-chain alkyl groups. Although the three macerals all contain certain amounts of aliphatic carbon, the distributions of oxygen-containing functional groups, such as ether, hydroxyl, carbonyl, carboxyl, and ester groups, are different: there are abundant ether links in vitrinite and exinite, whereas inertinite contains more carbonyl and carboxyl groups, and small amount of hydroxyl. Similarly, the major nitrogen-containing compounds are pyrroles and pyridines in vitrinite, amines and pyrroles in inertinite, and amines and pyridines in exinite. Sulfur is present mainly as sulfoxides (RSOR′) and sulfides.

Both oxygen-containing functional groups (such as C=O, C–O, –OH, and –O–) and aliphatic structures (such as $-CH_3$, $-CH_2$, and –CH) dissolve in pyridine to some extent, and the solubility is significantly influenced by the spatial positions of the functional groups and the electron donators/acceptors connected to the macromolecular network by intermolecular forces. The amount of isolated micromolecules influences the extraction rate, which is inversely proportional to the aromatic carbon content. The isolated micromolecules are analyzed in combination with a solvent extraction process, rather than by direct analysis as solid-state specimens. The order of the micromolecular content is exinite > vitrinite > inertinite.

3.3.2 *Micromolecules Structure of Pingshuo Coal*

3.3.2.1 Sample Preparation of Micromolecular Phase Sample

The dry pyridine extraction product (obtained in Sect. 3.1.2) was placed in a three-necked flask containing a benzene:methanol (3:1) mixture as the extraction solvent, with an extraction product to solvent ratio of 1:20. The mixture was continuously stirred at 40 °C for 48 h and then filtered. The residue was subjected to the same extraction procedure for 1 h. After filtration, the filtrates were combined and concentrated using a rotary evaporator, producing a benzene–methanol-soluble product, which was further dried in a vacuum oven. The FTIR spectra of the pyridine extracts were similar to those of the corresponding benzene–methanol insoluble products. Further analysis of the benzene–methanol-insoluble and soluble products focused on the vitrinite sample.

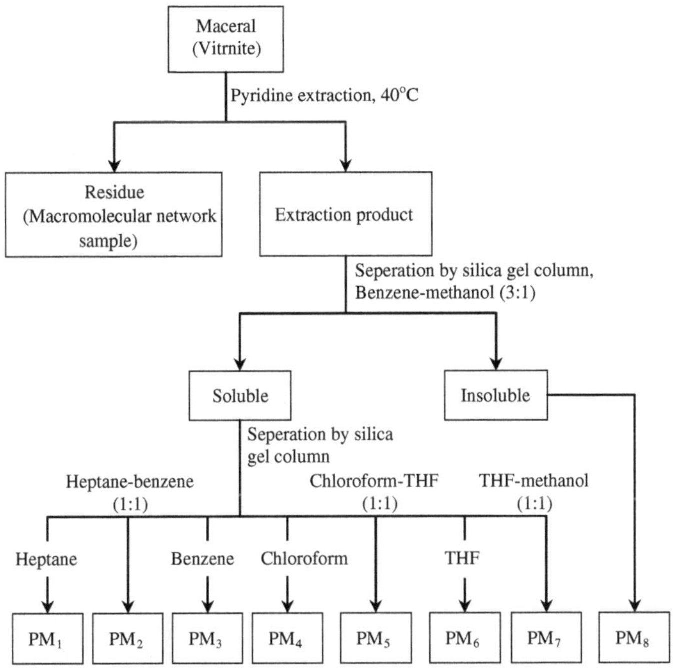

Fig. 3.10 Extraction and separation of micromolecules. Reprinted from Ref. [30], Copyright 1994, with permission from Taylor & Francis

Separation of vitrinite from the benzene–methanol-soluble product was carried out using silica-gel chromatography. The eluates were collected separately, and denoted by PM_1, PM_2, PM_3, PM_4, PM_5, PM_6, and PM_7. PM_8 represents an insoluble product. The procedure for micromolecular preparation is shown in Fig. 3.10, and the results of ultimate analysis of the macerals and the evaluated average molecular weights of PM_1–PM_8 are listed in the Table 3.17.

3.3.2.2 Micromolecular Phase Structure Characterization of Maceral

Analysis of Heptane Fractions of Pyridine Extracts by GC-MS

GC-MS, a tool for the rapid analysis of organic compounds, is widely used. In coal studies, it is useful for analyzing the constitutions and configurations of low-molecular-weight aromatic and aliphatic alkyl groups. The GC-MS results show there are at least 20 compounds in PM_1; they were identified as C_{14}–C_{34} *n*-alkanes by comparison of the acquired MS data with standard MS data. It should be noted that in a typical *n*-alkane mass spectrum, the ion intensity decays exponentially with increasing ion mass and the relative intensity of the molecular ion peak decreases with increasing length of the alkyl chain. The *n*-alkanes in the PM_1 fraction match

Table 3.17 Ultimate analysis of micromolecules and average molecular weights

Component	Ultimate Analysis (daf, %)					*H/C*	*O/H*	M^a
	C	*H*	*O*	*N*	*S*			
PM_1	85.67	14.33	0	0	0	2.00	0	357.26
PM_2	88.68	8.77	1.26	0	1.29	1.19	0.09	365.13
PM_3	86.12	9.20	2.45	0.92	1.31	1.28	0.02	373.66
PM_4	85.21	7.63	5.52	0.39	1.25	1.08	0.05	398.00
PM_5	79.05	6.49	41.64	1.12	1.70	0.99	0.01	587.67
PM_6	69.37	5.75	21.50	1.74	1.64	0.99	0.23	626.49
PM_7	46.82	4.13	43.77	4.86	0.42	1.06	0.66	267.99
PM_8	76.29	4.99	14.71	2.37	1.64	0.79	0.18	1024.9

[a]*M* is the average relative molecular mass, using VOP method, pyridine is the solvent

the features mentioned above; in particular, the mass spectrum of *n*-nonadecane (main peak) fitted the corresponding standard spectrum. Other results have shown that all coals contain series of *n*-alkanes. The major differences are the distribution of the carbon numbers, the carbon number of the main peak, and the odd–even preponderance (OEP). These factors vary with geological age and the coal-forming environment. For modern sedimentary rock, odd carbon atoms are predominant and the difference between odd and even atoms decreases with increasing sedimentation age. The absence of a preponderance of odd-carbon chains in Pingshuo bituminous coal, and major peaks with carbon numbers of C_{20}, C_{19}, and C_{22}, indicate that sedimentation occurred a long time ago. Moreover, the parameters such as the carbon preponderance index and OEP can be used to estimate the coal age:

$$\mathrm{CPI} = \frac{1}{2}\left[\frac{\sum_{i(\mathrm{odd})=25}^{33} C_i}{\sum_{i(\mathrm{even})=24}^{32} C_i} + \frac{\sum_{i(\mathrm{odd})=25}^{33} C_i}{\sum_{i(\mathrm{even})=26}^{34} C_i}\right] = 1.28,$$

$$\mathrm{OEP} = (-1)^{j+1}\frac{C_j + 6C_{j+2} + C_{j+4}}{4C_{j+1} + 4C_{j+3}},$$

j is the base carbon number for calculation of the OEP; for Pingshuo bituminous coal, $j = 25$.

Analysis of Fractions Using ^{1}H NMR Spectroscopy

The ^{1}H NMR spectra of fractions PM_1–PM_8 were obtained (Fig. 3.11). The hydrogen atom distributions were obtained from integration of the ^{1}H NMR spectra and accurate assignments of hydrogen atoms (as shown in Table 3.18).

The PM_1 fraction consists of aliphatic hydrocarbons, with long-chain saturated hydrocarbon compound as the major components. This finding is in accordance with the GC-MS results, and shows that the hydrogen atoms are mainly present in

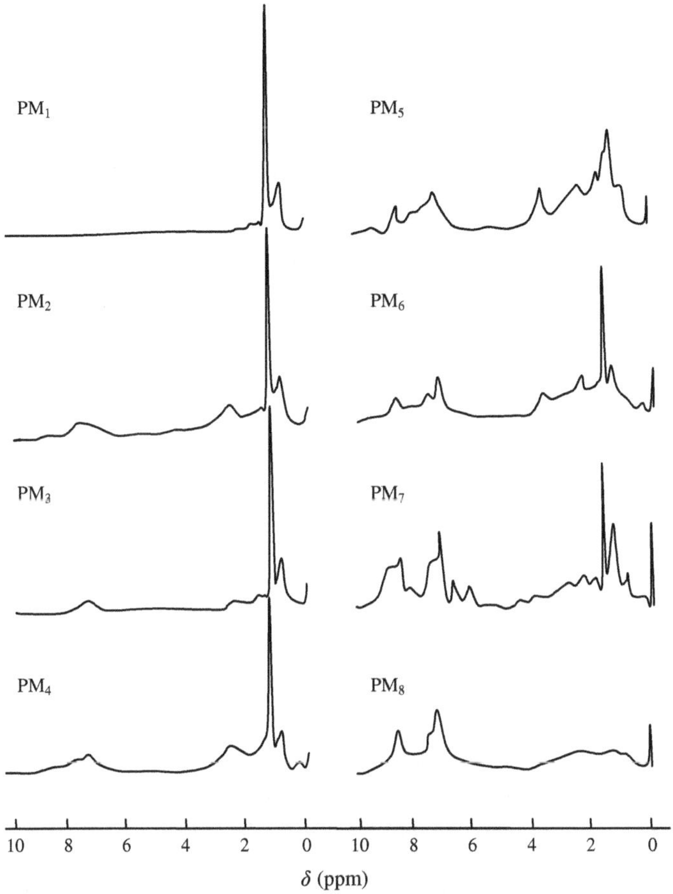

Fig. 3.11 ^{1}H NMR spectra of fractions of micromolecular phase

Table 3.18 Distribution of hydrogen atoms in fractions of micromolecules in Pingshuo Bituminous coal (%)

Type of H atom	^{1}H NMR chemical shift (ppm)	PM_1	PM_2	PM_3	PM_4	PM_5	PM_6	PM_7	PM_8
H_α	1.9–4.0	6.86	23.8	13.2	33.8	35.3	35.7	23.5	18.8
H_β	1.0–1.9	60.8	40.0	47.7	37.0	28.9	26.1	22.0	13.4
H_γ	0.5–1.0	32.4	15.0	23.2	10.5	10.6	6.2	5.9	3.6
H_a	6.0–9.0	0	21.8	15.9	19.3	25.4	32.0	49.0	64.3

the H_β region. For PM_2, more than 70 % of the hydrogen peaks originate from H_α, H_β, and H_γ. The broad peaks in the aromatic hydrogen region (6.5–8.5 ppm) indicate the presence of aromatic hydrogens in both single and condensed aromatic rings. In addition, the ratios of the three types of hydrogen indicate that in PM_2, aromatic rings substituted with alkyl groups containing more than three carbon atoms are enriched. For PM_3, the proportion of H_a decreases, and H_β accounts for 47.7 % of the total hydrogen, and signals from alicyclic rings are observed. In the aromatic hydrogen region (6.5–8.0 ppm), the proportion of H_a also decreases and small peaks originating from single, double, and triple rings are observed. In more detail, the PM_3 fraction mainly consists of single and double aromatic rings and a small amount of long-chain alkyl-substituted polyaromatic (more than three) rings. Similarly, PM_4 has aliphatic hydrogens in short alkyl-substituted side chains and aromatic hydrogens in condensed aromatics. For PM_5, the degree of aromatic condensation is much higher. The PM_5 fraction has shorter aliphatic side chains, more naphthenic structures, and complex substituents. PM_6 is similar to PM_5, with abundant cycloalkanes, and greater aromatic condensation. The ^1H NMR spectrum of PM_7 is even more complicated, because of possible hydrogen signals from CH_3–C–O as well as H_β from cycloalkanes. Aromatic hydrogens, representing 50 % of the total hydrogen atoms, are mainly contributed by structures such as

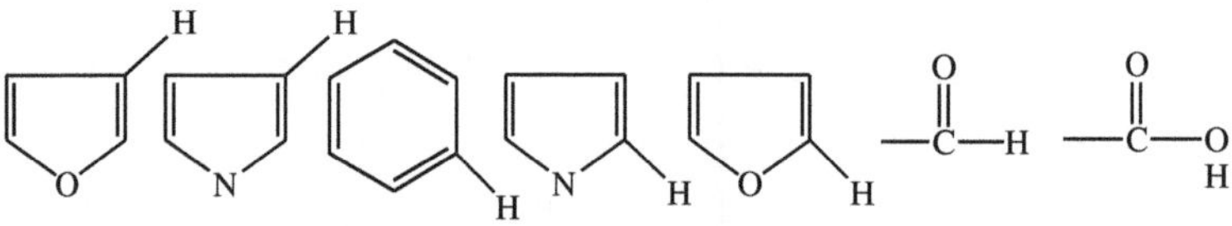

PM_8 contains benzene–methanol-insoluble aromatic structures, with the highest molecular weights. In more detail, the major hydrogen signal appears in the aromatic hydrogen range. The amount of phenolic hydroxyl groups is the highest in PM_8, with hydrogen signals at 4.5–6.0 ppm. There is a clear H_β peak representing aliphatic $(CH_2)_n$ chains. The weak H_α and H_γ peaks indicate that there are fewer alkyl-substituted groups in PM_8 than in the other fractions. PM_8 contains aromatic hydrocarbons with the highest degree of aromatic condensation, and the lowest amount of aliphatic hydrogens.

^{13}C NMR Analysis of Fractions PM_2–PM_7

It is not necessary to obtain the ^{13}C NMR spectrum of PM_1 because the structure has already been confirmed, as described in Sect. "Analysis of Heptane Fractions of Pyridine Extracts by GC-MS". In addition, because of the poor solubility of the PM_8 fraction, a high-resolution ^{13}C NMR spectrum cannot be obtained. The ^{13}C NMR spectra of PM_2–PM_7 were recorded using an FT-80A spectrometer, with a cumulative number of 8000–15 000; CS_2 was the solvent for PM_2, PM_3, and PM_4, and DMSO-d_6 was used for PM_5, PM_6, and PM_7. Based on the chemical shift assignments of the different types of carbons, the carbon distributions for fractions PM_2–PM_7 were calculated, and are listed in Table 3.19.

Table 3.19 Distribution of carbon atoms in micromolecules of Pingshuo Bituminous coal (%)

Type of carbon	^{13}C NMR chemical shifts (ppm)	PM_1	PM_2	PM_3	PM_4	PM_5	PM_6	PM_7
C_{al}	13–70	100	44.0	49.0	33.0	34.0	28.3	17.0
C_a	100–170	0	56.0	50.0	64.8	64.0	71.7	83.0

Table 3.20 Number of atoms in fractions of micromolecules of Pingshuo Bituminous coal

Fraction	C	C_a	H	H_α	H_β	H_γ	H_η	H_{OH}	H_a	S	N	O
PM_2	26.98	15.10	32.02	7.72	10.93	4.73	1.75	0	6.89	0.15	0	0.29
PM_3	26.2	13.41	34.38	4.47	14.71	7.90	1.79	0	5.51	0.15	0.25	0.57
PM_4	28.26	18.31	30.37	9.72	9.62	3.19	1.61	0.54	5.69	0.16	0.11	1.37
PM_5	38.71	24.77	38.14	12.93	7.36	4.04	3.60	0.53	9.68	0.31	0.46	4.27
PM_6	36.22	25.96	36.02	11.89	7.06	2.23	2.30	2.36	9.68	0.32	0.78	8.41
PM_7	10.46	8.96	11.07	2.31	2.00	0.65	0.43	1.10	4.68	0.04	0.93	7.33
PM_8	65.16	-	51.15	7.77	4.91	1.85	1.94	3.99	30.7	0.52	1.73	9.42

For PM_2, the aliphatic carbon signals occurred at 15.1, 20.7, 23.9, 30.8, 32.9, and 38.0 ppm, and were assigned to aliphatic alkyl and naphthenic groups. The main aromatic carbon peak appears at 128 ppm, and indicates the presence of phenylbenzene. The PM_3 fraction contains long-chain alkyl groups and phenylbenzene, similar to PM_4. For PM_5, the aliphatic carbon signal is weak and the signal at 66.1 ppm is ascribed to carbon in cycloalkyl alcohols. Carbon signals from aromatic carbons in condensed aromatic structures and signals from phenolic hydroxyl or ether groups are also observed. PM_6 contains aromatic structures with cycloalkyl alcohol substituents together with other oxygen-containing functional groups. For PM_7, the aliphatic signal is weak compared with the clear peaks in the aromatic range. In particular, the signal at 105 ppm represents aromatic carbons combined with oxygen or nitrogen. The other signals are assigned to protonated carbons, bridging carbons, aromatic carbons with various substituents, and carboxyl carbons. The numbers of atoms in the average molecular structures in PM_2–PM_8 were calculated (Table 3.20) based on the elemental analyses of fractions, the average molecular weights (listed in Table 3.17), and analysis of hydrogen and carbon distributions (listed in Tables 3.18 and 3.19).

3.3.2.3 Structural Parameters and Average Molecular Structures of Micromolecules

Integrated Analysis of PM_2–PM_7 Using ^1H NMR and ^{13}C NMR Spectroscopies

A versatile analytical equation (Table 3.21) was developed for calculating average molecular structural parameters, based on ^1H NMR and ^{13}C NMR data and the constituents of the respective pyridine extraction fractions. The calculation of the number of heteroatoms is calculated based on the amount of phenolic hydroxyl groups, and the carboxyl, carbonyl, saturated ether, and alcohol groups in the spectrum. The fractions contain only small amounts of aromatic ether links, so the distribution of oxygen atoms, besides in the form of phenolic oxygen, can be assumed to be one-third originating from aromatic ethers (denoted by O_{ae}), one-third from carboxyl and carbonyl groups, and others from alcohols. It is interesting to note that sulfur and nitrogen in PM_2–PM_7 are both present only in aromatic structures. Moreover, in the calculation of the number of bridging carbons in the aromatic rings or the number of aromatic rings, the carbons originating from carboxyl or carbonyl (denoted by C_o) should be subtracted, because they are not

Table 3.21 Definitions and calculations of structural parameters (PM_2–PM_7)

Structural parameter	Definition	Calculation
C_a	Number of aromatic carbon atoms	$C \times f_a$
x	Alkyl-substituted H/C	H_{al}/C_{al}
$C_{a(un)}$	Number of non-substituted aromatic carbon atoms	H_{al}
$C_{a(s)}$	Number of substituted aromatic carbon atoms	$\frac{H \times H_a^*}{x}$
C_{anb}	Number of aromatic non-bridging carbon atoms	$C_{a(un)} + C_{a(s)} + O_{ae} + N + S$
C_{ab}	Number of aromatic bridging carbon atoms	$C_a - C_{anb} - C_o$
M_{an}	Number of aromatic nuclei	$\frac{1}{3}C_{anb} - \frac{1}{6}C_a$
R_a	Number of aromatic rings	$\frac{1}{2}(C_a - C_{anb}) + M$
R	Number of ring structures	$\frac{1}{2}(2C + 2 - H - C_a)$
R_N	Number of aliphatic rings	$R - R_a$
C_N	Number of carbons in aliphatic rings	$3.5R_N$
C_γ	Number of γ-CH_3 on aromatic rings	$\frac{H \times H_r^*}{x}$
C_α	Number of α-CH_3, α-CH_2, and α-CH on aromatic ring	$C_{\alpha(s)}$
C_β	Number of β-CH_3 and β-CH_2 on aromatic ring	$C_{al} - C_\alpha - C_\gamma$

Table 3.22 Calculated average molecular structural parameters of Pingshuo Bituminous coal

Fraction	C_a	x	$C_{a(un)}$	$C_{a(s)}$	C_{anb}	C_{ab}	M_{an}
PM_2	15.10	2.11	6.89	3.61	10.75	4.35	1.06
PM_3	13.41	2.16	5.51	2.10	7.97	5.44	0.42
PM_4	18.31	2.48	5.69	4.46	11.66	7.89	0.84
PM_5	24.77	2.05	9.68	6.85	18.54	6.46	2.05
PM_6	25.96	2.39	9.68	7.53	20.31	7.65	2.44
PM_7	8.68	3.65	4.68	1.43	8.68	1.60	1.45
Fraction	R_a	R	R_N	C_N	C_α	C_β	C_γ
PM_2	3.18	4.40	1.22	4.9	3.61	6.68	1.58
PM_3	3.72	3.90	0.18	2.45	2.10	8.71	2.60
PM_4	4.17	4.90	0.73	1.22	3.92	4.97	1.06
PM_5	5.17	8.30	3.13	9.52	6.32	6.27	1.35
PM_6	5.27	6.80	1.53	2.63	5.17	4.34	0.74
PM_7	1.45	1.60	0.15	0.45	0.33	0.81	0.61

present in the skeletons of aromatic rings, although their peaks appear in the aromatic carbon region. The average molecular structural parameters based on the above mentioned calculation and adjustment methods are listed in Table 3.22.

Integrated Analysis of PM_8 Using ^1H NMR and FTIR Spectroscopies

The structural parameters of the PM_8 fraction were calculated (Table 3.23) based on ^1H NMR and FTIR spectroscopic data.

The average molecular structures (Fig. 3.12) were constructed according to the parameters listed above. This is useful information for the investigation of coal reactions. The micromolecules in PM_2–PM_8 are aromatic hydrocarbons (including heterocyclic compounds such as in PM_7) with different substituents (such as aliphatic alkyl, naphthenic, and oxygen-containing functional groups).

3.3.3 Summary

The approaches of integrating separation and extraction methods together with the physical analysis methods are powerful tool for understanding the basic chemical structure of coal, and linking these features to the physical structure. Information on the amount and type of functional groups present in a sample can be used to infer the structure of the micromolecular compounds that are present. The usefulness of this detailed understanding of coal structure and its relevance to the study of coal reactivity will be discussed further in the following chapters.

Table 3.23 Definitions and calculations of average molecular structural parameters (PM8)

Structural parameter	Definition	Calculation	Result
C	Number of carbon atoms	$\frac{\overline{M}\times C_{daf}}{12}$	65.15
x	Alkyl-substituted H/C	H_{al}/C_{al}	1.55[a]
f_a	Aromaticity	$\frac{C-(H_\alpha+H_\beta+H_\gamma)/x}{C}$	0.82
σ	Substitution index	$\frac{H_\alpha/x}{H_\alpha/x+H_\beta}$	0.13
C_P	Number of carbon atoms around aromatic nucleus	$\left(\frac{H_\alpha}{x}+H_\alpha\right)\times H_{daf}\times\overline{M}$	0.71
C_a	Number of aromatic carbon atoms	$C\times f_a$	53.42
C_P/C_a	Condensation index	$\frac{H_\alpha/x+H_\alpha}{C/H-\left(H_\alpha+H_\beta+H_\gamma\right)/x}$	32.72
R_a	Number of aromatic rings	$\frac{1}{2}\left(C_a-C_p\right)+1$	11.35
R	Number of ring structures	$\frac{1}{2}C\times(2-H/C-f_a)+1$	13.87
R_N	Number of aliphatic rings	$R-R_a$	2.52
n	Number of alkyl-substituted aromatic nuclei	$\frac{H_\alpha}{x}\times H_{daf}\times\overline{M}$	5.14

[a]The IR-related data were calculated using the following equation:

$$\frac{H_{CH_3}}{H_{al}}=k_1\frac{A_{1380}}{A_{2920}},\quad \text{where } k_1=2.4$$

$$\frac{H_{CH_3}+H_{CH_2}}{C_{CH_3}+C_{CH_2}}=\frac{\frac{1}{5}+\frac{1}{7}k_2\frac{A_{2920}}{A_{1380}}}{\frac{1}{15}+\frac{1}{14}k_2\frac{A_{2920}}{A_{1380}}},\quad \text{where } k_2=0.3966$$

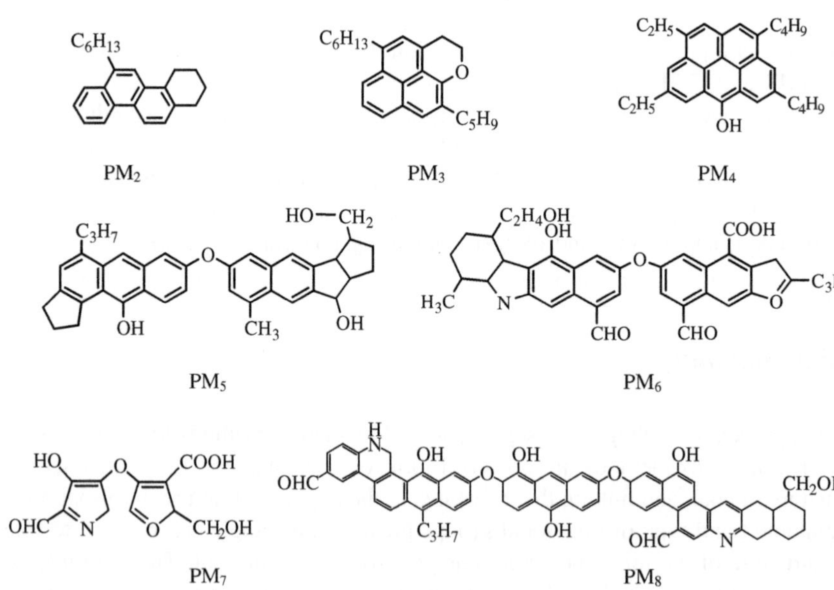

Fig. 3.12 Average molecular structure of micromolecular phase in various fractions

References

1. He XM (2010) Coal chemistry. Metallurgical Industry Press, Beijing, p 42
2. Niekerka DV, Pugmire RJ, Solum MS et al (2008) Int J Coal Geol 76:290–300
3. Zhu PZ, Gao JS (1984) Coal chemistry. Shanghai Science and Technology Press, Shanghai, p 58
4. Erdenetsogt B-O, Lee I, Lee SK et al (2010) Int J Coal Geol 82:37–44
5. Thomas KM (1997) Fuel 76:457–473
6. Li F, Zhang YF, Xie KC (1993) Fuel Sci Technol Int 11(8):1113–1131
7. Chou CL (2012) Int J Coal Geol 100:1–13
8. Yu JS (2000) Coal chemistry. Metallurgical Industry Press, Beijing, p 61
9. Petersen HI, Rosenberg P, Nytoft HP (2008) Int J Coal Geol 74:93–113
10. Blom LL, Edelhausen L, van Krevelen DW (1957) Fuel 36:135–153
11. Xie KC (1992) Coal Convers 15(1):24–30
12. Botto RE, Cody GD (1994) ACS Div Fuel Chem Prepr 39(1):205
13. Lafferty CJ, Eloi C, Anderson PK et al (1994) ACS Div Fuel Chem Prepr 39(3):810
14. Alan AH, Kandiyoti R (1995) Fuel 74(5):784–786
15. Krzton A, Cagniant D, Gruber R (1995) Fuel 74(2):217–225
16. Cody GD, French DC, Botto RE (1994) ACS Div Fuel Chem Prepr 39(1):59–63
17. Zhu PZ, Gao JS (1984) Coal chemistry. Shanghai Science and Technology Press, Shanghai, p 94
18. He XM (2010) Coal chemistry. Metallurgical Industry Press, Beijing, p 97
19. Spiro GL (1981) Fuel 60:1121
20. Given PH (1960) Fuel 39:147
21. Devidson RM (1982) Coal science. Academic Press, New York, p 5
22. Solomon PR (1981) ACS Symp Ser 169:61
23. Heredy LA et al (1980) ACS Div Fuel Chem Prepr 25(4):38
24. Carlson GA et al (1989) ACS Div Fuel Chem Prepr 34(3):780
25. Shinn JH (1984) Fuel 63:1187
26. Carlson GA et al (1991) ACS Div Fuel Chem Prepr 36(1):398
27. White WE et al (1990) Adsorpt Sci Tech 7:108
28. Faulon JL et al (1993) Fuel Process Tech 43:277
29. Xie KC, Li WY, Zhu SY (1994) Fuel Sci Technol Int 12(9):1159–1168
30. Li F, Xie KC et al (1994) Fuel Sci Technol Int 12(1):151–162

References

Chapter 4
Coal Pyrolysis Reactions

Abstract Pyrolysis processes are among the most important of coal's reactions and a necessary process for the production of coke used in multiple industries. Additionally, the study of pyrolysis processes gives insight into the structure and composition of different types of coal. Owing to the complexity of these chemical and physical decomposition processes, they remain poorly understood. In this chapter, we discuss some of the general physical and chemical models that are used to describe the changes that various types of coal undergo during thermal decomposition in the absence of oxygen. The distribution of chemical products, kinetics and thermodynamic parameters of pyrolysis reactions are introduced and related to specific experimental methods and techniques. In this context, the findings of studies into various types of Chinese coal are presented. This includes discussion of the stages that occur at different pyrolysis temperatures and how the pyrolysis behavior is affected by coal type. The results we present are rationalized in terms of the structural properties of the coal.

4.1 Introduction to Coal Pyrolysis

The word "pyrolysis", derived from the Greek words "pyr" (fire) and "lysis" (decomposition), refers to the chemical processes that change a material into another or several other materials by heat energy. Coal pyrolysis is a series of chemical and physical reactions that occur when coal is heated continuously in an air-free or inert atmosphere without a catalyst. The breaking of chemical bonds is the fundamental process in pyrolysis. Coal pyrolysis is the first step in coal gasification and other chemical processes, and is the basic process in clean coal technologies. Coal pyrolysis is closely related to coal composition and structure. The molecular structure of coal can be clarified based on pyrolysis studies. Coal pyrolysis is an artificial carbonizing process, which is similar to the natural coal-forming process. An in-depth knowledge of pyrolysis is therefore helpful in coalification studies. The mechanism of coal pyrolysis and the nature and distribution of the products are

K.-C. Xie, *Structure and Reactivity of Coal*, DOI 10.1007/978-3-662-47337-5_4

significantly affected by specific conditions, including the coal quality, heating rate, heat transfer, and pyrolysis atmosphere. Thermal processing is currently the most important method of coal processing. The large-scale coking industry is a typical example of the thermal processing of coal. The study of coal pyrolysis has a very close technical relationship with the thermal processing of coal. In the coking industry, results of pyrolysis studies help in the selection of the coal feed, expansion of coking methods, determination of the optimum processing conditions, and improving the product quality. It can also guide the development of new technologies for coal thermal processing such as high-temperature rapid pyrolysis, hydropyrolysis, and plasma pyrolysis.

4.1.1 Pyrolysis Processes

4.1.1.1 Changes to Typical Bituminite During Heating

When coal is heated under air-free conditions, the organic matter undergoes a series of changes as the temperature increases, forming gaseous (coal gas), liquid (tar), and solid (semi-coke or coke) products. The changes that occur in a typical bitumite during heating are shown in Fig. 4.1.

Coal pyrolysis can be roughly divided into three phases as shown in Fig. 4.1.

The first phase, RT (RT to T_d (300 °C, i.e., active thermolysis temperature) is referred to as the dry degassing phase. The coal shape remains basically unchanged during this phase. Lignite decarboxylates above 200 °C, and pyrolyzes at about 300 °C. Anthracite and bitumite remain generally unchanged. Dehydration occurs below 120 °C. The removal of gases such as CH_4, CO_2, and N_2, completes at approximately 200 °C.

The second phase occurs at around 600 °C. Depolymerization and decomposition occur mainly in this phase. Large amounts of volatiles (gas and tar) are generated and discharged. The maximum amount of tar is discharged at about 450 °C, and the maximum amount of gas is released at 450–600 °C. The gas is mainly composed of hydrocarbons, CO_2, and CO, and has a higher calorific value. The tar

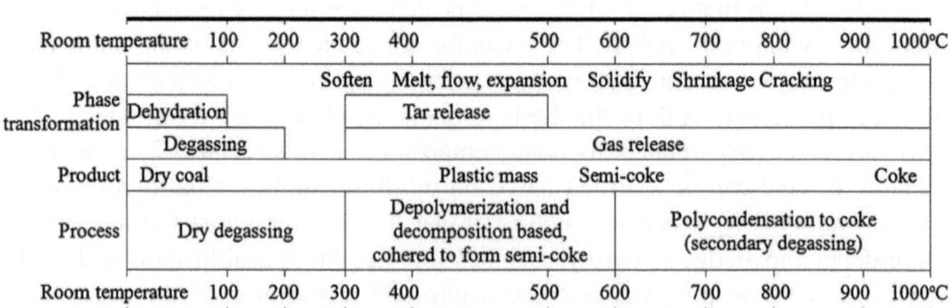

Fig. 4.1 Pyrolysis process for typical bitumite

is mainly composed of complex aromatic and fused ring aromatic compounds. Bitumite begins to soften at approximately 350 °C, and then melts and coheres, before forming semi-coke at 600 °C. Some of the physical indicators of semi-coke such as the average size of the aromatic layers and the helium density, do not change significantly compared with those of raw coal, which indicates that little polycondensation occurs during semi-coke formation. In this phase, bitumite (especially bituminite of medium metamorphic grade) softens, melts, flows, and expands, until it is solidified. A series of specific phenomena are observed, and the plastic mass coexists in three phases (i.e., gas, liquid, and solid). Liquid crystals (intermediate phase) are present in the liquid phase. The quantity and quality of the plastic mass determine the degree of cohesion and the coking properties.

The third phase occurs at 600–1000 °C. Semi-coke is converted to coke, and the main reaction is polycondensation. The amount of tar released is minimal and the volatiles are mainly gases, so it is also known as the secondary degassing phase. The gas consists mainly of H_2 and some CH_4. During conversion of semi-coke to coke, a large amount of gas is produced, the coke density increases, and the volume decreases, leading to generation of many cracks and fragment formation. The coke lumpiness and strength are directly related to the contraction conditions. If the final temperature is above 1500 °C, there is a graphitizing phase, which is used for the production of graphite carbon products.

4.1.1.2 Coal Pyrolysis Research

Coal has a complex structure, and coal pyrolysis involves three phases (i.e., coke, tar, and gas), therefore a quantitative description of the process is difficult, and thus the key is the rationality of the theoretical models. Early theoretical studies established the basic chemical reactions in the pyrolysis process, determined the gaseous products, and described the heat and mass transfer processes. With the development of research tools, some new theoretical models appeared, e.g., the single first-order chemical kinetic model, a set of first-order chemical kinetic models, and a kinetic model for a group of chemical reactions. As increasing numbers of pyrolysis reactions were proposed, these models became increasingly sophisticated. Many different pyrolysis theories have been proposed, and these can be broadly divided into two types: the classical theory and the recently proposed depolymerization theory. The classical theory is represented by the pioneering work of Howard [1]. Suuberg [2, 3] made a detailed summary in a comprehensive review of work in this area. The classical theory proposes the release of volatiles during the secondary polycondensation of the residue. In the depolymerization theory proposed by Gavalas et al. [4, 5] coal is regarded as a crosslinked macromolecular solid, and pyrolysis is regarded as a depolymerization process; the model compound is regarded as representing the coal structure and composition.

The two pyrolysis theories are currently studied using simple research approaches, e.g., thermogravimetric (TG) online chromatography [6], which determines product and weight losses to deduce the reaction process, and the model compound method [7], which identifies the pyrolysis mechanism using model compounds. Many factors affect pyrolytic activity: coal type, coal rank, particle size, pyrolysis medium, heating rate, final pyrolysis temperature, and the reactor. Pyrolysis reactor types include fixed-bed, fluidized-bed, ebullated-bed, wire-mesh, and Curie-point reactors. Previous studies have been restricted to raw coal or vitrinite, and the methods have been simple, resulting in defects or limitations in the studies of pyrolysis.

The early theoretical studies mainly established the basic chemical reactions during thermolysis, and the heat and mass transfer processes. Based on chemical measurements, Van Krevelen [8] proposed a series of chemical reactions for pulverized coal decomposition and calculated their characteristic coefficients. Nusselt [9] presented a simple sphere heat transfer model, which only describes the energy changes in coal particles during thermolysis. Badzioch [10] studied the chemical kinetics of pyrolysis and determined the main gaseous products in the thermolysis of pulverized coal. Improved experimental methods enabled new insights into coal pyrolysis, and new theoretical models have emerged. The weight loss curve during coal pyrolysis was calculated using a single first-order chemical kinetic equation by Kabayashi et al. [11], Pitt [12] described the pyrolysis process using a set of parallel or independent chemical kinetic equations. The activation energies of the steps in the Pitt model were calculated, using Gaussian functions, by Anthong et al. [13] Suuberg [14] suggested that coal pyrolysis may be primarily a radical process, which is caused by weak bond fractures within the structure. Based on his coal model structure, Given [15] proposed that coal pyrolysis involves the following four steps: removal of hydroxyl groups at low temperature (400–500 °C), dehydrogenation of some hydrogenated aromatic structures, molecular fracture at methylene bridges, and alicyclic fracture. A series of pyrolysis processes were proposed by Wiser et al. [16] starting with two free radicals generated from pyrolysis of aromatic cluster bonds. These free radicals are stabilized by rearrangement within the fragment or by collisions with other radicals.

Suuberg [17] deduced, based on the bond energy, that the activation energy of coal pyrolysis is 146,293 J mol^{-1}, and the pre-exponential factor values are in the range 1010–1013 s^{-1}. In the models developed by Suuberg and others, it is assumed that coal pyrolysis products are derived from solid structural decomposition of two or more types of coal. In the Pitt–Anthong model [12], the reaction rate is calculated using a set of first-order chemical kinetic equations. Gavalas et al. [18] made comparatively detailed studies of the chemical reaction mechanism in coal pyrolysis. Their experiments showed that coal pyrolysis consists of 14 chemical reaction groups and over 50 reaction equations. A complex chemical kinetic model has been proposed, and is by far the most complex and detailed model.

Currently, some active research approaches are being pursued. One approach is to detect the products in the pyrolysis process from the corresponding weight losses, and to determine the reaction process by deducing the products formed and the reaction extent. This research is performed using TG online chromatography. The advantage of this method is that it is simple and practical, and it is easy to analyze the products. The disadvantage is that many assumptions are made.

The limitations of the thermal analysis equipment and heat and mass transfers cause complications, and sometimes false conclusions are reached. The other approach is to regard the model compound as the pyrolysate, and this can accurately determine the coal pyrolysis mechanism. This approach of establishing and correcting pyrolysis models cannot be replaced by the former approach. However, because of the non-linearity of coal in a complex system, the individual processes cannot be explained separately, and can only be fed back into the original system to reflect the coal pyrolysis. This approach therefore has significant limitations. Accurate quantitative description of the pyrolysis process is currently an important issue in coal pyrolysis research. Despite considerable efforts, the reaction mechanisms of coal pyrolysis remain unclear. In prior studies, there has clearly been inadequate consideration of the maceral compositions and the effects of their interactions in coal pyrolysis. The effects of additives such as CaO or minerals in coal on the pyrolysis reactivity of macerals have a direct impact on clean coal conversion and use. The pyrolysis reaction capacities of functional groups in coal are directly related to the pyrolysis reactivity of the entire coal. It may be possible to avoid complex reaction mechanisms and obtain a relatively complete understanding of coal pyrolysis activity using an online response analysis.

4.1.1.3 Factors That Affect Coal Pyrolysis

The impact of coalification degree: The coalification degree is one of the most important factors in coal pyrolysis. Table 4.1 shows that the initial pyrolysis temperature gradually increases with increasing coalification degree. The composition of the pyrolysis products and pyrolysis reactivity are also related to the coalification degree. In general, the gas and tar yields and pyrolysis reactivity of young coal are higher than those of old coal.

The impact of coal particle size: If coal particle pyrolysis is controlled by chemical reactions, the pyrolysis rate will not be related to the particle size or particle pore structure. Badzioch [19] has identified two particle sizes, 20 and 60 μm, with the same pyrolysis rates. Howard et al. [1] used pulverized coal samples with common particle sizes (50 wt% < 35 μm, 80 wt% < 74 μm) and compared their pyrolysis rates, but no impact of particle size was found. Similarly, Wiser [20] used coal samples of particle sizes 60–74 and 246–417 μm in weight loss experiments; no differences between the weight loss curves were observed.

The impact of heat transfer: If heat transfer resistance mainly occurs between the particle and its surroundings, the temperature of the particle during heating is consistent, and the heating rate decreases with increasing particle size. Under these conditions, the pyrolysis rate during the heating process increases with decreasing

Table 4.1 Initial pyrolysis temperatures (°C) of various coal types

Coal type	Peat	Lignite	Bitumite	Anthracite
Initial temperature for pyrolysis	190–200	230–260	300–390	390–400

Table 4.2 Relationship between particle size and heating rate

Heating rate (°C s^{-1})	100	1000	10,000
Critical diameter (μm)	2000	500	200

particle size, but when the heating rate exceeds a certain value, the reaction extent during the heating process can be neglected. For coal samples smaller than a certain particle size, the pyrolysis rate is chemically controlled, and does not depend on the heating rate and particle size. Badzioch's [10] calculation shows that the turning point is at a particle size of about 100 μm. Under extreme conditions, the heat transfer rate is controlled completely by the particle interior, and the switch from a chemically controlled pyrolysis rate to heat transfer control is related to the particle size. When the particle size is smaller than the particle size at this turning point, the temperature at the particle center is approximately equal to its surface temperature. Koch et al. [21] showed that the change with critical particle size of control of the devolatilization rate from first-order reaction control to heat transfer control is a function of the heating rate (Table 4.2).

In addition, the heating rate, the final temperature, the reaction pressure, and chemical processing of coal samples have significant impacts on coal pyrolysis. These will be discussed in subsequent sections together with specific pyrolysis processes.

4.1.1.4 Changes in Surface Structure During Pyrolysis

Coal pyrolysis occurs before coal ignition. Changes in the surface structure during coal pyrolysis directly affect the coal combustion status. Maria [22] studied the changes in surface area at low temperatures during coal pyrolysis. The results are shown in Table 4.3. The table shows that the surface areas measured by both N_2 and CO_2 physisorption methods increased in varying degrees with increasing temperature and pyrolysis time. However, the range of the increase in the surface area measured by CO_2 physisorption is far larger than that in the surface area measured by N_2 physisorption. Below 500 °C, the surface area measured by CO_2 physisorption increases continuously, whereas the surface area measured by N_2 physisorption is almost

Table 4.3 Variations in specific surface area during coal pyrolysis

Temperature (°C)	Weight loss ratio (%)	Specific surface area (m^2 g^{-1})		
		N_2 (77 K)	CO_2 (195 K)	CO_2 (298 K)
350	4.6	<1	21	118
375	8.2	<1	22	127
400	13.0	<1	27	147
500	22.4	2.4	233	304
600	24.9	10	206	383

Reprinted from Ref. [22], Copyright 1983, with permission Elsevier unchanged. Only N_2 adsorption on mesopores and macropores can be detected, whereas with CO_2, the microporous structure of the coal can be detected. This shows that large amounts of microporous structures are formed as a result of volatile release when the temperature starts to rise in pyrolysis, but the macropores and mesopores are mainly unchanged. When the temperature rises further, volatile emission increases. This not only creates a new microporous structure, but also further changes the original micropores into mesopore and macropores, so the surface areas measured by N_2 and CO_2 physisorption both increase. Maria also studied the pyrolysis of high-volatile coal samples. At the beginning of pyrolysis, the surface area of CO_2 increases continuously, whereas the surface area of N_2 decreases. Maria suggests that this phenomenon is the result of severe reactions at the beginning of pyrolysis, caused by the high volatile content. When micropores are formed, the macropores collapse, resulting in different trends in the surface areas measured using the two methods.

Studies of the pore structure changes during pyrolysis of four types of Chinese coal have been reported [23]. Below 500 °C, the pore volume and pore surface area change little, consistent with Maria's conclusions. At 500–700 °C, large amounts of volatiles in the coal are emitted, such as C_nH_{2n}, CO, and CO_2, which mainly come from inside the coal particles, resulting in increased pore volume and area. Above 700 °C, some of the pores are blocked, and the number of pores is reduced because of the coal plasticity and tar release, decreasing the pore surface area and volume. Above 800 °C, the volatiles are all lighter substances such as H_2, and their release leaves many pores, so the surface area rapidly increases.

As volatiles are released during pyrolysis, the coal pore system develops, forming microporous structures, and increasing the internal surface and roughness. Studies of the changes in the fractal dimension of Victoria lignite during pyrolysis showed that the fractal dimension change varies with different atmospheres and processing conditions [24]. Typically the fractal dimension in the low-temperature range does changes little, but increases significantly with temperature. In a N_2 atmosphere, the increase in the fractal dimension is less noticeable. For acid-washed coal samples, the fractal dimension is always stable, at around 2, which explains why acid washing produces coal with a smooth surface, forming a Euclidean plane.

4.1.2 Chemical Reactions of Coal During Pyrolysis

The complexity of the chemical reactions during coal pyrolysis prevents a complete understanding of the chemistry. Coal pyrolysis involves changes in the elemental composition, and chemical and physical properties over different stages during coal pyrolysis. Chemical reactions during coal pyrolysis can be divided into two major types: cracking and polycondensation. These reactions involve many paths such as the cracking of organic matter in coal, volatilization of low- molecular-weight pyrolysis products, polycondensation of cracking residues, decomposition and combination of volatile products during emission, and further decomposition and repolycondensation of the polycondensation products. The molecular structure of coal suggests that pyrolysis involves continual cracking of thermally unstable elements, including the side chains, bridging bonds, and the functional groups around the basic structure. Low-molecular-weight compounds are formed and then escape. The condensed aromatic nuclei of the basic structural element remain heat stable, and mutual polycondensation results in a solid product (i.e., semi-coke or coke).

4.1.2.1 Laws Governing Pyrolysis of Organic Compounds

The general laws governing pyrolysis of organic compounds are presented to explain coal pyrolysis. The thermal stability of organic compounds depends mainly on the molecules' chemical bonds. The order of the thermal stabilities of hydrocarbons is generally condensed arene > arene > cyclane > alkene > alkane. The longer the side chain on the aromatic ring, the more unstable the side chain; the more aromatic rings there are, the more unstable the side chains; and the more condensed the polycyclic aromatic hydrocarbons are, the higher the thermal stability.

Coal pyrolysis products are extremely complex compounds. The main products are cyclic aromatics. These compounds contain from a single ring to over 20 rings, and there are also oxygen-, nitrogen-, and sulfur-containing heterocyclic compounds. There is a small amount of aliphatic hydrogen and partially saturated hydrogenated aromatic structures in the ring systems. Generally, there is only one heteroatom at most. Aromatic and heterocyclic structures are present in two forms: unsubstituted and substituted. The main substituents are methyl, ethyl, and hydroxyl. Small amounts of aliphatic chains are present in the low-boiling fraction, but they tend to disappear in the high-boiling fraction. Molecules containing more than four rings are generally condensed, but can also have chain structures, with incomplete condensation.

4.1.2.2 Coal Pyrolysis Reactions

Coal pyrolysis is governed by the general laws for pyrolysis of organic compounds. Based on the reactions and phases involved, pyrolysis reactions are generally classified into coal-cracking, secondary, and polycondensation reactions.

Cracking Reactions in Coal Pyrolysis

When the coal temperature reaches a certain value, the chemical bonds in its structure break. This decomposition reaction directly affects coal molecules, and is the first reaction during coal pyrolysis; therefore it is often called primary pyrolysis and includes the following reaction types: free-radical generation by cracking of bridging bonds which are the weakest links in coal structure. The concentration of free radicals in coal increases with increasing temperature; cracking of aliphatic side chains during heating generates gaseous hydrocarbons such as CH_4, C_2H_6, and

C_2H_4; cracking of oxygen-containing functional groups occurs where the order of thermal stabilities of oxygen-containing functional groups in coal is –OH > C=O > –COOH > –OCH_3. Hydroxyl is not easily removed, and water can be generated in the presence of a large amount of hydrogen at 700–800 °C or above. The cracking of carbonyl groups at about 400 °C generates CO. At 200 °C or above, carboxyl groups decompose to CO_2. Oxygen-containing heterocyclic rings are broken and cracked, generating CO, at 500 °C and above; cracking of low-molecular-weight compounds such as aliphatics occurs with decomposition to gaseous hydrocarbons.

Secondary Reactions in Coal Pyrolysis

The volatile components of the primary pyrolysis product continue to decompose and create secondary cracking reactions during releasing at higher temperatures (e.g., in a coking oven). The main secondary cracking reactions are as follows.

Direct cracking:

$$C_2H_6 \xrightarrow{\text{£-}H_2} C_2H_4 \xrightarrow{\text{£-}CH_4} C$$

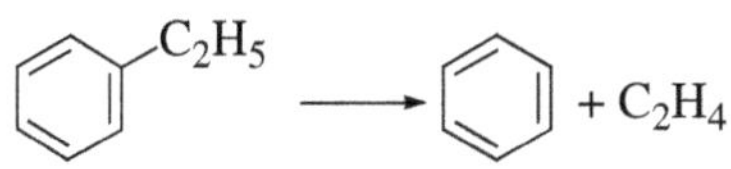

Aromatization:

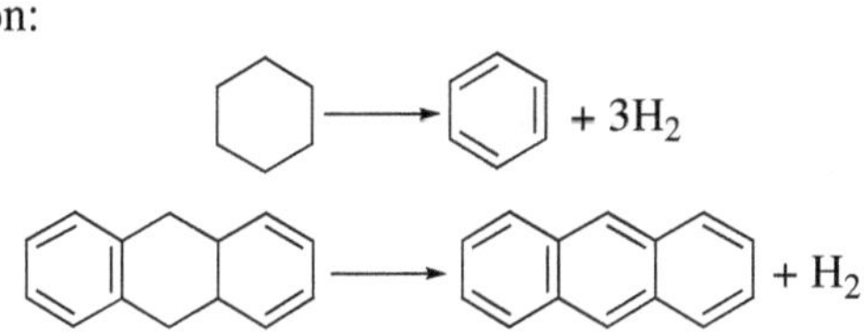

Hydrogenation:

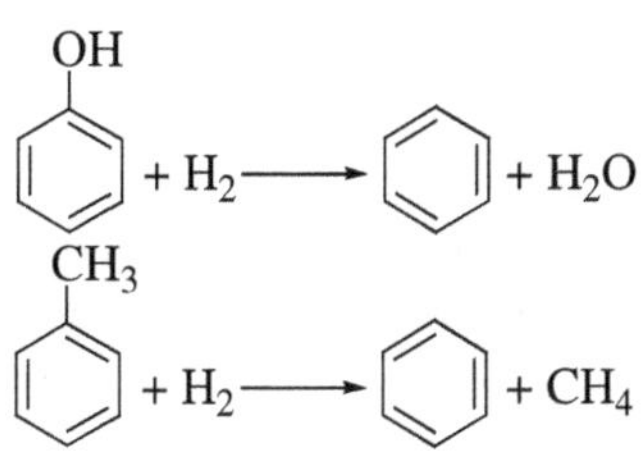

Condensation:

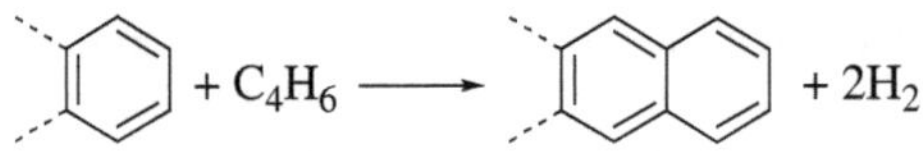

Polycondensation Reactions in Coal Pyrolysis

Cracking reactions dominate the early phase of coal pyrolysis and the latter phase is mainly polycondensation reactions. Polycondensation occurs in the plastic mass during maturation. It involves the combination of free radicals generated by pyrolysis, intermolecular polycondensation of liquid products, polycondensation of liquid and solid phases, and internal polycondensation of the solid phase. These reactions are mostly completed below 550–600 °C, generating semi-coke. Polycondensation converts semi-coke to coke. The reaction is characterized by dehydrogenation condensation of aromatic structures, and the number of aromatic layers increases. This may include condensation of fused ring aromatic structures and small molecules such as benzene, naphthalene, biphenyl, and ethylene. It may also involve condensation between polycyclic aromatic hydrocarbons. During the change from semi-coke to coke, the indexes of various physical properties of coal, such as density, reflectivity, conductivity, characteristic X-ray diffraction (XRD) peak intensities, and aromatic nucleus size, increase slightly at 500–600 °C, but the changes are not significant. These indexes change significantly at about 700 °C, and they further increase with increasing temperature.

Table 4.4 lists the changes in aromatic nucleus size L_a [25]. The increases in L_a and other physical indexes are the results of polycondensation reactions.

4.1.3 Kinetics of Coal Pyrolysis

In coal pyrolysis kinetic studies, the reaction types, mechanisms, products, rates, controlling factors, and kinetic constants are investigated. Such studies not only form the basis of coal science, but also the basis of clean coal use. The study of coal pyrolysis kinetics covers two aspects: the reaction kinetics of the plastic mass and the devolatilization kinetics.

Table 4.4 Changes in aromatic nucleus size (nm) during pyrolysis

Temperature (°C)	Room temperature	300	400	500	600	700	800	1100
Gas coal	17	17	21	24	26	–	38	46
Coking coal	16	16	19	22	24	30	35	37

4.1.3.1 Reaction Kinetics of Plastic Mass

Van Krevelen et al. [26] proposed the plastic mass (metaplast) theory, based on division of coal pyrolysis phases; they also obtained large amounts of experimental quantitative data. In the theory, it is first assumed that coke formation is represented by three successive related reactions:

$$\underset{\text{Coking coal}}{\mathrm{P}} \underset{E_1}{\overset{k_1}{\longrightarrow}} \underset{\text{Plastic mass}}{\mathrm{M}} \tag{4.1}$$

$$\mathrm{M} \underset{E_2}{\overset{k_2}{\longrightarrow}} \underset{\text{Semi-coke}}{\mathrm{R}} + \mathrm{G_1} \tag{4.2}$$

$$\mathrm{R} \underset{E_3}{\overset{k_3}{\longrightarrow}} \underset{\text{Coke}}{\mathrm{S}} + \mathrm{G_2} \tag{4.3}$$

where k_1 is the reaction rate constant (s^{-1});
$E_{1\text{–}3}$ are the activation energies ($\mathrm{kJ\ mol}^{-1}$).

Reaction 4.1 is depolymerization, which generates an unstable intermediate phase, namely the plastic mass. Reaction 4.2 is polycondensation, and during this reaction, the tar is released, non-aromatic groups are desquamated, and semi-coke is formed. Reaction 4.3 is polycondensation degassing, and during this reaction, the semi-coke volume shrinks, and cracks are generated.

The depolymerization and cracking reactions are generally first-order reactions. It can therefore be assumed that reactions 4.1 and 4.2 are first-order reactions. Reaction 4.3 is much more complicated than reactions 4.1 and 4.2, but it is still assumed to be a first-order reaction. The three reactions can be described by the following kinetic equations:

$$-\frac{\mathrm{d[P]}}{\mathrm{dt}} = k_1[\mathrm{P}],$$
$$\frac{\mathrm{d[M]}}{\mathrm{dt}} = k_1[\mathrm{P}] - k_2[\mathrm{M}],$$
$$\frac{\mathrm{d}[G]}{\mathrm{dt}} = \frac{\mathrm{d[G_1]}}{\mathrm{dt}} + \frac{\mathrm{d[G_2]}}{\mathrm{dt}} = k_2[\mathrm{M}] + k_3[\mathrm{R}].$$

The experimental data show that k_1 and k_2 are almost equal during coking. It can therefore be assumed that $k_1 = k_2 = k$. Introduction of the boundary condition $t = 0$ and some empirical approximations gives

$$[P] = [P]_0 e^{-\bar{k}t},$$
$$[M] = [P]_0 \bar{k} e^{-\bar{k}t},$$
$$[G] \approx [P]_0 \left[1 - (\bar{k}t + 1)e^{-\bar{k}t}\right],$$

where $\bar{k}$ is the corrected rate constant k.

The kinetic theory is consistent with some of the phenomena observed experimentally during the heating of coking coal. The activation energy E can be used in the Arrhenius equation to give

$$\ln k = -\frac{E}{RT} + b,$$

where k is the chemical reaction rate constant (s^{-1}); E is the activation energy ($kJ\ mol^{-1}$); R is the gas constant ($8.314\ J\ mol^{-1}\ K^{-1}$); T is the absolute temperature (K); and b equals ln A, where A is the pre-exponential factor. The obtained coal pyrolysis activation energy, E_1, is 209–251 $kJ\ mol^{-1}$, which is similar to the activation energy of cracking for polymers such as polypropylene and polystyrene. It is roughly equivalent to the $-CH_2-CH_2-$. bond energy. Generally, the E value in the initial phase of coal pyrolysis is small, and the k value is large. With increasing temperature and deepening pyrolysis, E increases and k decreases. Reactions 4.1, 4.2, and 4.3 are successive related reactions, and the reaction rate order is $k_1 > k_2 \gg k_3$. The average apparent activation energy of coal pyrolysis increases with increasing coalification. The normal gas coal activation energy is 148 $kJ\ mol^{-1}$ and the activation energy of coking coal is 224 $kJ\ mol^{-1}$.

4.1.3.2 Devolatilization Kinetics

The study of the devolatilization rate using thermogravimetry is an important aspect of coal pyrolysis kinetics. A thermobalance is used, which can continuously measure changes in a heated sample at a certain heating rate. When coal is subjected to thermolysis, volatiles are released and leave the reaction system. The mass loss caused by pyrolysis of a coal sample can be measured using the thermobalance. The reaction weight loss can be used to study the kinetics of devolatilization in coal pyrolysis.

Isothermal Pyrolysis

The coal is heated as quickly as possible to a predetermined temperature T, which is then maintained and the weight loss is measured. The values of $-dW/dt$ and the constant weight can be calculated from tangential lines at various points in the weight loss curve. The final weight loss ($-\Delta W_e$) at temperature T is generally measured several hours after the weight loss has stabilized. Typical weight loss curves at different temperatures and the total weight losses are shown in Fig. 4.2. The corresponding temperatures of the three weight loss curves in the figure

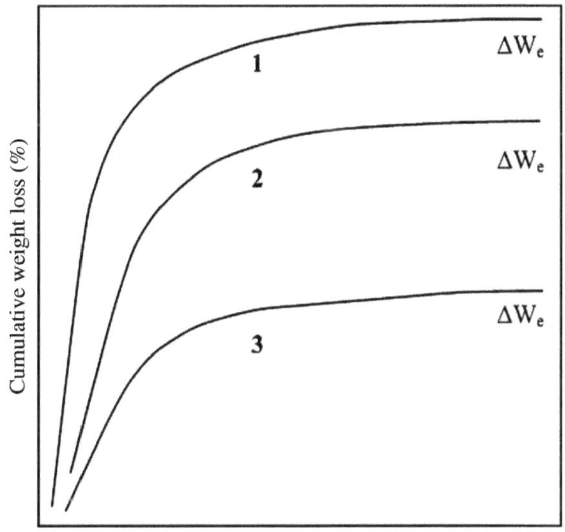

Fig. 4.2 Isothermal weight loss *curves* at different temperatures: *1* T_1; *2* T_2; and *3* T_3; $T_1 > T_2 > T_3$

decrease from top to bottom. At the beginning of the reaction, the cumulative weight loss and the time show a linear relationship. After a transition period, equilibrium is gradually reached. The equilibrium value (ΔW_e) is related to the coal type and the heating temperature. The time to reach equilibrium is generally 20 − 25 h. It must first be assumed that the decomposition rate is equivalent to the volatile releasing rate. Based on the shape of the ΔW–t curves, these reactions can be treated as first-order reactions. The reaction rate constant is

$$k = \frac{1}{t}\ln\frac{1}{1-x},$$

where x is the ratio W_t/W_e, i.e., the ratio of the weight loss at reaction time t and the maximum weight loss.

The apparent activation energy obtained using a first-order reaction calculation is only about 20 kJ mol^{-1}. This is because the coal particles heat up rapidly in the initial stage of the reaction and a temporary pressure gradient is generated in the microporous system within the coal particles. The process is controlled by diffusion rather than by reaction rate. The activation energy at this time is the diffusion activation energy. This shows that the pyrolysis rate (i.e., the reaction rate) and the devolatilization rate (i.e., the overall rate of reaction and diffusion) are not the same concepts. During isothermal pyrolysis, many reactions occur simultaneously. For coal pyrolysis, this can cause overlap of the primary pyrolysis degassing and the secondary polycondensation degassing. It is therefore difficult to create a kinetic equation system on the basis of gas removed. It is still unclear whether the isothermal devolatilization process is controlled by diffusion or by volatiles. However, large amounts of data show that either process may be the main devolatilization mechanism, depending on the environment.

Temperature-Programmed Pyrolysis

During the experiments using isothermal methods, differences among experimental results for the same sample inevitably affect the accuracy. The values reflect the average for the selected temperature range, which does not completely reflect the entire reaction process. Non-isothermal methods have many advantages over isothermal methods. First, the experimental time is shorter, and constant data for various temperature points within the reaction temperature range can be obtained from a single experiment. Second, the experimental data are obtained from the same sample, which can eliminate experimental errors due to sample differences. Additionally the results reflect the situation across the entire reaction temperature range, which helps in determining the temperature range for calculating the activation energy. This avoids incompatibility of experimental data as a result of inappropriate choice of temperature range. Similarly problems occurring with the time needed for the sample temperature to reach the required temperature T are avoided. In theory, all the kinetic parameters can be calculated from a single weight loss rate curve, using a temperature-programmed method. This simplifies the measurement method and gives results as reliable as those obtained using isothermal methods. Temperature-programmed pyrolysis avoids much of the inconvenience of pyrolysis under isothermal conditions. The method using linear temperature programming is therefore generally adopted for experiments. In this method, it is also assumed that the decomposition rate is equivalent to the volatile releasing rate. For certain reactions or reaction sequences, the relationship between gas releasing rate and concentration is as follows:

$$\frac{\mathrm{d}x}{\mathrm{d}t} = Ae^{-\frac{E}{RT}}(1-x)^n,$$

$$x = \frac{M_0 - M}{M_0 - M_f} = \frac{\Delta M}{\Delta M_f},$$

where

x is the coal pyrolysis conversion (%);
n is the reaction order;
E is the activation energy (kJ mol^{-1});
R is the gas constant (kJ mol^{-1} K^{-1});
A is the pre-exponential factor (s^{-1});
M_0 is the initial mass of the sample (g);
M, ΔM are the quantity and weight loss of the sample at a given time in the pyrolysis process (g);
M_f, ΔM_f are the residual quantity and weight loss of the sample at the termination of pyrolysis (g).

There have been many studies of the reaction order n. The coal weight loss and devolatilization rate vary with the coal type, heating rate, pressure, atmosphere, and other conditions. There is no unified kinetic equation. Based on the linear temperature programming method, Coats-Redfern [27] suggested that the temperature T and time t have a linear relationship,

$$T = T_0 + \lambda t,$$

where λ is the heating rate (K s^{-1}).

The following approximate solutions can be obtained:

$$\ln\left[-\frac{\ln(1-x)}{T^2}\right] = \ln\left[\frac{AR}{\lambda E}\cdot\left(1-\frac{2RT}{E}\right)\right] - \frac{E}{RT} \quad \text{when } n = 1$$

$$\ln\left[\frac{1-(1-x)^{1-n}}{T^2(1-n)}\right] = \ln\left[\frac{AR}{\lambda E}\cdot\left(1-\frac{2RT}{E}\right)\right] - \frac{E}{RT} \quad \text{when } n \neq 1$$

E is very large, therefore $2RT/E$ can be taken as zero. If the reaction order obtained is correct, plots of the left-hand sides of the above equations versus reciprocal temperature $1/T$ are straight lines. The activation energy E and the pre-exponential factor A can be obtained from the slope and intercept, respectively, of the resulting straight line.

4.2 Study of Coal Pyrolysis

4.2.1 Study of Coal Pyrolysis by PyGC Technique

The use of pyrolysis-gas chromatography (PyGC) enables in-depth studies of various factors affecting cracking reactions. The combination of a multifunctional pyrolyzer, concentration techniques, PyGC, and nuclear magnetic resonance (NMR) spectroscopy, together with developments in PyGC-mass spectrometry (MS), PyGC-Fourier-transform infrared spectroscopy (FTIR)-MS, and other techniques has increased the amount of information obtainable by PyGC. High-resolution PyGC has become a powerful analytical tool for studying macromolecular microstructures. Fast pyrolysis can greatly improve the coal tar yield and facilitate research on the determination of products and their distributions. The results of such research expand knowledge of coal structure and pyrolysis theory, and also provide the basis and standard for accurately testing models of pyrolysis and coal structure. The similarities and differences among different coal pyrolysis products have direct impacts on the characteristics of coal combustion. The accurate determination of pyrolysis products can be used to optimize the combustion system, and to adapt equipment operation depending on the coal combustion characteristics, to improve the combustion efficiency. Tar products obtained from conventional pyrolysis have higher polycondensation degrees, and thus the accuracy of information on the coal structure based on tar from conventional pyrolysis is low. To avoid this, we used an advanced PyGC coupling technique. The heating rate can reach 20,000 °C s^{-1} and a temperature lag in the pyrolysis process can be avoided.

Coal samples from lignite to anthracite coal, their proximate and ultimate analysis data are listed in Table 2.4 of Chap. 2, were pyrolyzed in a pyrolysis reactor (CDS2000) coupled a gas chromatograph (PE-8500). The atmosphere was N_2 and the particles size was about 50 μm. To determine the sample composition in different pyrolysis phases, the pyrolysis process was divided into the following three phases. In the first phase, the temperature was increased from RT to 1100 °C at a rate of 10^4 °C s^{-1}, and the temperature was maintained for 0.5 min. In the second phase, the temperature was increased from RT to 450 °C at the same heating rate, maintained for 0.5 min, and then reduced to RT. In the third phase, the temperature of the residue from the second phase was increased from RT to 750 °C, using the same heating rate. The temperature limits were determined based on work by Marzec [28], Van Heeks [29].

The pyrolysis products were identified by comparing the retention times of all chromatographic peaks with those of standard normal hydrocarbons, and converted to retention indexes, using the following formula:

$$I = 100 \times \frac{\log t_{\mathrm{RI}} - \log t_{\mathrm{R}_z}}{\log t_{\mathrm{R}_{(Z+1)}} - \log t_{\mathrm{R}_z}} + 100z,$$

where

I is the retention index of the substance in the specified column (stationary phase);

t_{Rz} is the carbon number of an *n*-alkane adjusted to the retention time for a standard homolog *z*;

$t_{\mathrm{R}(z+1)}$ refers to the carbon number of an *n*-alkane adjusted to the retention time for another standard homolog *z* +1.

t_{RI} is the adjusted retention time for the sample.

4.2.1.1 Effects of Coal Rank and Temperature on Product Distribution

Product Distribution From TR to 450 °C

The compounds at around C_5 in the pyrolysis products of high-rank coals such as anthracite, meager coal, and lean coal account for only 80–90 % of the total pyrolysis products. Based on analysis of the carbon numbers, compounds with such structures can only be derivatives of C–H chain compounds, which indicates that high-rank coal has an ordered structure. The interactions of small-molecule phases and aromatic rings are relatively weak, and they are easily released from the coal structure at 450 °C. As a result of the imperfect development of the aromatic ring system, a considerable proportion of ring clusters in the structure of low-rank coal at 450 °C are broken and escape. At the same time, because of the heating rate, the effusion of larger molecules is also seen in the chromatographic peak. The structural units of long-flame coal are the smallest, and the maximum carbon number that can be desorbed is about C_7–C_8. It can be reasonably assumed that benzenoid derivatives and alicyclic compounds are mainly in the tar. With increasing coal rank, degree of polymerization of the aromatic ring system increases along with the number of structural units that can be desorbed also gradually increases. Lignite tar (hard lignite) consists mainly of C_{10}, and fat coal tar consists mainly of C_{20} and C_{26}. The sharp rise in the carbon number is related to fat coal, which has a volatile content as high as 46 %. The pyrolysis products of coking coal are mainly C_{25} compounds. Peat is an exception to this rule, and has more structures with 15–16 carbons than those of long-flame coal. This may suggest that instead of aromatic rings there are alicyclic compounds present in the structure. In recent years, studies of tar have also shown that the biggest unknown factor is the form of aliphatic carbon in tar. As the majority of aliphatic carbon is in ring systems, the presence of long-chain aliphatic hydrocarbons becomes less likely. Stereoscopic analysis has shown that tetracyclic aromatic carbons can be neglected, and the NMR data also confirm that no tetracyclic aromatic carbons are present.

Product Distribution From 450 to 750 °C

After the small-molecule phases in the coal is removed before 450 °C, new products will be generated if these types of semi-coke are heated further to 750 °C. Peat has a loose structure in this process. The carbon number of a specific product increases slightly compared with that in the first phase, reaching about C_{30}, which indicates that the system has been completely destroyed. After the first-phase pyrolysis of lignite, large amounts of C_6 compounds remain in the pyrolysis products. This indicates that a temperature limit of 450 °C is not appropriate for lignite, and its products at 750 °C are still those from the unfinished process in the previous phase. Long-flame coal is quite different and its pyrolysis products not only continue for the previous phase C_6–C_7, but also volatilize at high temperatures after the condensation of aromatic rings in the C_{16}–C_{18} coal structure. After the removal of large amounts of aliphatic and aromatic hydrocarbons from fat coal below 450 °C, its semi-coke structure has a high degree of condensation and only a small amount of C_5 aliphatic hydrocarbon is removed under this condition because the heat at 750 °C cannot further crack the condensed aromatic rings. These results show that fat coal and long-flame coal with high volatile contents should mainly be processed at low temperatures. The products from coking coal not only continue to release C_{25} compounds from the previous process, but also generate C_{20} depolymerization products of aromatic rings with increasing temperature. However, depolymerization does not have a significant advantage during the competitive processes of depolymerization and condensation. For meager coal, the condensation extent is much larger than that of depolymerization. The C_{26} product does not appear in the previous phase, which indicates that the structural units of meager coal have higher activities and tend to condense into larger clusters. For meager coal and anthracite, the condensed aromatic ring systems start to depolymerize at 750 °C. There are three peaks, namely C_5, C_{10}, and C_{15}, in the anthracite products, and C_7, C_{10}, and C_{15} in meager coal. This indicates that the depolymerization of condensed rings is based on single-unit aromatic rings.

Product Distribution from RT to 1100 °C

For a sample for which the temperature was raised directly from RT to 1100 °C, the tar generated mainly consisted of C_{12}–C_{16} compounds, except in the case of peat. The pyrolysis products of peat are mainly linear-chain aliphatic hydrocarbons (C_4–C_5), and this result is basically consistent with long-term studies of woody plants and peat. As peat has not undergone coal petrographic metamorphism, it can be assumed that the C_{12}–C_{16} compounds reflect the structure of small molecules in the coal structure. Vulava [30] found that monocyclic arene and dicyclic polyaromatic hydrocarbons are present in coal tar. They are mainly the alkylation derivatives of benzene and naphthalene, accounting for nearly 70 % of the total fraction. Studies have also shown that there are significant differences among coal tar yields and compositions. These differences may be related to the nature of the raw coal. It can be concluded that the majority of structural units exist in the "monomeric" form during primary cracking of coal and coal polymers. The smaller molecules in the tar are generated by secondary cracking of such structural units. If a monocyclic saturation of four is used, the C_{12}–C_{16} compounds mainly consist of two to three cyclic arene derivatives, and their content accounts for about 70 % of the total pyrolysis products.

The volatiles of meager coal, with a relatively high degree of metamorphism, only account for 12.22 %, which is one-half to one-third of the amounts in high-volatile coals such as fat coal and coking coal. However, their pyrolysis rates are roughly equal, which indicates that the energy provided under rapid pyrolysis conditions (using 12.54 J K^{-1} g^{-1} as the specific heat of coal, a heating rate of 10^4 °C s^{-1} is equivalent to release of 125.4 kJ of energy in 1 ms) is sufficient to meet the requirements of small-molecule phases in coal to overcome mass transfer, and the releasing rate is too rapid for the macromolecular structure in the coal to be captured and condensed before escaping from the solid phase surface of the coal. Moreover, because of the higher degree of metamorphism of meager coal, the carbon number of the tar components is about four times those of other coal types. As a result of the high condensation degree and graphite-like structure of anthracite, the pyrolysis yield is only about 30 %, even under such severe reaction conditions. It is worth noting that the pyrolysis products of anthracite are not the same as those of meager coal, which can generate higher carbon number products. As for other coal types, the carbon numbers are concentrated in the range C_{12}–C_{13}. This shows that the pyrolysis products of meager coal may contain some extremely active components, and relatively serious polycondensation occurs in the process of transfer from the interior to the exterior of coal particles. However, almost all the active groups in anthracite have been lost.

When the heating temperature is increased directly from RT to 1100 °C, the pyrolysis products are not simply the sum of those in the first two phases. Peat has a loose structure, but depolymerization, which has a higher activation energy, occurs at high temperatures; only two peaks occur in the C_4–C_6 range, which indicates that almost no polycondensation occurs. The case for lignite is slightly different and C_6–C_7 products were detected. The amount of products around C_{14} accounts for over 90 % of the total products, and this shows that the structure can be condensed. For instance, there are relatively large amounts of unsaturated alicyclic and diene structures, and heating can easily condense them into compact aromatic or naphthalene ring structures, which are difficult to crack. This is probably why lignite is suitable for hydrotreatment, which saturates its structure rapidly. The situation for long-flame coal is similar. Its products are also mainly C_{14} that appear in different stages. For fat coal, the trend in products is clearer and also concentrated in the range C_{12}–C_{14}. This indicates that the stabilities of structures with more than two rings are relatively poor under these conditions (1100 °C, with rapid heating and ignoring mass transfer resistance), although polycondensation occurs readily while polymerization is disfavored. The equilibrium point in the competing processes is in the vicinity of bicyclic rings and their derivatives. At the same time, the cracking fragments in the macromolecular cluster system should remain at C_{12}–C_{14}, which is the most favorable size. Otherwise, they are either recaptured and condensed in the system as a result of the slow mass transfer process, or they are cracked further because of poor stability. When the coal structure is reassessed based on the fragments of pyrolysis products, it can be deduced that the presence of polycyclic structures is less likely. This is because if there is such a structure, it should have already been transformed into a more stable monocyclic or bicyclic ring, or an aromatic ring network linked by tricyclic-based structures, in the early stages of the long-term coal-forming process.

4.2.1.2 Pyrolysis Products of Coal Extracts

Studying the pyrolysis reactions of coal extracts is based on the following considerations. If the solvent extract can be regarded as gaseous products and a non-polycondensation tar of pyrolysis products in the raw coal, then the small-molecule structures escaping from the coal matrix without condensation may be present in complete or quasi-complete forms under the same pyrolysis conditions. The aim of this work is to help us to conclude whether the coal activity is determined by the structure of the small-molecule phase, or by the degree of polycondensation of the coal macromolecular network, or related to active structures in the hot-working process. The precise definition of pyrolysis here is thermal degradation, and it is used to degrade complex compounds before analysis. We therefore used 450 °C as the pyrolysis temperature to avoid excessive cracking of the extracts.

Using the extraction method described in Sect. 3.2, various extracted products were obtained from raw coal samples with the solvent *N*-methyl-2-pyrrolidone. A concentrated solution of the extract was pyrolyzed using the Py-GC technique at 450 °C. The C_{12} compounds of all the products below 450 °C account for over 80 % of the total, which is exactly the same as the product distribution obtained by increasing the temperature directly from RT to 1100 °C. This means that small molecules in various coal types have similar structures, i.e., phenyl (probably bicyclic) derivatives with carbon numbers of around 12.

4.2.2 Study of Coal Pyrolysis by Py-FTIR Techniques

In the past two decades, there have been many studies of coal pyrolysis mechanisms and the various types of reactors used in the experiments have been compared. The Solomon FG-DVC model [31] has been used in in-depth studies of coal chemical structures; it has been used to study functional groups and pyrolysis products simultaneously, leading to progress in understanding the role of functional groups in coal pyrolysis. In recent years, with the development of FTIR, Py-FTIR has become available. Although the quantitative accuracy of this coupling technique is currently no better than that of PyGC-MS, this coupling technique has an important role in research on molecular structures.

Different coal rank samples (listed in Table 2.4 of Chap. 2) were studied in a pyrolysis reactor (Pyroprob CDS2000) coupled an infrared spectrometer (Bio-Rad FTS165). The interface temperature of ZnAs was 220 °C, the flow rate was 40 mL min^{-1}, and heating rates of 20, 100, and 1200 °C min^{-1} were used. The changes in the activities of the functional groups in the primitive materials were compared through their vibrational peak intensities of the pyrolysis reaction products at various heating rates.

4.2.2.1 Functional Group Released During Peat Pyrolysis

Peat pyrolysis at 20 °C min^{-1} can be divided into three phases. The first phase is the prereaction phase. The concentrations of most functional groups in this phase are basically the same, but the concentration of saturated C–H bonds is increasing. This indicates that the bond energy of the linear chain structure in peat is lower than those of other functional groups, and they begin to crack at about 300 °C. The intensity of the vibration at 2178 cm^{-1}, representing diene or alkyne moieties, also increases, which indicates that cracking of the side groups on the macromolecular chains may occur during pyrolysis. The side chains undergo elimination reactions, generating small-molecule compounds, and the main chain forms polyene structures; the main chain is probably the side group of a larger main chain.

The absorption at 2978 cm^{-1}, attributed to CH_2, is derived from the main chain. When the temperature rises to about 570 °C, all the vibrational intensities increase significantly; the intensity of the band attributed to diene structures, at around 2100 cm^{-1}, increases fastest. This indicates that the depolymerization reactions of bonds in peat are very intense under these conditions, but cyclization of the corresponding main chain is relatively weak. This is confirmed by the increase in the intensity of the aromatic ring vibration at 3065 cm^{-1}, which is only about half those of the other products. Other evidence for the competing reactions is the change in the C–H vibrational intensity of the product. At 580 °C, polycondensation results in a slight decrease in the intensity of the saturated C–H vibration, and then the intensity increases continuously with breakage of increasing amounts of side chains in the peat. The reaction then enters the third phase and the product concentrations are generally constant. The main functional groups are carbonyl (C=O) at 1700 cm^{-1} and ether (C–O–C) at 1200 cm^{-1}; during pyrolysis, these functional groups are released as CO_2.

At heating rates such as 100 and 1200 °C min^{-1}, the segmented process that occurs at a slow heating rate, does not occur during rapid heating, and the changes become exponential. All the changes in the vibrational intensities of the functional groups are similar, and the only differences are in the change range and reaction rate.

4.2.2.2 Functional Group Released During Lignite Pyrolysis

In coal structural analysis, the structural changes during peat pyrolysis can be used to investigate the coal pyrolysis mechanism, but lignite pyrolysis more closely reflects the actual process. A a heating rate of 20 °C min^{-1} the main change to lignite functional groups is for CO_2, at 2360 cm^{-1}, The changes in CO_2 concentration reflect the condensation of functional groups such as C=O and COOH in the coal structural unit. There are three distinct peaks (i.e., at 170, 350, and 850 °C) in the entire pyrolysis process. The strong peak at 170 °C can be attributed to free CO_2 adsorbed inside the sample. The second CO_2 peak occurs at 340–350 °C, and marks the beginning of lignite pyrolysis. At this temperature, C–C and C–O single bonds connected to the main chain start cracking and breaking off because lignite has a high pyrolysis reactivity and its chemical bonds are easily broken. Additionally, because the lignite activity can be represented by ether oxygen bonds and the number of side chains, this shows that lignite has not attained the maximum hydrogenation capacity during hydrogenation. With increasing pyrolysis temperature, two smaller CO_2 peaks occur, at 460 and 540 °C. The results show that formation of these peaks is accompanied by an increase in $-CH_3$ concentration, indicating crosslinking of the lignite. This early crosslinking occurs in low-rank coal at low temperatures and suppresses formation of large fragments in the pyrolysis products. Between 800 and 850 °C, another CO_2 peak occurs, indicating another reaction in the lignite structure. The number of unsaturated aliphatic hydrocarbons in lignite (2602 cm^{-1}) in the same temperature range clearly decreases, whereas the amount of arene groups (3065 cm^{-1}) abruptly increases. This indicates that pyrolysis fragments extract hydrogen from aliphatic hydrocarbons and hydrogenated aromatic rings to form large amounts of aromatic compounds, i.e., that the products of lignite at 750 °C are mainly C_7–C_8 compounds, but large amounts of C_{13} compounds are formed at 1100 °C. This is the result of condensation of carbonyl and carboxyl groups in the lignite. The detailed process is still unknown, but it is certain that up to 21 % of oxygen in lignite plays an important role in lignite pyrolysis activity, the pyrolysis mechanism, and product composition.

The maximum absorption intensity of CO_2 occurs at 930 °C, and then the intensity remains unchanged until the end of the reaction, as the heating rate increases up to 1200 °C min^{-1}. The absorption intensity shows a maximum at 1000 °C, and then it is unchanged until the end of the reaction. Under fast pyrolysis conditions, the reacting coal particles have a larger temperature gradient because of the impact of heat transfer.

4.2.2.3 Functional Groups Released During Long Flame Coal Pyrolysis

At 1200 °C min^{-1} the changes during pyrolysis of long-flame coal are not as great as those for peat and lignite, but the overall pyrolysis rate is higher. During pyrolysis, the main changes in the products from long-flame coal are for arenes (3056 cm^{-1}) and aliphatic hydrocarbons (2890 cm^{-1}). The changes are basically the same as those for lignite. However, the intensity of the peak at 2358 cm^{-1}, attributed to C=O and C–O–C, decreases in the constant-temperature section in a short period. The decrease in concentration over such a short period of time indicates a lack of oxygen functional groups in long-flame coal and shows that the trend in the changes in the light hydrocarbons in the coal is unchanged after the loss of active oxygen functional groups. This shows that in the fast pyrolysis of coal, the activation energy of the macromolecular network or side-chain cracking has an important effect on the kinetics of coal pyrolysis, but this rapid heating effect becomes less obvious at lower heating rates (100 and 20 °C min^{-1}). The condensation of ether bonds occurs throughout the reaction, and the trend in the release of light hydrocarbons and aliphatic hydrocarbons is exponential. This further indicates that variations in the heating rate change the distribution of products during any time interval and that the pyrolysis mechanism also changes. The progressive cracking of side-chain functional groups leads to collapse of the entire system at some point.

4.2.2.4 Functional Group Released During Fat Coal Pyrolysis

During pyrolysis at 1200 °C min^{-1}, the thermal lag of fat coal is significantly higher than those of several other coal types, and the time at which the maximum product concentration appears is delayed by nearly 50 s, compared with peat, lignite, and long flame coal. A reasonable explanation is that fat coal is in a disordered to ordered structural transition and the connections between aromatic rings in the structure are relatively loose. Many larger aromatic rings and long-chain aliphatic hydrocarbons are found in the extracts. The energy required for further condensation to a macromolecular network is relatively low. Crosslinking is prone to occur at 300 °C, and C_{25} compounds are found in the pyrolysis chromatography products at 450 °C. Macromolecular tar formation is difficult. Fat coal has the thickest plastic layer and its impermeability and higher thermal stability make pyrolysis products hard to release. This is shown by the condensation of ether bonds, which is higher than for the previous three coal types, and lasts longer. This further indicates that although in many ways fat coal should show a high degree of pyrolysis, its overall performance does not result in higher pyrolysis yields and rates because its polycondensation energy is much lower than its depolymerization energy. The properties of various coal components and the overall ring condensation degree after coal formation should therefore all be taken into account in studies of coal pyrolysis reaction rates.

4.2.2.5 Functional Groups Released During Coking Coal Pyrolysis

At a heating rate of 20 °C min^{-1}, the amounts of arenes and aliphatic hydrocarbons slowly increase throughout pyrolysis. The reaction rate slows down at 550 °C; this temperature is the same as that for coking coal softening, indicating enhancement of coal plasticity and increased crosslinking in the system at this point. At the same time, the CH_2 concentration in the product decreases correspondingly, which indicates that coking coal softening is closely related to consumption of coal side chains. The maximum concentration of CO_2 in the product occurs at 490 °C, which indicates the start of crosslinking of the coal structure. The entire process therefore involves condensation of oxygen functional groups in the coal by structural crosslinking and consumption of hydrogen-containing side chains. Side chains decrease the molecular weight of the crosslinked structure, preventing loss of mobility; however, a larger number of CH_2 groups does not necessarily improve the mobility. The cause of the poor mobility of low-rank coal is therefore a mismatch between its crosslinking capacity and the hydrogen supply.

4.2.2.6 Functional Group Released During Lean Coal Pyrolysis

At a heating rate 20 °C min^{-1}, a large amount of CO_2 appears at 650 °C and two relatively strong peaks appear at 850 and 970 °C. Normally, C–O bonds in coal are weak, and are completely released in the form of CO_2 or CO below 600 °C. Ultimate and proximate analyses of lean coal show that its volatile and oxygen contents are the highest among all high-rank coals, except for peat and lignite. However, strong C–O vibrations were not found in lean coal during experiments on ether bonds. The oxygen in lean coal is therefore present in heterocyclic rings or other structures with single covalent oxygen bonds. Such structures are highly stable. Lean coal has the highest yield of pyrolysis products among all types of coal and also has the maximum amount of hydrogen; therefore its CH_2 chain lengths are shorter than those of other coal types. The structure of lean coal during depolymerization of aromatic clusters in the late stage of pyrolysis provides highly reactive oxygen and promotes cracking of aromatic clusters. This is another reason why the lean coal pyrolysis products with the highest carbon numbers are found during GC analysis of the products at 750 °C. At heating rates of 100 and 1200 °C min^{-1}, the characteristics of lean coal pyrolysis are the same as those of the coals discussed previously; the tar yield is greatly improved and the reaction rate is accelerated.

4.2.2.7 Functional Group Released During Meager Coal and Anthracite Pyrolysis

Because the concentrations of functional groups are below the limit of detection, conclusive information about pyrolysis of meager coal and anthracite at slow heating rates is difficult to obtain. However, the trends in changes of the concentrations of functional groups with increasing heating rate are similar to the previous ones.

4.2.2.8 Functional Group Activity

It is clear that oxygen functional groups (2358 cm^{-1}) are always the primary products, in terms of quantity or rate of formation after the reaction begins. This shows that oxygen functional groups have high activities and low activation energies. The main effect of the heating rate on pyrolysis is to change the pyrolysis mechanism.

The reactions of hydroxyl and carbonyl groups are in line with the first-order reaction in the pyrolysis model. The predicted values for highly active functional groups such as methylene bridges and ether bonds are much lower than the experimental values. This positive deviation indicates that their reaction order is higher than first order. The release rates and the amounts of products related to reactions of methylene and ether oxygen rapidly increase with increasing temperature. The impact of these components with high activation energies on the whole system cannot be ignored when developing coal pyrolysis models. The differences among coal reactivities are shown to be mainly caused by differences among activation energies of the same functional groups in various types of coal during pyrolysis. Most functional groups in coal undergo first-order reactions during pyrolysis, but the reaction orders of functional groups generating CO_2 are higher than first order.

During low-rank coal (such as peat and lignite) pyrolysis, changes in the oxygen content affect the coal reactivity and the product composition, and the proportion of aromatic rings and side chains in high-volatile bitumites (such as long-flame coal, fat coal, and coking coal) determines the coal reactivity. Changes in the pyrolysis rate change the distribution ratio of each component in the product and affect the pyrolysis reaction mechanism. Fast pyrolysis is a reliable approach for studying coal pyrolysis activity and dynamics. The experimentally observed phenomenon of zero tar retention indicates that in reversible reactions in high-volatile young coal pyrolysis under certain reaction conditions, the forward reaction predominates.

4.2.3 Study of Coal Pyrolysis by Thermogravimetry

Thermogravimetry is used to investigate the chemical kinetics of pyrolysis and the relationship between product quantity and temperature during coal pyrolysis. The advantage of this approach is that the coal pyrolysis process can be studied directly. The disadvantage is that heterogeneous reactions occur in coal pyrolysis, and a large number of assumptions and omissions are used in processing the chemical kinetic data. Great care should be taken with results obtained using thermogravimetry.

A thermal analyzer (SETARAMTGA92) was used as the thermogravimetric reactor. The amount of sample was about 10 mg, the heating rates were 10 and 20 °C min^{-1}, the final temperature was 1100 °C, a N_2 gas atmosphere was used, and the gas flow rate was 90 mL min^{-1}. The TG kinetic parameters and TG characteristic parameters were calculated using the methods reported in the literature [32, 33].

4.2.3.1 Coal Pyrolysis Activity

Weight Loss Temperature

Initial weight loss temperature. The initial weight loss temperatures of eight coal samples at two heating rate are listed in Table 4.5. The initial weight loss temperature reflects the initial coal decomposition, and is closely related to the coal combustion properties. In recent years, many researchers have used this value as a substitute for activation energy to express coal pyrolysis reactivity. The data in the table show that the weight loss temperatures at various heating rates are similar. The temperature gradually increases with increasing coal rank. However, lean coal is an exception and its initial weight loss temperature is similar to that of lignite.

Table 4.5 Initial weight loss temperatures (°C) of coal samples at various heating rates

Heating rate (°C min^{-1})	Peat	Lignite	Long flame coal	Fat coal	Coking coal	Lean coal	Meager coal	Anthracite
20	253	386	435	434	429	325	347	558
10	259	368	422	440	420	371	501	545

Table 4.6 Maximum weight loss temperatures (°C) of coal samples at various heating rates

Heating rate (°C $\min^{-1}$)	Peat	Lignite	Long flame coal	Fat coal	Coking coal	Lean coal	Meager coal	Anthracite
20	300	447	476	491	476	453	480	762
10	412	436	468	477	459	463	563	618

Maximum weight loss temperature. The maximum weight loss temperature represents the average stability of a macromolecular structure in the entire coal. A higher temperature indicates that the system is more closely combined and the entire network structure is not easily destroyed during pyrolysis. It also indicates higher coal activity and that each part is prone to cracking during heating and participating in a reaction. A comparison of the data in Table 4.6 shows that the heating rate has a large impact on the maximum weight loss temperature of the coal. At a rate of 20 °C $\min^{-1}$, the maximum weight loss temperatures for coal of different ranks differ significantly. With increasing coal metamorphic degree, the maximum weight loss temperature rapidly increases, with only slight pauses in the coking and lean coal phases. It continues to increase, and the overall temperature difference is nearly 460 ° C. At a heating rate of 10 °C $\min^{-1}$, the differences among different types of coal are significantly less; the difference between the maximum and minimum values decreases to about 200 °C. This indicates that during slow pyrolysis, the macromolecular structure of coal has sufficient time to adjust and coordinate the energy absorbed from heating the system, and the coal macromolecular structure shows "thermal inertia". However, at faster heating rates, such coordination is weaker.

Activation Energy

The non-isothermal method is used to calculate the activation energy of coal pyrolysis. Based on TG curves, the residual fraction at any point in the reaction is as follows:

$$C = \frac{m_0 - m_t}{m_0 - m_f},$$

where

m_0 is the initial mass of the sample (g);
m_t is the mass of the sample at any time t (g);
m_f is the mass of the sample after pyrolysis (g).

The weight loss rate is

$$-\frac{dC}{dt} = k \cdot f(C).$$

The specific form of $f(C)$ depends on the reaction mechanism. To simplify the calculation, in accordance with usual understanding, coal pyrolysis is considered to be a first-order reaction, and the Arrhenius equation also applies:

$$-\frac{\mathrm{d}C}{\mathrm{d}t} = A\exp\left(-\frac{E}{RT}\right)\cdot C,$$

A constant heating rate is maintained:

$$\lambda = \frac{\mathrm{d}T}{\mathrm{d}t},$$

Therefore,

$$-\frac{\mathrm{d}C}{\mathrm{d}t} = \frac{A}{\lambda}\exp\left(-\frac{E}{RT}\right)\cdot C.$$

By using integration, E is obtained by least-squares fitting. The results are shown in Table 4.7. The data in the table show that with increasing of coal rank, the coal pyrolysis activation energy increases. Except for meager coal and anthracite, which have the highest coal ranks, the activation energies of the coal samples at 20 °C min^{-1} are higher than those at 10 °C min^{-1}. Coal pyrolysis involves depolymerization and polycondensation processes. The activation energy of forward polycondensation should be slightly lower than that of reverse depolymerization. For the forward reaction, the activation energy indicates the potential barrier ΔE between the initial state and the activated state. However, because of the participation of polycondensation processes, the entire pyrolysis process consists of several overlapping reactions. The apparent activation energy is often used to represent this non-elementary reaction. This makes the corresponding relationship between the activation energy and the pyrolysis reaction unbalanced, i.e., a high activation energy should correspond to poor reactivity, but deviations may be caused by structural changes or the pyrolysis reaction mechanism. The data in Table 4.7 show these deviations. All high-volatile coal types have high activation energies under given reaction conditions, except for meager coal and anthracite, which have decreased activation energies. If we exclude the impact of calculation accuracy, we can assume that the activation energy in high-rank coals is different from the true coal reactivity. To date, there have been few reports verifying experimental values with kinetic parameters calculated from reaction mechanism

Table 4.7 Activation energies (kJ mol^{-1}) of pyrolysis of various coal types

Heating rate (°C min^{-1})	Peat	Lignite	Long fame coal	Fat coal	Coking coal	Lean coal	Meager coal	Anthracite
20	19.3	32.0	28.6	33.2	38.6	35.0	18.2	17.9
10	13.8	18.1	16.5	25.0	21.1	16.4	23.0	22.6

measurements. The majority of results are based on apparent kinetics without any correlation with microscopic processes.

Conversion Time

Based on the general form of the kinetic equation,

$$\frac{dx}{dt} = A\exp\left(-\frac{E}{RT}\right)f(x),$$

where x is the conversion fraction of the response at time t.

Integrating this gives

$$\int_0^x \frac{dx}{f(x)} = \int_0^t A\exp\left(-\frac{E}{RT}\right)dt.$$

The left-hand side of the equation is a function of x, so we make it $F(x)$. The right-hand side of the equation has a time dimension, so a new variable is introduced, i.e., the conversion time θ:

$$\theta = \int_0^t A\exp\left(-\frac{E}{RT}\right)dt,$$

$$F(x) = \int_0^x \frac{dx}{f(x)}.$$

This gives

$$F(x) = A \cdot \theta.$$

This equation shows that the conversion times for a reaction at constant temperature and under heating conditions are equal, when the rate of change of x is the same. This means that the conversion time also represents the common dimension of the reaction degree. However, the magnitude of the conversion time is not equal to the coal reactivity. Plots of conversion times versus conversion rates for various coals are shown in Fig. 4.3. The non-linear transformation of the time axis provides a more direct impression of the relationship between the extent of reaction and reactant concentrations. It also eliminates data unreliability arising from errors in temperature measurements related to the pre-exponential factor, which have been discovered in sensitivity analysis of pre-exponential factors [34]. Moreover, the experimental curve for a coal sample directly corresponds to its reactivity. Figure 4.3 shows that peat can be converted to the maximum amounts of reactants, with the same reactivity

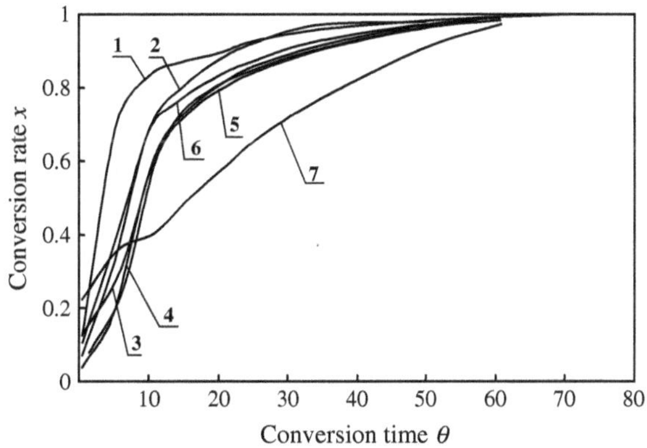

Fig. 4.3 Relationship between conversion rate and conversion time at heating rate of 10 °C min^{-1}: *1* peat; *2* lignite; *3* long flame coal; *4* fat coal; *5* coking coal; *6* lean coal; and *7* anthracite

as anthracite, in the early phase of the reaction, but has the lowest conversion fraction after 510 °C. This phenomenon indicates that anthracite products after polycondensation have much higher stabilities than the polycondensates of other coal types. The variations among the conversion rates of various coal types for conversion fractions with the same reactivities at a heating rate of 10 °C min^{-1} are larger than the variations at 20 °C min^{-1}. This indicates that the impact of heat and mass transfers in a thermogravimetric reactor cannot be ignored. There is a certain lag during product transfer from coal particles to the atmosphere. If the heating rate is decreased by only 10 °C min^{-1}, the components detected in the products vary. As the reaction progresses, even if the necessary energy is provided to give various types of coal the same reactivities in the late phase of pyrolysis, the reaction capacities will not differ much. This shows that the activation energies of cracking of polycondensation products are constantly converging in coal pyrolysis processes.

4.2.3.2 Coal Maceral Pyrolysis Activity

The pyrolysis activities of macerals and their pyridine extracts and residues from pyrolysis reactions were tested in the same reactor (SETARAMTGA92 thermal analysis system). The experimental samples were Pingshuo bitumite. The experimental conditions were heating rate 20 °C min^{-1}, final temperature 1250 °C, and reaction atmosphere CO_2 at normal pressure.

Pyrolysis Reactions of Original Maceral Samples

Figure 4.4 shows differential TG (DTG) curves for pyrolysis of three types of maceral in CO_2. The maximum devolatilization rates of the three macerals all occur at about 470 °C. However, different macerals have different maximum devolatilization rates; their order is exinite (−20 % min^{-1}) ≫ vitrinite (−3 % min^{-1}) > inertinite (−1.8 % min^{-1}). Chemical analyses of the three macerals show that the

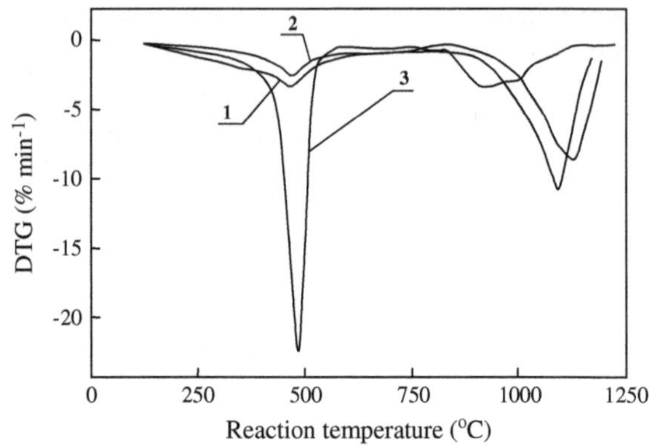

Fig. 4.4 DTG curves of three macerals in CO_2 atmosphere in pyrolysis and gasification reactions: *1* vitrinite; *2* inertinite; and *3* exinite

Table 4.8 Kinetic parameters of maceral pyrolysis in CO_2 atmosphere

Maceral	E(kJ mol^{-1})	ln$A$	Correlation coefficient
Vitrinite	99.26	14.71	0.8885
Inertinite	63.3	8.54	0.9422
Exinite	97.37	14.63	0.9070

aromaticity increases progressively in the order exinite (0.39), vitrinite (0.67), and inertinite (0.75). The values of $CH_3/(CH + CH_2)$ also follow this order (0.52, 0.71, and 1.10), but the H/C atomic ratios show the reverse order (0.1025, 0.0678, and 0.0467). It can be deduced that the weakest C–C bonds, in aliphatic hydrocarbon bridges and C–C bonds between alkyl and aromatic structures, are cracked to a greater extent in exinite than in the other macerals. The primary reaction is cracking of weak bonds, leading to the formation of various fragments during pyrolysis. Aromatic C–C bonds are very strong, so the aromatic ring system in inertinite is relatively stable during pyrolysis.

The kinetic parameters of the three macerals in the pyrolysis reaction can be obtained based on the first-order reaction model. The results are listed in Table 4.8.

Pyridine Extracts of Macerals and Their Extraction Residue Pyrolysis Characteristics

TG curves for macerals and pyridine extraction residues of macerals show that there is little difference between the residue and original maceral for vitrinite and exinite, but there is a clear difference in the case of inertinite (see Fig. 4.5). It is known that the macromolecular structure does not change significantly after pyridine extraction at 40 °C. The weight loss value of the extraction residue should be less than that of the corresponding original maceral. This is because small molecules in the original maceral phase have already been extracted. The extraction residue TG curves should be significantly different from those of the original macerals, but this has not been observed, and their TG curves do not differ significantly. It can therefore be

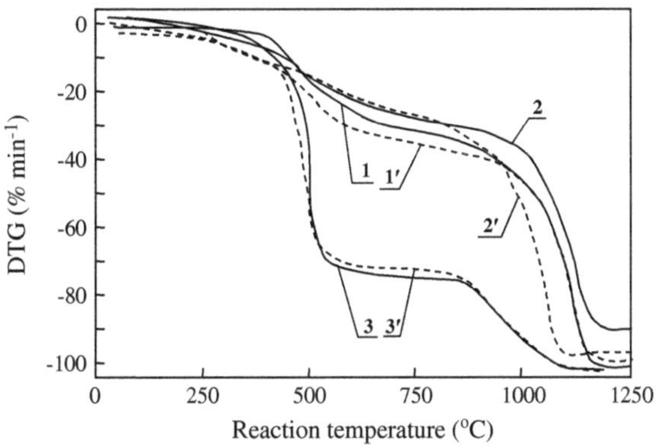

Fig. 4.5 Pyrolysis characteristics of three macerals and pyridine extraction residues: *1* vitrinite; *1′* vitrinite extraction residue; *2* inertinite; *2′* inertinite extraction residue; *3* exinite; and *3′* exinite extraction residue

deduced that pyrolysis mainly involves degradation of the macromolecular structure. It is worth noting that the temperature of the maximum devolatilization rate of the extraction residue is the same as that of the original maceral, but the maximum devolatilization rates differ significantly, for example, for the original exinite the rate is $-20\ \%\ min^{-1}$, but that for the extraction residue is only $-4.2\ \%\ min^{-1}$.

Compared with the original macerals, weight loss of the pyrolysis extracts occurs earlier and it gradually increases between 100 and 500 °C. It can therefore be deduced that the extract contains mobile phases embedded in gaps in a macromolecular network. When the temperature is high enough to allow free diffusion, these substances can be distilled and moved away from the particle surface by diffusion. There are also some relatively large fragments of the molecular network in the mobile phase, and their stabilities are higher than those of the coal macromolecular structures with weak C C weak bonds, but lower than that of the condensed aromatic structures.

The reaction kinetic parameters of the pyridine extracts and pyrolysis residues of the macerals, obtained using the first-order reaction model, are listed in Table 4.9.

Table 4.9 Pyrolysis kinetic parameters of pyridine extracts of macerals and extraction residues in CO_2 atmosphere

Sample	Maceral	E(kJ mol^{-1})	ln$A$	Correlation coefficient
Extract	Vitrinite	62.43	8.45	0.9094
	Inertinite	56.08	7.55	0.8280
	Exinite	49.37	6.89	0.9178
Residue	Vitrinite	92.86	14.29	0.9683
	Inertinite	59.86	7.85	0.8616
	Exinite	165.82	26.01	0.9767

4.2.3.3 Pyrolysis Characteristics of Coal Maceral Mixture

The interactions occurring among macerals during pyrolysis were investigated to clarify the coal pyrolysis mechanism. Nine maceral mixtures (i.e., simulated coals) containing three individual macerals in different amounts, based on various lithofacies of Pingshuo bitumite, were prepared. The compositions of the simulated coals are listed in Table 4.10.

We used non-isothermal TGA to study the simulated coal pyrolysis kinetics. Samples of 100–200-mg were heated using a PRT-1 thermal analyzer. The experimental conditions were N_2 atmosphere at normal pressure, flow rate 65 mL min^{-1}, heating rate 10 °C min^{-1}, and final temperature 900 °C.

The experimental pyrolysis conversion rates are listed in Table 4.11.

The calculated conversion rate in Table 4.11 were determined as follows:

$$x_{cal} = \sum x_i h_i,$$

where x_i is the pyrolysis conversion rate of individual maceral i (%) and h_i is the mass percentage of individual maceral i in the simulated coal (%).

Table 4.11 shows that below 600 °C, the calculated values, x_{cal}, are greater than the experimental values, x_{obs}; the difference, Δx, (= $x_{cal} - x_{obs}$) in the range 350–450 °C is relatively large. This shows that the interactions among macerals in the temperature range 350–450 °C are relatively strong. There are two possible explanations based on the chemical structures. One is that small polar groups generated from various macerals in the pyrolysis process agglomerate. The other is that the bridged aromatic clusters in the macerals polymerize. The data for samples 1, 4, and 7 show that the Δx values of maceral mixtures that contain little inertinite and a large amount of exinite are high. This also means that the interactions are strong. If the effects of interactions are considered, x_{cal} can be corrected as follows:

$$x_{cal} = f \sum x_i h_i,$$

where f is the maceral interaction coefficients.

Table 4.10 Compositions of Simulated Coals (wt%)

Simulated coal no.	Vitrinite	Inertinite	Exinite
1.	26.67	6.67	66.66
2.	17.29	21.74	60.87
3.	12.95	29.08	57.97
4.	34.84	4.34	60.82
5.	25.79	16.11	58.09
6.	29.63	33.37	30.00
7.	38.75	3.22	58.03
8.	44.45	18.53	37.03
9.	34.28	25.72	40.00

Table 4.11 Pyrolysis conversion rates for nine simulated coals at various temperatures (%)

Sample	Data Source	Temperature (°C)							
		200	300	400	500	600	700	800	900
1.	Calculated	0	6.16	22.90	61.29	76.04	85.44	91.16	100.00
	Experimental	0	3.63	13.86	53.03	71.09	83.29	90.98	100.00
2.	Calculated	0	6.09	27.69	66.95	78.69	86.90	91.88	100.00
	Experimental	0	50.21	21.49	69.05	82.41	89.58	94.87	100.00
3.	Calculated	0	6.11	29.35	68.71	79.49	78.47	92.13	100.00
	Experimental	0	2.77	17.13	68.53	80.56	88.43	92.61	100.00
4.	Calculated	0	6.28	21.94	59.31	74.73	84.64	90.72	100.00
	Experimental	0	4.06	14.86	48.65	68.91	81.70	87.84	100.00
5.	Calculated	0	6.21	26.23	64.77	77.40	86.13	91.48	99.97
	Experimental	0	4.68	18.37	63.23	77.83	86.49	92.43	100.00
6.	Calculated	0	6.23	30.59	53.34	78.80	86.77	91.76	100.00
	Experimental	0	5.00	24.58	70.43	80.42	87.10	92.09	100.00
7.	Calculated	0	6.36	21.45	58.27	74.05	84.19	90.51	100.00
	Experimental	0	3.68	13.10	50.33	69.12	79.88	88.60	100.00
8.	Calculated	0	6.48	27.29	63.93	76.05	85.12	90.89	100.00
	Experimental	0	4.37	16.95	61.41	77.19	87.13	94.16	100.00
9.	Calculated	0	6.35	29.02	66.83	77.77	86.18	91.47	100.00
	Experimental	0	4.80	51.32	68.61	78.76	87.76	92.56	100.00

Maceral mixture pyrolysis studies also show that the pyrolysis temperature reaches a maximum at about 400 °C and tends to be constant above 600 °C. This implies that there is a correlation between f and temperature. The x values of individual macerals and Δx values of mixed macerals both show such a trend; f can be calculated using the empirical equation

$$f = 1.94\exp\left(\frac{-687.73}{T + 273}\right).$$

The pyrolysis conversion rate x at any temperature can be therefore be calculated from the f value at any temperature.

Similarly, the pyrolysis activation energies of maceral mixtures in any temperature range can be obtained using the following equation:

$$x_{\text{cal}} = \sum E_i h_i,$$

where E_i is the activation energy of maceral i in a certain temperature range (kJ mol^{-1}).

Table 4.12 shows the calculated and experimental pyrolysis activation energies of the maceral mixtures in various pyrolysis temperature ranges (see Table 4.13).

Table 4.12 Pyrolysis activation energies (kJ mol^{-1}) of maceral mixtures in various temperature ranges

Sample	Data source	Activation energy		
		First phase	Second phase	Third phase
1.	Calculated values	20.38	64.47	14.17
	Experimental values	20.62	64.21	16.74
2.	Calculated values	–	70.14	–
	Experimental values	–	67.88	–
3.	Calculated values	23.03	56.54	11.53
	Experimental values	18.47	61.87	15.77
4.	Calculated values	27.47	65.79	16.68
	Experimental values	20.29	68.16	14.28

Table 4.13 Temperature ranges (°C) for three phases of pyrolysis

Maceral	First phase	Second phase	Third phase
Vitrinite	200–325	325–350	350–600
Inertinite	200–350	350–440	440–700
Exinite	200–330	330–450	450–700

The data in the table show that the calculated values and experimental values are in agreement.

In conclusion, interactions are shown to occur among macerals in simulated coal pyrolysis. The conversion rate and pyrolysis activation energy of simulated coal consisting of a maceral mixture can be calculated by summation of the pyrolysis conversion rates of the individual macerals. Deviations in the calculated values can be corrected by the introduction of temperature-related interaction coefficients. The discovery of the interactions among macerals during pyrolysis and the quantitative description of this process are important in understanding coal pyrolysis.

4.3 Pyrolysis Simulation Applied to Coalification Case Study

Coalification can be simulated based on the time–temperature compensation effects described by the Arrhenius equation; coalification with pyrolysis under laboratory conditions can be simulated based on the theory of the thermal evolution of organic matter. Based on this concept, coalification can be considered as a natural pyrolysis process. This was proposed as early as 1924 by Roberts [35], and by Gillet [36] in 1946, based on their experimental results. The experimental results obtained by Dulhunty et al. [37] in 1953 proved that all types of coal can be completely devolatilized, similar to the natural coalification process in the absence of expansion and metamorphic conditions, regardless of their metamorphic grade, if the heating rate is sufficiently slow. This provides further evidence for the possibility of reproducing the natural coalification process under laboratory conditions. In 1985, Saxby et al. [38] simulated 6 years of thermal evolution. Lignite was used as the raw coal sample in the experiment and the heating rate was increased by 1 °C per week. After 300 weeks, the temperature increased from 100 to 400 °C. The research showed that with a sufficiently low heating rate and long enough experimental period, the coal rank gradually increased; comprehensive data for the evolution from lignite to coal types of high metamorphic grade were obtained. The experiment further confirmed that coalification is a natural pyrolysis process, and provided a solid theoretical foundation for simulation experiments. The greatest difficulty in simulation of the thermal evolution of organic matter under laboratory conditions is that it is impossible to reproduce millions of years of geological time. Time–temperature compensation can be used to simulate the long coalification process and to shorten the time for pyrolysis experiments, by increasing the experimental temperature and accelerating the reaction rate.

Some conclusions on natural coalification and pyrolysis simulation have been reached, based on extensive research. A large amount of CO_2 gas is generated in the early coalification stage and in the low-temperature phase of pyrolysis simulation. Large amounts of CH_4 are generated by thermodynamic activity in the mid to late stage (>50 °C) of coalification, and by methanation of carbon and aliphatic decomposition in coal during pyrolysis simulation in the high-temperature range (>400 °C). In the late stage of coalification, a large amount of hydrogen is generated in the high-temperature phase of pyrolysis simulation, and anthracite is formed. The release of large amounts of CO_2, CH_4, and hydrogen means that the overall processes in coalification and pyrolysis simulation are carbon enrichment, dehydrogenation, and deoxygenation. The O/C and H/C atomic ratios decrease with increasing coalification degree. It can therefore be seen that coalification can basically be simulated by pyrolysis. The only difference is that coal pyrolysis requires a higher temperature to shorten the experimental time.

The CH_4 content of coal bed gas depends on two factors, namely cumulative generation and diffusion loss of CH_4 during coalification. The prediction of the theoretical coal bed gas production rate and simulation of the natural coalification process under laboratory conditions are therefore important in coal bed CH_4 research. Pyrolysis is currently used to simulate natural coalification. Although temperature, pressure, and time differences between the natural and artificial coalification processes may lead to deviations in coal pyrolysis simulation experiments, this can still be regarded as a useful approach to evaluating the potential capacity for hydrocarbon generation of a coal sample. These data can be adopted for estimation of the CH_4 generation rate in the coal bed.

4.3.1 Main Factors Affecting Simulation Experiments

In pyrolysis simulation experiments, the temperature, time, pressure, water, and many other factors play important roles; temperature and time have the most significant effects.

4.3.1.1 Effect of Temperature

In the thermal evolution of coal, large amounts of hydrocarbon gases are generated and processed in sediments up to several kilometers thick, in the temperature range of 50–200 °C, at pressures up to several hundred atmospheres, and in the absence of microorganism activities. This is a cracking process, driven by thermodynamic activity and thermal catalysis. However, it is not a simple cracking reaction, but a combination of a series of parallel reactions and consecutive reactions. The temperature plays a very important role in this process. Temperature is the main factor affecting the experimental results, so minimal variations in the temperature affect the accuracy and comparability of simulation results. The reactor should therefore be designed to minimize the temperature gradient within the reactor, to make the temperature consistent. Appropriate temperature measurement positions, and accurate and sensitive temperature measurement and control instruments should be used. This improves the reproducibility of parallel experimental results and comparability of data from different laboratories under the same conditions.

4.3.1.2 Effect of Time

The thermal evolution of organic matter has occurred over millions of years of geological time, so the influence of time cannot be ignored. Within a limited period under laboratory conditions, it is very difficult to make organic matter reach the anthracitic stage, i.e., a very high degree of metamorphism, at 250–300 °C, but under the influence of long geological time periods, the appropriate degree of metamorphism can be reached even if the temperature is relatively low. Peters and Rohrback [39], Saxby et al. [38] performed experiments lasting 625 days and 6 years, respectively. Their results show that the influence of time is related to the temperature. In particular, when the experimental heating time is relatively short, its influence is more obvious.

4.3.1.3 Combined Effects of Temperature and Time

Simulation experiments are performed based on the Arrhenius time–temperature compensation effect in chemical reaction kinetics to reproduce the natural

Table 4.14 Vitrinite reflectances (%) of coal samples for different heating temperatures and reaction times (%)

Experimental condition	Temperature (°C)	100	150	200
	Time (h)	5000	25	1
Vitrinite Reflectance (%)		0.25	0.25	0.25

coalification process by pyrolysis under laboratory conditions. This compensation effect has been confirmed experimentally (see Table 4.14; [39]).

Connan [40] found that the equation for the time–temperature relationship based on the Arrhenius equation is as follows:

$$\ln t = \frac{E}{TR} - A$$

where
A is the frequency constant;
T is the temperature (K);
t is time (s);
E is the activation energy (J mol^{-1}).

Karweil [41] proposed that medium-metamorphic grade coal formed at 150 °C over 2000 years would not be formed at 50–60 °C, even over a very long geological time period (two hundred million years). Some researchers have suggested that the temperature only has a significant influence when it reaches a certain value. Otherwise, even if the time is long enough, it has no influence on the metamorphic grade of the coal; for example, lignite in the Moscow basin was formed during the Paleozoic Early Carboniferous epoch, but it only reaches the metamorphic grade of lignite because of shallow burial and a low temperature. All these results indicate that the time compensation effect has limits. When the temperature is too low (<50 °C), the influence of time is very small. Only when the temperature exceeds a certain value is the influence of time observed.

4.3.1.4 Effect of Pressure

Currently, there is no consistent view on the effect of pressure. Most researchers believe that pressure is the main cause of variations in the physical structure of coal. For coal of low metamorphic grade, decreased porosity, reduction in the amount of water, and an increase in density are mainly caused by pressure. Moreover, the high static pressure in the stratum facilitates the regular arrangement of parallel layers of aromatic fused rings. The extrusion force and shear force, in particular, significantly promote formation of an anthracite lattice, transforming it to a graphite lattice. The presence of static pressure in simulation experiments influences the mechanism of product formation, and product quantity and composition. Theoretically, high pressure is not conducive to cracking reactions, which cause volume expansion. An increase in the reaction pressure therefore inhibits cracking of macromolecular liquid hydrocarbons to small-molecule gaseous hydrocarbons (i.e., it restrains secondary pyrolysis reactions), decreasing hydrocarbon gas production.

4.3.1.5 Effect of Hot Water

The experimental results show that under the same conditions, without a load pressure, the composition of gaseous products obtained during "dry pyrolysis" is very different from that of a natural sample, whereas the agreement between the experimental results of "hydrous pyrolysis", i.e., with addition of excess water to the system, and those for a natural sample can be greatly improved [42, 43]. This indicates that the effect of water cannot be ignored in pyrolysis simulation without a load pressure. However, the addition of a small amount of water, or no water, has no significant effect on the pyrolysis results with an external load pressure. The characteristics of the solid products obtained during pyrolysis are basically unaffected by water [44]. It can therefore be deduced that changes in the time and mode of changes of each element in the reaction process are basically the same. The composition of the gaseous products is also basically the same. In terms of the reaction conditions, hydrous pyrolysis requires temperatures below 374 °C, and a large excess of water to ensure that the coal samples are always submerged in water during pyrolysis. Such reaction conditions indicate that the hydrogen in the reaction process must be derived from water. In confined-system pyrolysis above 374 °C, hydrogen is mainly derived from aromatization reactions, when only a small amount of water or no water is added. However, if a small amount of water is added before the reaction, it can accelerate decarboxylation, enabling oxygen to escape in the form of CO_2 in the early phase of pyrolysis, as in natural coalification. If no water is added, the oxygen in coal escapes in the form of water [45]. The effect of this portion of water in the late phase of pyrolysis is equivalent to that of water added before the reaction. The differences between confined-system pyrolyses with and without water are therefore not obvious. In terms of the effect of pressure and water in different systems on the experimental results, data in the literature show that a load pressure greatly improves the simulation. Simulation experiments are affected not only by the presence or absence of water, but also by the quantity.

4.3.1.6 Effect of Mineral Matrix

Organic matter is dispersed, or exists, in a mineral matrix; this matrix affects the thermal evolution of organic matter. Metal halide catalysis increases the production of CH_4 from lignite by 60–70 %. Some studies [46] have shown that under slow heating, mixtures of clay and organic matter decarboxylate and deaminate to form low-molecular-weight alkanes, cycloalkanes, and arenes. Experiments have shown that catalysis with clay minerals affects the reaction mechanism. In catalysis without clay minerals, C–C bond cracking is a free-radical reaction and the original linear chains form linear hydrocarbons. In catalysis with clay minerals, carbocations are generated, and these form branched hydrocarbons by rearrangement of the carbon skeletons. The effect of the mineral matrix on pyrolysis experiments is therefore mainly observed as catalysis of pyrolysis and adsorption of the pyrolysis products.

After several decades of research, it has been found that pyrolysis simulation is an effective method for determining the properties and processes of hydrocarbon generation of whole coal and macerals in different basins, different coal-bearing strata, and different geological ages. It can also provide a theoretical basis for resource assessment of coal being transformed into hydrocarbon gases, and selection of exploration directions. However, for obvious reasons, simulation of the natural coalification process by pyrolysis under laboratory conditions has limitations. Experimental research using pyrolysis to simulate coalification should therefore be based on actual data and assessments for further exploration of the details and fundamental processes of pyrolysis and natural coalification. The experimental conditions should also be refined to make the pyrolysis simulation process very similar to the natural coalification process.

4.3.2 Simulation Study of Coalification

Using a modified a closed pyrolysis system with the "slow rate of heating + long constant temperature" method. We selected four metamorphic coal grades (Shenmu coal, Datong coal, Dongshan coal, and Yangcheng coal) from bitumite to anthracite phases for pyrolysis simulation experiments of gas generation, to obtain the yields, compositions, and trends in the evolution of gas, liquid, and solid products in different phases of thermal evolution of organic matter during coal formation. Proximate and ultimate analysis data for the coal samples are shown in Table 4.15.

The coal samples were crushed to the desired particle sizes before the experiments and dried at low temperature. A weighed amount of sample was then loaded in the autoclave. The autoclave body was sealed and pressurized with argon gas. The autoclave was flushed two or three times to remove air. Finally, the gas in the autoclave was evacuated and the valve was closed. The autoclave was placed in a heating furnace. Heating was performed under automatic temperature control and the power was cut off after the temperature had been constant for a long time, and then the autoclave was cooled naturally. The needle valve was opened, and the gaseous products passed into the condenser for condensation of water and oil

Table 4.15 Proximate and ultimate analytical data for coal samples (%)

Coal sample	Proximate analysis				Ultimate analysis				
	M_{ad}	A_d	V_d	FC_d	C_d	H_d	O_d	N_d	S_{td}
Shenmu coal	2.37	5.56	36.19	63.81	81.65	4.95	12.09	1.00	0.32
Datong coal	2.51	5.58	26.43	73.57	75.52	4.63	16.40	0.79	2.65
Dongshan coal	0.66	10.32	15.61	84.39	90.16	4.24	3.74	1.21	0.66
Yangcheng coal	2.31	9.32	6.98	93.02	90.11	2.81	5.42	1.34	0.33

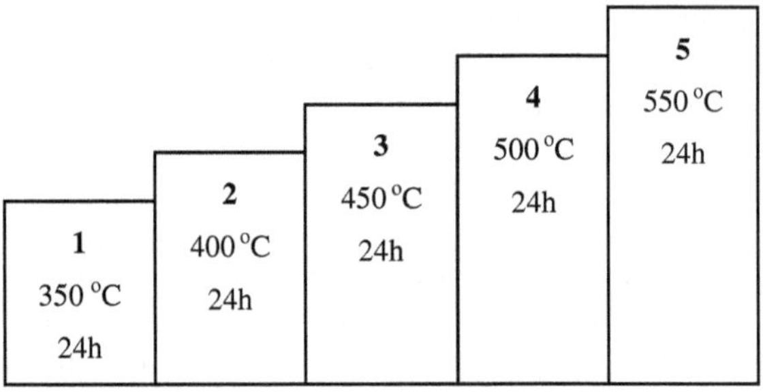

Fig. 4.6 Staged heating program

vapors. The water-displacement method was used to collect the gaseous products. The product was then sampled and its composition was determined. Finally, the device was opened and the solid pyrolysis residue was removed and weighed. After solvent extraction with trichloromethane, the extracts were analyzed using column chromatography. The dry residue after extraction was analyzed using techniques such as vitrinite reflectance measurements, organic elemental analysis, and IR spectroscopy.

Staged heating to the final temperature was used (Fig. 4.6), after various final pyrolysis temperatures had been reached and maintained for 24 h, the experiments were terminated and the products were analyzed.

4.3.2.1 Features and Composition of Gaseous Products from Pyrolysis Simulation

Under the same reaction conditions as above, the compositions and quantities of gaseous, liquid, and solid products obtained at different temperatures were investigated. The analytical results for the total amounts of gaseous pyrolysis products and main component compositions of four types of coal in different experimental temperature ranges are shown in Table 4.16. The data in the table show that the proportions of CO_2 and H_2 in the gaseous products were high compared with the situation in nature. This is because of the small size of the H_2 molecule, which allows it to diffuse easily so that it does not participate easily in natural aggregation. CO_2 gas is highly soluble in water (436 mL L^{-1} at 50 °C, 0.1 MPa), so its adsorption on surrounding rocks is relatively strong and acidic gases are difficult to preserve in alkalescent environments [47].

The general trends in the changes in pyrolysis gas production for the four types of coal at different temperatures are consistent, i.e., the evolution degree of the coal samples gradually increases and total gas production increases continuously with increasing pyrolysis temperature within the experimental temperature ranges; however, the magnitude of the total gas production depends on the coal rank: the higher the coal rank is, the less gas is produced.

Table 4.16 Compositions of gaseous products obtained from four coal samples during pyrolysis

Coal sample	Temperature (°C)	Volume of gaseous products (mL g^{-1})[a]	Mass percentage of gaseous products				
			CH_4	H_2	CO_2	CO	C_{2+}
Shenmu coal	350	30.03	42.8	11.3	35.5	3.64	6.73
	400	57.79	54.8	9.30	27.8	3.13	5.09
	450	87.67	64.9	6.56	22.5	2.61	3.46
	500	118.73	67.3	7.80	20.1	2.34	2.51
	550	136.05	64.4	12.8	17.4	2.65	2.83
Datong coal	350	21.29	38.0	13.3	38.6	4.13	6.03
	400	47.19	48.2	10.9	32.3	3.71	4.87
	450	72.81	57.4	8.82	28.3	2.19	3.31
	500	92.42	64.4	7.95	23.1	2.44	2.14
	550	111.57	65.9	9.46	20.2	2.74	1.65
Dongshan coal	400	16.34	53.6	24.6	17.1	0.77	3.83
	450	45.00	78.1	11.8	7.55	0.29	1.92
	500	68.85	80.9	9.7	7.03	1.29	1.05
	550	83.14	80.5	13.8	3.38	0.60	0.68
Yangcheng coal	450	14.66	45.8	10.3	37.5	2.60	3.74
	500	21.18	62.0	7.16	28.5	0.49	1.92
	550	32.72	60.2	15.0	22.6	1.29	1.07

[a]Volume conversion is based on the volume at standard temperature and pressure. Coal sample product amounts were calculated on dry ash-free basis

Hydrocarbon Gases

The main gaseous hydrocarbon component is CH_4. The total yield of C_{2+} hydrocarbons gradually decreases with increasing pyrolysis temperature. In the temperature range 400–500 °C, the volume fraction of CH_4 gradually increases with increasing pyrolysis temperature. When the final pyrolysis temperature reaches 550 °C, the amount decreases slightly. The main methods of CH_4 generation form coal samples during thermal evolution are as follows [48].

Direct generation of CH_4 from coal by primary pyrolysis with active hydrogen, H*:

$$\text{Coal—CH}_3 + \text{H}^* \rightarrow \text{Coal} + \text{CH}_4,$$

Hydrogenation of solid products generated in pyrolysis:

$$\text{C(solid)} + 2\text{H}_2 \rightarrow \text{CH}_4.$$

Secondary pyrolysis of liquid products generated in the early pyrolysis phase also generates CH_4. During the experiment, the maturity of the coal sample increases with gradual increases in the final pyrolysis temperature, and the aliphatic content of the coal gradually decreases because of generation of gaseous products during pyrolysis. The longer constant-temperature time and other experimental conditions used slow down the release of liquid hydrocarbons from the coal pores, which promotes cracking of the early pyrolysis products within the particles. High final pyrolysis temperature and long constant-temperature time are conducive to the CH_4 content gradually increasing with increasing final pyrolysis temperature. When the temperature reaches 550–600 °C, the solid pyrolysis residue contains only a small amount of non-aromatic carbon, and the amount of liquid products gradually reduce to zero. The pyrolysis of aliphatics and cracking of the early pyrolysis products within the particles are weakened, so CH_4 production decreases.

Hydrogen

All four types of coal show similar variations in H_2 generation, i.e., the volume fraction of H_2 initially gradually decreases and then begins to increase at 550 °C. The main sources of H_2 during pyrolysis are as follows.

Hydrogen generation from active hydrogen atoms (H*) in the system:

$$H^* + H^* \rightarrow H_2.$$

Vapor-phase reaction of CO and water:

$$CO + H_2O \rightarrow CO_2 + H_2.$$

Water-gas reaction of carbonic organic matter at high temperature:

$$C + H_2O \rightarrow CO + H_2.$$

The condensation of organic matter and cyclization and aromatization of hydrocarbons are also sources of H_2. Large amounts of active hydrogen atoms are generated during pyrolysis, and these readily react to form H_2. Secondary cracking of liquid hydrocarbons generated in the early phase of pyrolysis also consumes active hydrogens, but the coal residue after the active decomposition phase is completed aromatized. The liquid products disappear and hydrogen consumption decreases. The amount of H_2 generated therefore varies.

CO_2 and CO

The oxygenic groups in coal can generate CO, CO_2, and water during pyrolysis. Competition occurs among the various reactions. If hydrogen is present in the system, the oxygenic groups preferentially react with hydrogen to generate water. Furthermore, the stabilization energy of carboxyl-generated CO_2 is lower than that of carbonyl-generated CO. A large amount of CO_2 is therefore generated in the initial pyrolysis phase.

Previous pyrolysis experiments have shown that in the absence of external hydrogen, hydrogen is only generated gradually during the pyrolysis process. The compositions of the gaseous pyrolysis products (see Table 4.16) show that the proportion of CO_2 in the product gases generated from various coal samples is larger when the pyrolysis temperature is lower; for example, the amounts are Shenmu coal, 35.54 % (350 °C); Datong coal, 38.56 % (350 °C); Dongshan coal, 17.12 % (400 °C); and Yangcheng coal, 37.54 % (450 °C). This indicates that the desorption of carboxyl, which is an oxygenic functional group, from the coal sample mainly occurs at a relatively early phase of evolution, and the characteristics of the coal sample determine the volume fraction of CO_2 in the total gaseous products. CO_2 production initially decreases with increasing pyrolysis temperature, but then increases slightly at 550 °C. This is because oxygenic groups in coal are deoxygenated and generate CO_2 or CO during thermal evolution. Only carboxyl groups of poor thermal stability are released in the early phase of pyrolysis, and other oxygen sources (such as ether, quinone, and heteroring oxygens) are only released after active decomposition in the secondary pyrolysis phase (about 550–900 °C).

Relationship Between Coal Sample and Composition of Gaseous Pyrolysis Products

The distributions of the three main elements, namely C, H, and O, in gaseous pyrolysis products generated from four types of coal at different temperatures were compared with the results for ultimate analysis of raw coal. The results are shown in Figs. 4.7 and 4.8. The order of the oxygen contents of the four types of coal is Dongshan coal (3.744 %) < Yangcheng coal (5.417 %) < Shenmu coal (12.097 %) < Datong coal (16.405 %). The results in Fig. 4.7 show that the oxygen content of the gaseous products obtained during coal pyrolysis has a positive

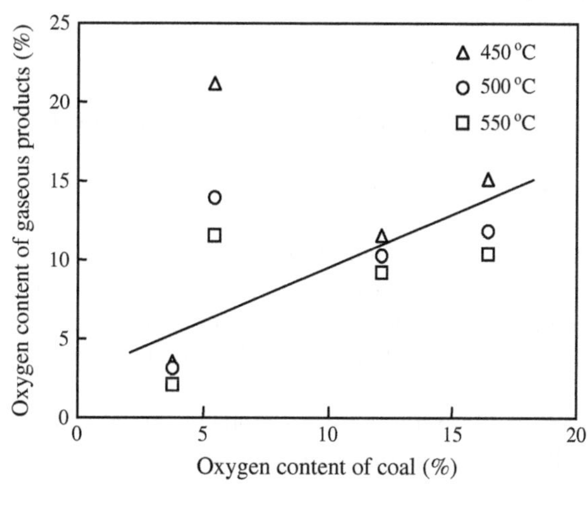

Fig. 4.7 Relationship between oxygen content of gaseous pyrolysis products and oxygen content of raw coal sample

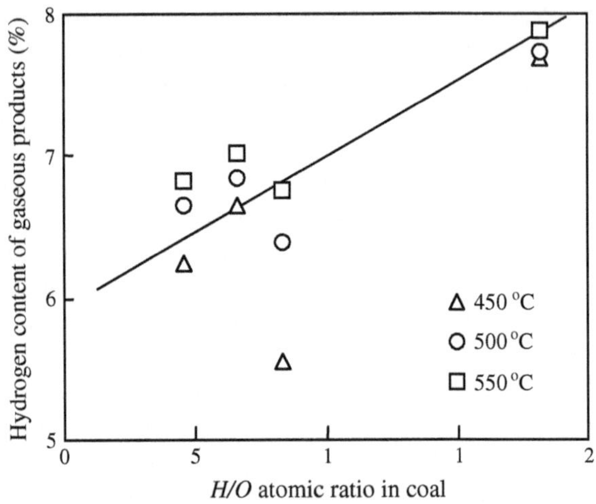

Fig. 4.8 Relationship between hydrogen content of gaseous pyrolysis products and H/O atomic ratio in raw coal sample

correlation with the oxygen content of the raw coal sample, for the same final pyrolysis temperature. The order of the H/O ratios of the four types of coal is Datong (4.518) < Shenmu coal (6.553) < Yangcheng coal (8.317) < Dongshan coal (18.110). The results in Fig. 4.8 show that the hydrogen content of the gaseous products obtained during coal pyrolysis also has a positive correlation with the *H/O* of the raw coal sample.

The highest final pyrolysis temperature in the experiments was 550 °C, which is significantly higher than the decomposition temperature observed during TGA of Shenmu, Datong, and Dongshan coals, but lower than the peak temperature (678 °C) corresponding to the maximum weight loss rate of Yangcheng anthracite. Shenmu, Datong, and Dongshan coals therefore undergo significant pyrolysis and cracking reactions at this temperature, and release large amounts of volatiles, whereas the high metamorphic grade Yangcheng anthracite only undergoes weak decomposition and cracking reactions. The trend for Yangcheng coal was therefore different from those for the other coal samples.

4.3.2.2 Analysis of Solid Products from Pyrolysis Simulation

Proximate and Ultimate Analyses

The proximate analytical results for the solid products obtained during pyrolysis at different final temperatures are shown in Table 4.17. The analytical data for the solid residues show changes with increasing pyrolysis temperature. At a final pyrolysis temperature of 350–550 °C, the coal porosity is reduced and the equilibrium moisture (M_{ad}) decreases with increasing pyrolysis temperature. The ash content (A_d) reflects the non-combustible part of the coal, and volatiles are gradually released during pyrolysis; therefore, the ash content increases. The volatile

Table 4.17 Proximate analytical data for solid pyrolysis products (%)

Coal sample	Pyrolysis conditions (final temperature, °C/retention time, h)	M_{ad}	A_d	V_{daf}	FC_{daf}
Shenmu coal	350/24	1.98	6.03	33.61	66.39
	400/24	1.72	6.58	28.27	71.73
	450/24	1.36	7.05	24.44	75.56
	500/24	1.07	7.81	18.46	81.54
	550/24	0.77	8.28	15.52	84.48
Datong coal	350/24	2.03	5.93	22.95	77.05
	400/24	1.88	6.72	20.01	79.99
	450/24	1.83	7.18	17.87	82.13
	500/24	1.25	6.64	13.29	85.71
	550/24	1.03	8.33	10.76	89.24
Dongshan coal	400/24	0.53	10.90	12.60	87.40
	450/24	0.49	11.21	10.62	89.38
	500/24	0.47	12.63	9.17	90.83
	550/24	0.85	14.90	6.35	93.65
Yangcheng coal	450/24	1.76	9.82	6.93	93.07
	500/24	1.82	10.09	6.50	93.50
	550/24	1.50	11.31	5.35	94.65

content (V_{daf}) gradually decreases, but the fixed carbon content (FC_{daf}) gradually increases. This is easily explained by the chemical nature of pyrolysis.

Coalification and pyrolysis both involve dehydrogenation, carbon enrichment, and deoxygenation. During thermal evolution, the H/C and O/C atomic ratios change regularly as the metamorphic grade gradually increases; Van Krevelen evolution (H/C–O/C) diagrams are the standard method for evaluating the maturity of a coal sample. We therefore calculated H/C and O/C atomic ratios from the elemental compositions of four raw coal samples and their solid products at various final pyrolysis temperatures. Van Krevelen evolution diagrams for the samples were compiled (Fig. 4.9) to determine the maturity of the solid pyrolysis products and to evaluate the authenticity of the simulation process.

The two curves in Fig. 4.9 respectively represent the regional boundaries of possible O/C and H/C distributions for natural coal samples. The O/C and H/C values for the raw coal samples fall in the region between the two curves. The figure shows that a series of solid pyrolysis products with gradually increasing maturity are obtained by pyrolysis of the raw coal samples with gradually increasing final temperature. The experimental values for the corresponding O/C–H/C diagrams mostly fall between the two boundary curves [49, 50]. The distribution ranges of the sequential solid pyrolysis products of the coal samples vary with coal metamorphic grade. The lower the metamorphic grade of the raw coal sample, the wider the distribution range. For Yangcheng anthracite, because of its high metamorphic

Fig. 4.9 Van Krevelen diagrams for solid pyrolysis products for four types of coal and raw coal: *1* Shenmu coal; *2* Datong coal; *3* Dongshan coal; and *4* Yangcheng coal

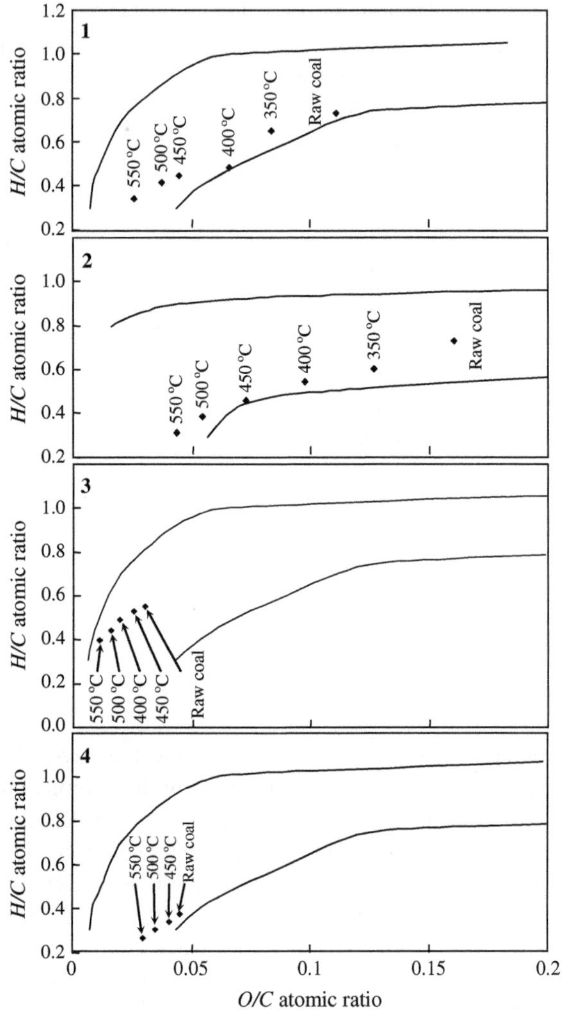

grade, the O/C–H/C distribution range is particularly narrow. This indicates that it is feasible to use the modified closed-system pyrolysis experimental apparatus and experimental conditions in coal sample pyrolysis simulation, especially medium-metamorphic bitumite pyrolysis simulation.

FTIR Spectroscopy

The IR spectra of coal change during coalification as small molecules are released and the degree of aromatization increases. The intensity of the absorption peaks at 2920 and 2960 cm^{-1} have a good positive correlation with the metamorphic grade of the kerogen or coal. Isolated C=C, C≡C, and other unsaturated bonds are not found in coal, but the content of condensed aromatic ring structures with different degrees of substitution increases with increasing coal rank. The intensity of the absorption peak at 1600 cm^{-1} has a negative correlation with the degree of coalification because of the strong influence of the chemical environment and the structure of the aromatic carbon skeleton. The changes in the absorption peak at 1700 cm^{-1} reflect carbonyl group changes, but not changes in other oxygenic groups such as hydroxyl, ether, and methoxy groups. Non-structural factors such as the purity and weight of samples, instrument performance, and operating process cause variations in the absolute intensities of IR absorption peaks; this affects quantitative analysis. If the relative ratio of the absorption intensity of a characteristic peak is used as the analytical parameter, this eliminates the effects of such factors to some extent. The following analytical parameters are usually used [51, 52]; (2920 + 2860) cm^{-1}/1600 cm^{-1} and (1380 + 1460) cm^{-1}/1600 cm^{-1} are used to express the ratio between aromatic and aliphatic groups in coal; 1700 cm^{-1}/1600 cm^{-1} is used to express variations in the O/C atomic ratio or oxygen functional groups; 1380 cm^{-1}/2920 cm^{-1} and 1380 cm^{-1}/1460 cm^{-1} are used to express the enrichment degree of CH_3 in the residual aliphatic series, i.e., the degree of methylation.

We used a WIN-IR spectrometer in the wavenumber range 400–4000 cm^{-1} to obtain IR spectra of KBr/coal sample tablets. IR spectra of the solid products from pyrolysis of four types of raw coal at different final temperatures and durations were obtained. The analytical parameters were obtained from the intensities of the absorption peaks. Correlated analytical data for Shenmu coal and Yangcheng coal are shown in Table 4.18.

The data in the table show that with increasing final pyrolysis temperature, (2920 + 2860) cm^{-1}/1600 cm^{-1} and (1380 + 1460) cm^{-1}/1600 cm^{-1} decrease,

Table 4.18 Relative intensities of characteristic IR absorption peaks of solid pyrolysis products and raw coal samples

Sample		Analytical parameter[a]				
		A	B	C	D	E
Shenmu coal	Raw coal	3.051	1.743	0.788	0.948	0.528
	350 °C/24 h	2.774	1.628	0.738	0.968	0.535
	400 °C/24 h	2.740	1.497	0.713	0.965	0.520
	450 °C/24 h	2.481	1.399	0.672	0.983	0.552
	500 °C/24 h	2.260	1.318	0.615	1.010	0.582
	550 °C/24 h	2.143	1.222	0.560	1.041	0.574
Yangcheng coal	Raw coal	2.109	2.339	0.706	1.012	1.034
	450 °C/24 h	2.056	2.267	0.643	1.035	1.077
	500 °C/24 h	2.011	2.185	0.587	1.091	1.103
	550 °C/24 h	1.898	2.166	0.521	1.132	1.154

[a]A = (2920 + 2860) cm^{-1}/1600 cm^{-1}; B = (1380 + 1460) cm^{-1}/1600 cm^{-1}; C = 1700 cm^{-1}/1600 cm^{-1}; D = 1380 cm^{-1}/1460 cm^{-1}; and E = 1380 cm^{-1}/2920 cm^{-1}

which indicates that the maturity of the coal sample is increasing, the amount of condensed aromatic rings is increasing, and the amount of aromatic compounds is gradually increasing during pyrolysis. The increases in 1380 cm^{-1}/1460 cm^{-1} and 1380 cm^{-1}/2920 cm^{-1} indicate enrichment of CH_3 in the aliphatic series in residual solid products. To some extent, the gradual decrease in 1600 cm^{-1}/1700 cm^{-1} shows that the O/C atomic ratio in coal decreases with deepening pyrolysis. This further proves that pyrolysis involves dehydrogenation, deoxygenation, and carbon enrichment. The IR characteristic parameters of the other types of coal follow similar trends; the only difference is the magnitude of the variation.

X-Ray Diffraction

Coal is an amorphous material with short-range order and long-range disorder. Incompletely mature microlitic structures similar to graphite are present in the coal structure, and the arrangement rules for their size and orientation vary with the metamorphic grade of the coal. The (002) diffraction peak in the XRD pattern reflects the extent of parallel orientation of aromatic lamellae in the aromatic nucleus, and the (100) diffraction peak reflects the size of the aromatic lamella. The thickness, L_c, in the vertical direction and the length, L_a, in the parallel direction between the aromatic lamella and a microlite, and the distance d between aromatic lamellae can be calculated using the Bragg equations.

The microlitic parameters determined for the solid pyrolysis products of various coal samples, including d, L_a, and L_c values are shown in Table 4.19.

The data in the table show that the stack height, L_c, of the aromatic nucleus lamella increased, and the distance, d, between the lamellae decreased with increasing final pyrolysis temperature. The calculated values of L_c and d are consistent with the values for the coke microlitic structure reported in the literature.

Table 4.19 XRD analysis of raw coal samples and solid pyrolysis products

Sample		Characteristic parameter of diffraction peak (°)				Microlitic structure parameter (nm)		
		(002)	(002)	(100)	(100)	d	L_c	L_a
Datong coal	Raw coal	8.14	24.2	–	–	0.368	1.044	–
	350 °C/24 h	8.60	25.1	–	–	0.355	0.990	–
	400 °C/24 h	8.14	25.2	–	–	0.354	1.046	–
	450 °C/24 h	7.91	25.2	–	–	0.353	1.076	–
	500 °C/24 h	7.41	25.1	–	–	0.354	1.151	–
	550 °C/24 h	6.98	25.4	–	–	0.351	1.218	–
Dongshan coal	Raw coal	5.81	25.4	–	–	0.350	1.471	–
	400 °C/24 h	5.67	25.5	–	–	0.349	1.501	–
	450 °C/24 h	5.58	25.5	–	–	0.349	1.532	–
	500 °C/24 h	5.39	25.1	–	–	0.355	1.579	–
	550 °C/24 h	5.58	26.1	5.58	43.7	0.342	1.534	3.151

Analysis of Distribution of Gas, Liquid, and Solid Products

The products of the experiments performed under the experimental conditions determined by us are divided into three states, namely gas, liquid, and solid. The total amounts of products in each state are shown in Table 4.20. The data in the table show that the final pyrolysis temperature is an important factor in the composition of the product states for the same type of coal at similar heating rates, pressures, constant temperatures, and times, i.e., the degrees of coal sample pyrolysis and cracking increase, the total amounts of small-molecule gaseous products generated increase, and the amounts of solid products decrease with increasing final pyrolysis temperature. The trends in the changes for liquid products such as tar, pyrolysis water, and liquid hydrocarbons with changes in final pyrolysis temperature are less obvious. There are large differences in terms of the main type of reaction and the extent of reaction for coals with different metamorphic degrees under the same final pyrolysis temperature conditions because of variations in the physical and chemical properties of the coals. For lower metamorphic grade Datong and Shenmu coals, the total amount of gaseous products increases significantly and the maturity of solid residual products increases significantly for a final pyrolysis temperature range of 350–550 °C; however, for high metamorphic grade Dongshan coal and Yangcheng anthracite, the corresponding variations are much smaller.

Table 4.20 Mass percentages of gaseous, liquid, and solid pyrolysis products obtained from four types of coal

Coal sample	Pyrolysis conditions (°C h^{-1})	Gaseous product	Solid product	Liquid product
Shenmu coal	350/24	3.84	89.83	6.33
	400/24	6.61	85.18	8.21
	450/24	9.72	82.93	7.35
	500/24	14.37	79.09	6.54
	550/24	15.51	75.90	8.59
Datong coal	350/24	2.74	92.33	4.93
	400/24	5.88	89.10	5.02
	450/24	9.34	85.83	4.83
	500/24	11.64	81.19	7.17
	550/24	12.94	78.26	8.80
Dongshan coal	400/24	2.01	95.82	2.17
	450/24	3.49	93.84	2.67
	500/24	5.21	92.70	2.09
	550/24	6.20	91.96	1.84
Yangcheng coal	450/24	1.79	97.62	0.59
	500/24	2.36	96.70	0.94
	550/24	3.05	95.35	1.60

4.4 Structure–Reactivity Relationships in Coal Pyrolysis

4.4.1 Application of Pyrolysis Methods in Study of Coal Structure–Reactivity Relationships

Coal pyrolysis can be used as a non-isothermal method for studying intermolecular interactions. Structure–reactivity relationship studies using pyrolysis provide information on the activities of the structural units in the coal, by analyzing the changes in coal structure and related pyrolysis products during pyrolysis. First, no heat transfer is considered in rapid pyrolysis, and the most active small molecules in the structural units will be the first to break from the bulk. If mass transfer is not considered, the broken small molecules escape as volatile matter, therefore the larger the amount of active small molecules in the coal, the more volatiles it produces. If we can prevent condensation of volatiles, we can tell from the end groups of the small-molecule fragments how they are connected to the bulk. After release of the active components from the outermost side of the coal particles, the next inner layer of the coal particle cannot receive enough energy to quickly break the small-molecule fragments, and condensation and rearrangement occur between neighboring structural units or inside the unit. If proton or electron pairs can be provided at this time, sites that can easily give or receive electrons will be the most active sites in this structural unit. Such sites are identified using molecule probes in catalysis studies, or molecular modeling using quantum chemistry calculations. If mass transfer is considered, some of the small molecules in the volatile matter will be captured by the polymeric network, forming a relatively stable structure under this condition. Such structures play important roles in coal reactivity [53]. In slow pyrolysis, the effect of heat transfer can be virtually neglected, but because the active groups or sites cannot receive enough energy in the required time to break the active fragments, reactions with activation energies lower than that of active bond breaking will occur instead, and the coal structure tends to build an intermittent stable form by intramolecular restructuring, therefore the broken molecular fragments are not able to represent the active sites of the coal structure. Under slow pyrolysis conditions, actions can be taken to obtain more precise information on the coal structure and reactivity. Hydrogen can be provided during pyrolysis to stabilize the structure after bond breakage, and to identify the active sites by changing the structure before and after pyrolysis; hydrogen atoms stabilize the coal structure during pyrolysis, by providing easy electron-exchange sites. Crosslinks can be weakened by solvent swelling before pyrolysis, to expose more active sites to reactants. These methods can also increase the amount of volatiles from active groups in fast pyrolysis.

Hyphenated pyrolysis techniques with precise temperature control are useful tools for studying coal structure and reactivity. TGA-FTIR [54] and PyGC/FIMS (field ion mass spectroscopy) [53, 55] are popular techniques. The former is usually used to study the relationship between pyrolysis rate and coal functional groups under slow heating conditions, and the latter is used in studies of the structure of free radicals and small molecules and their reactivities under fast heating conditions.

In-depth studies of coal structure and reactivity were possible in the 20th century using hyphenated techniques, and Py-FIMS enabled structure and reactivity studies to progress from indirect estimations to real-time online stages, and became an important method for molecular-level studies of coal structure and reactivity. In this method, the sample is placed directly in a plasma field in a high vacuum for pyrolysis; this reduces the changes caused by post-pyrolysis secondary condensation reactions, and represents the coal composition more accurately. The use of field ionization technology significantly reduces the number of molecular fragment peaks in MS, and forms high density molecular ions with a wide range of masses and polarities, which can be more precisely analyzed. Currently, there are two MS groups performing related work, i.e., the Fachhochschule Fresenius in Germany, and Stanford Research Institute in the USA, involving coals from lignite to high-metamorphic bituminous coals [53]. One weakness of this method is that if the active and inactive structures have the same mass, all measurements related to coal reactivity will be independent of the FIMS signal strength.

Solomon et al. [56] discovered when analyzing Illinois#6, using TG-FTIR, that the weight loss at about 500 °C is the result of breaking of weak bridges, and the temperature of this weight loss peak increases as the coal rank increases; this is in agreement with the consensus that lower rank coals have higher reactivities. At low temperatures, the shoulder peak representing tar release correlates with "object" molecules built by non-covalent bonding. Marzec et al. [57] measured the rate of crosslinking during pyrolysis based on the rate of CH_4 release, using TG-FTIR, and determined the bond-breaking rate from the tar volatile data. Both these studies show the effects of enhancement and inhibition of the coal structure on the reactivity. Solomon suggested that as a result of hydrogen bond breakage at 200–400 °C, volatile non-covalent-bonded small molecules form CO_2 and water. The amounts of hydrogen bonds and carboxyl groups can therefore be preliminarily estimated from the amounts of CO_2 and H_2O. For low-rank coals, the release of CO_2 and H_2O also depends on low-temperature crosslinking, which in turns depends on the coal metamorphism. As the metamorphism increases, the extent of low-temperature crosslinking increases. Low-rank coal is more reactive than high-rank coal because of these highly active groups, which can crosslink at low temperature; for example, CO_2 release from lignite indicates the start of crosslinking, whereas the release of CH_4 indicates the start of crosslinking in bituminous coal. This represents the differences between the coal structures and the changes in active sites for these two coals. Based on the impregnation of metal ions [58], it has been suggested that the release of CO_2 at this point may be related to carboxyl groups attached to base metal ions. This result provides a method for predicting the catalysis of coal reactions by primary mineral materials. In one pyrolysis process, the weakest C–C and C–O bonds start to break and generate molecular fragments; the starting temperature range of this process was identified as 340–390 °C, using pyrolysis-MS [53]. These two types of component and the amount and distribution of chemical bonds significantly affect the initial activity. Hydrogen extraction from breaking molecules increases the aromatic hydrogen concentration. These aromatic hydrogens cannot be directly measured at present, but the amount may be related to the hydrogen-donating ability of the coal. The amount of aromatic hydrogen has a significant effect on the coal reactivity; components with such aromatic hydrogens can generate highly active cyclohexadiene free radicals [59]. In terms of coal thermophysical properties, this temperature range is that of the initial softening point of coal, at which point the Van der Waals forces and hydrogen bonds are not strong enough to hold all the molecular units together. Coal thermoplastic studies have also shown that the more inactive the component, the longer it takes to reach softening, and the higher the related Gieseler temperature [60]. Structural analysis of the volatile matter in this stage, using pyrolysis-MS, shows that the active structure consists of large amounts of non-substituted aromatics or substituted aromatics with three to seven rings, most of which are ethyl and propyl type, as well as alkyl-substituted phenolic components with one to three aromatic rings, alkyl-substituted naphthalenes, fluorenes, pyrenes, anthracenes, and acenaphthenes with chains of less than three carbon atoms, alkylated aromatic hydrocarbons with four rings and C_5 substitution and below (e.g., pyrenes), and alkylated aromatic hydrocarbons with six rings (pyrene diphenylurea). If the fragments are small enough to eliminate mass transfer effects, and no mid-temperature crosslinking occurs, so that they are able to pass through the particle and form tar oil, then the yield of tar under the same conditions can represent the amount of small molecules that the coal sample could generate. The amount of such volatiles to some extent represents the coal reactivity. As the temperature increases further, crosslinking occurs inside the coal network. This process mainly involves condensation of the components that cannot be easily volatilized. As a result, the volume increases, the structure is stabilized, and the reactivity decreases. The rate of mid-temperature crosslinking is slightly lower than that of the breaking of bridging bonds. The connectivity among large molecules (chain or network) also affects the reactivity; researchers [56] have used percolation theory to estimate the average number of bridges in each structural unit cluster to determine the effect of networks on coal structure and pyrolysis. It has been shown that a chain structure represent highly volatile bituminous coal, whereas a network structure represents lignite and sub-bituminous coals. This conclusion seems counter-intuitive, but it shows that the chemical bonds in highly volatile bituminous coals are much stronger than the ionic bonds in low-rank coals. The increase in crosslinking does not greatly affect the rate of bond breaking, but it can change the product distribution to a large extent, i.e., there are more monomers in low-rank coals. When the hydrogen generated from hydrogenated aromatic rings and aliphatic chains has been consumed, primary pyrolysis stops, which also indicates termination of the effects of hydrogen on coal reactivity.

Chauvin and Deelder [61] assumed that pyrolysis consists of the following processes: the breaking of weak bonds between elemental structural units and the formation of free radicals; the evaporation of some elemental structural units, if the temperature is high enough; the condensation of some monomers into polymers, which remain in the char; hydrogen transfer reactions, which enable condensation between elemental structural units via free radicals; in the vapor phase, the stabilization of vaporized elemental units via hydrogen transfer; and secondary cracking in the solid and liquid phases. They suggest that short-time, quick pyrolysis techniques that minimize secondary cracking favor entry of monomers into the vapor phase, and inhibit condensation and hydrogen transfer reactions. Consequently, the composition of the primary tar represents the coal elemental structural units stabilized by acquiring protons or via oligomerization. The results of hydrogenation experiments show that it can be assumed that the bridges between elemental structural units of the oligomerized tar resins are oxygen atoms.

4.4.2 Application of Solvent Extraction in Research on Coal Structure and Reactivity

In extraction, small molecules in the coal are released via the electron-donating/receiving abilities of solvents. Staged extraction offers a solid experimental basis for the establishment and validation of coal structural models and analysis of the soluble and insoluble components under different solvent conditions. In addition, because the amount of small molecules in the extract to some extent represents the strength of the coal reactivity, information on the reactivity can be acquired. The generally accepted models such as association, host–guest, and two-phase models were first established by studying extracts. Currently, it is agreed that large three-dimensional crosslinked molecular networks form the bulk of the coal structure; the networks are connected by covalent bonds; non-covalent secondary bonds (bond energy <62.7 kJ mol^{-1}such) such as hydrogen bonds, Van der Waals forces, and weak complex forces are also important. Small molecules are incorporated into the large molecular network. Differences among coal types are determined by the variations in the small-molecule phase and variations in the bonds between small molecules and the coal crosslinked network. It is the secondary bonds or bridging bonds of crosslinked networks that are broken during extraction.

4.4.2.1 Relationship Between Soluble Components and Reactivity

According to van Krevelen, extracts are important components of the metaplast generated from pyrolysis, with molecular weights higher than 600–700. This range also represents the main contribution of the small-molecule phase, defined by Derbyshire et al. [62]. The definition of extracted soluble components is small molecules connected to large molecular networks via non-covalent bonds, which are released by physical interactions with solvents, but excluding fragments depolymerized from the large molecular network. The main difference between the soluble and insoluble components is that the former are connected to the large molecular network via non-covalent bonds. Some extraction solvents degrade the large molecules, and the presence of degraded products in the extracts complicates their study, so reactions between solvents and coal should be avoided. The molecular weights of the extracts range from 20 to 3000, with the peak at around 500–1500 [63]. The molecular-weight distribution is one of the most important properties of the extract. Components with relatively low molecular weights are volatile in the thermal treatment of coal, and increase the coal reactivity, whereas components with higher molecular weights remain in the coal and undergo polymerization because of the tar vapor pressure at the coal particle surface.

Currently, the composition of the small-molecule phase is not well understood. Studies [64–68] of extracts using MS and solid ^{1}H NMR spectroscopy have been reported in the literature, and it was shown that small molecules with molecular weights less than 500 account for 30–40 % of the coal composition, which is much higher than that has been assumed in the past. However, the large-molecule phase does not represent the organic biolith completely. Studies have been performed [53] to determine the reactivity of the small-molecule phase, using tetralin as the hydrogen-donor solvent, and comparing the rate of extraction into the tetralin after the hydrogenation of different coals. FIMS analysis of the products showed that aromatics with four and five rings show high activities in liquefaction and pyrolysis.

4.4.2.2 Application of Insoluble Residues in Research on Coal Structure and Reactivity

Strictly speaking, the insoluble components during extraction consist of large molecular networks with covalent bonds. The insoluble components mainly affect the coal reactivity by influencing the strengths of different types of connections in the molecular network; these effects are generally analyzed using swelling measurements. This method has developed into an important qualitative method for studying large molecules in coal, and offers a simple method for estimating the strength of crosslinking bonds inside the coal. The more the sample swells, the weaker the crosslinks become. The more hydrogen there is in the soluble components of the extract, the lower the softening point. These effects could be considered as expansion of the agglomerated large-molecule network as a result of solvent interactions, which, to some extent, transform it back into the state before coal formation. If this assumption is correct, this provides a feasible method for studying coal structure and increasing the activity of the large molecular network. It was found during a ^{1}H NMR spectroscopic study of pyridine-insoluble residues [69, 70] that there are two types of molecular structure in the pyridine-insoluble components: one is completely rigid, without any variable conformations, and the other contains rotatable structures. From the chemical point of view, this result suggests that polymerized components are dominant in the large-molecule phase, and there are some other components attached to the polymerized rings as side chains by single bonds, which enable them to rotate. Literature reports [53] on studies of insoluble components using Py-FIMS and NMR spectroscopy suggest that such side-chain radicals have molecular weights of 50–850, with the peak in the range 150–650.

4.4.3 *Model Compounds in Studies of Coal Structure and Reactivity*

Model compounds are used to acquire kinetic data for coal liquefaction and pyrolysis, and to explore the reaction mechanism, based on an understanding of the coal structure. Reference data for coal structure and reactivity studies can be obtained by studying the relationships between model compounds and reactivity. Model compounds also provide the necessary preconditions for the use of quantum chemistry and molecular mechanics. In the early stages, the model compounds were usually simple molecules in homogeneous phases, with the focus on microscopic studies of bond breakage during liquefaction and pyrolysis, and the processing conditions that can enhance this. Modern models focus on mesoscopic studies of large molecules in multiphase systems during coal conversion; typical model compounds and the corresponding processes studied are listed in Table 4.21.

4.4.4 *Models for Pyrolysis Reactivity*

The significance of establishing models for coal structure and reactivity is the prediction of the coal reactivity when it is heated, based on the studies of the common behavior of different types of coal with different ranks, and also the

Table 4.21 Typical model compounds in coal reactivity studies

Model compound	Study process
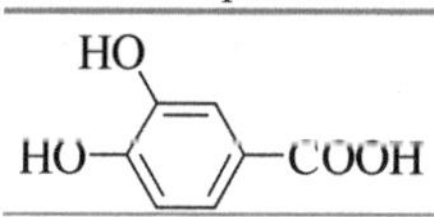	Decarboxylation to form H_2O and CO_2 [71] during heating
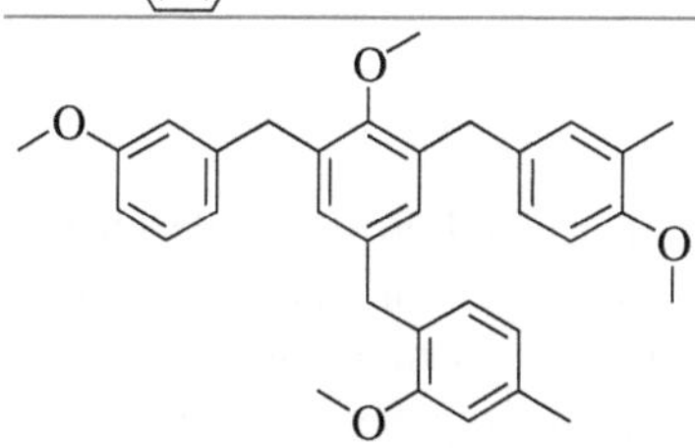	Formation of CH_4, CO, and benzene in coal [29]
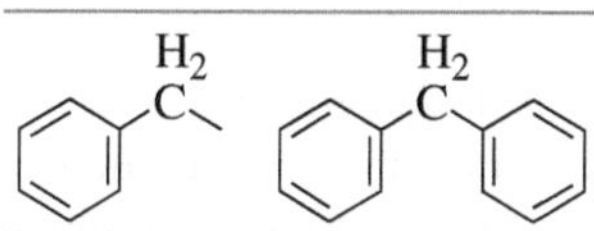	Hydrogen transfer in coal, including free-radical H and reversible free-radical imbalance [72]
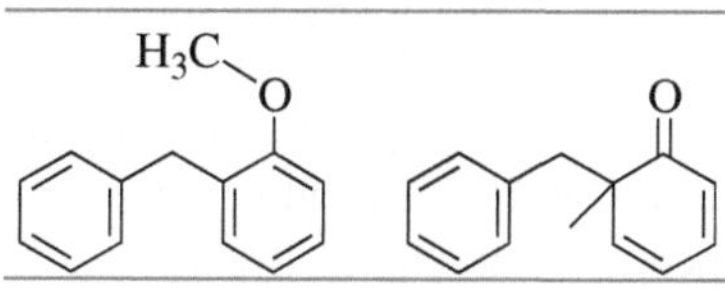	Breaking of ether bonds in pyrolysis [73]

analysis such as proximate, ultimate analysis and chemical structure analysis. The reactivity of coal depends on the coal rank, types, physical structure and chemical structure, as well as the reaction condition and environment. It is with no doubt the most comprehensive consideration to establish a model for coal structure and reactivity with respect to all the above parameters. But not only because all the parameters are coupled with each other, but also that some of those are not quite understood yet, in this section, the heat and mass transfer, and the effect of mineral components are not considered, and we only take consideration of the main parameters such as coal proximate analysis, ultimate analysis, and the reactivity of structural groups, to establish the relationship of coal structure and reactivity, and use it to predict the coal reactivity.

4.4.4.1 Selection of Parameters that Affects Coal Reactivity

Coal structure parameters can be roughly divided into two types: one is the fundamental data such as the proximate and ultimate analysis; there other is those represent the individual uniqueness, such as the type and quantity of functional groups, and their concentration, the degree of polymerization of coal, average molecular weight, etc. the former determines the common characteristics between the same coal rank, and the latter determines the properties and the product distribution for pyrolysis. As summarizing the existing common understanding, the following relevant parameters can be considered: the proximate and ultimate analysis; the quantity of oxygen containing functional groups and its activation energy in the pyrolysis of different coal types; the ratio of aromatic and aliphatic hydrogen; the structure parameters of the ring clusters; the pyrolysis condition, including the heating rate and final temperature; the amount of hydration. Such parameters are listed in *Table*. The selection of coal type is same with those in Table 2.4 of Chap. 2.

4.4.4.2 Parameter Independent Analysis

Because the parameters in Table 4.22 may have interlinked cross couple interactions, the first step to establish a model is to do an independent analysis of all the above parameters, and find a set of variables that contains minimal and linearly independent parameters. The author uses a method of progressive regression of main component to do the independent analysis. The principle of such method is: by the calculation of the covariance matrix of the actual measured results of different variable, progressively extract the component with maximal contribution to the variance, to accomplish the reduction of the variables, and finally get a minimal set of independent variables. As shown in Table 4.23, considering the pyrolysis activation energy E of coal samples under 20 °C min^{-1} heating rate, and 40 min hydration amount H_{40} as related variables, two groups of independent variables can be obtained by using the above mentioned independent variable analysis.

Table 4.22 Analysis of coal structure and reactivity parameters

Parameters[a]	Mud coal	Lignite	Long flame coal	Fat coal	Coking coal	Lean coal	Meager coal	Anthracite
E_{10} (kJ mol^{-1})	13.77	18.11	16.51	25.00	21.14	23.06	16.36	22.62
E_{20} (kJ mol^{-1})	19.33	32.02	28.66	33.24	38.59	37.62	34.95	17.86
H_{40} (mol 100 g coal^{-1})	3.50	4.97	1.63	0.49	3.82	4.52	0.51	1.02
C_d	26.40	66.44	68.08	70.69	76.25	75.85	72.72	77.95
H_d	2.97	4.57	4.27	4.45	4.67	4.85	3.36	2.93
O_d	19.15	21.90	11.61	7.33	9.55	7.56	5.04	0.57
N_d	1.86	1.08	0.91	1.13	1.05	1.45	1.05	0.94
S_{td}	0.24	0.29	0.89	1.88	0.52	1.01	0.29	0.36
M_{ad}	8.02	13.49	2.36	0.81	1.25	0.79	0.87	0.72
A_d	49.38	5.72	14.24	14.52	7.96	6.82	17.54	14.52
V_d	40.11	43.68	32.27	29.52	33.60	11.05	12.22	7.93
M_δ	119.3	53.2	51.4	47.4	38.0	38.7	44.1	35.4
M_r	520.1	370.5	373.1	389.6	329.2	401.8	454.6	462.4
$\sigma + 1$	2.716	0.629	0.524	0.415	0.500	0.434	0.388	0.404
p_0	3.365	4.695	5.441	5.515	5.192	5.082	5.602	5.420
H_{ar}/H_{al}	0.19	0.30	0.475	0.49	0.522	0.523	0.87	1.10
CH_2	2.17	1.26	0.69	0.61	0.85	0.25	0.92	0.15
OH	26.50	39.00	32.72	17.76	10.16	15.34	14.87	7.00

[a] E_{10} is the activation energy in the heating rate of 10 °C min^{-1}, E_{20} is the activation energy in the heating rate of 20 °C min^{-1}, H_{40} is the hydration amount under the conditions of 40 min and 100 g coal, M_δ and M_r are the average molecular weight of side chain and clusters respectively, $\sigma + 1$ is the reaction sites that one group cluster can offer, p_0 is the quantity of breakable bond. The calculation method for all these parameters are from Genetti and Fletcher [74]

Table 4.23 Calculation results for parameter independent analysis

Pyrolysis activation energy for 20 °C min^{-1} heating rate (E)		40 min hydration amount (H_{40})	
Independent variables	Correlation coefficient (R^2)	Variables	Correlation coefficient (R^2)
H	0.654328	O	0.436105
OH	0.715858	OH	0.566804
CH_2	0.744942	CH_2	0.862920
H_{ar}/H_{al}	0.885887	V_d	0.954602
M_δ	0.967469	S_{td}	0.963736

From the contribution of the coals structure parameter to reactivity (E) variance, H, OH, CH_2, H_{ar}/H_{al} and M_δ has 96.74 % impaction to the reactivity model, while the other parameters are eliminated because they are correlated to the above 5 parameters or between each other. Both considers that O, OH and CH_2 have significant effects in the coal reactivity model; but the difference is that the volatile and sulfur content are not considered in the reactivity measured from thermal analysis, instead, the ratio of aromatics to aliphatics and the average molecular weight of side chains are used. Such differences may result from that the reaction rate in thermal analysis is determined by the release rate, instead of amount, of the volatiles, while the reaction rate measured from solvent hydration represents the average reaction active sites that the coal structure can provide.

4.4.4.3 Establishment of the Coal Structure and Reactivity Model

It can be seen by comparing the selected two groups of variables to the reactivity they represent that, the properties of H, CH_2, and H_{ar}/H_{al} represent the contribution of coal rank to the activation energy characterized reactivity, while the average molecular weight of side chains M_δ and OH contributes to the coal structure-reactivity model by effecting the behavior of particular coal types in the pyrolysis. When the other independent variables are not considered, the variation of any of the independent variable to the coal pyrolysis reactivity represents the how much such variables affect the reactivity, in which H, CH_2 and H_{ar}/H_{al} are positivity related to E, while M_δ and OH are negatively related.

The contributions of O, V_d, CH_2 and OH to the hydration reactivity are represented via the change of coal rank, where all parameters except for O are negatively related; the effect of sulfur in the hydration is to provide electron pair, this makes that the effects of hetero-atoms on hydration plays an non-negligible role, which has not been considered, and will be a topic for further study.

The contributions of the independent variables to the variance of the regressed equation are different, which is surmised in Table 4.24. In the model, the contributions of H, M_δ, H_{ar}/H_{al}, CH_2 and OH to the model successively decrease, this implies that, the reactivity in the coal pyrolysis mainly depends on the hydrogen

Table 4.24 Contributions of variables to the variance of the regression

Activation energy under 20 °C min^{-1} heating rate (E)		40 min hydration amount (H_{40})	
Variables	Contribution to variance	Variables	Contribution to variance
H	0.980604	O	0.804533
M_δ	0.982084	S_{td}	0.805782
H_{ar}/H_{al}	0.982674	V_d	0.815775
CH_2	0.988185	CH_2	0.831224
OH	0.991579	OH	0.862411

amount and the average molecular weight of side chain, while the other effects play a secondary role to increase the reactivity. The analysis on the parameters that affect the hydration suggests that, the active sites during the hydration reaction are main provided by O and hetero sulfur atoms.

From Fig. 4.10, the prediction of the structure reactivity model for the activation energy agrees well with experimental data, with the correlation coefficient of 0.99. Figure 4.11 compares the model prediction of the coal hydration reactivity to the experimental data, with a correlation coefficient of 0.86, suggesting a poor predictability of the pyrolysis activation energy.

The reactivity of coal for pyrolysis can be relatively precisely predicted by the coal analysis data. A relationship between coal structure and reactivity can be established according to the above analysis.

Coal pyrolysis reactivity model:

$$E = a_1 H + a_2 M_\delta + a_3 \frac{H_{ar}}{H_{al}} + a_4 \mathrm{CH_2} + a_5 \mathrm{OH},$$

$$a_1 = 8641.77, \quad a_2 = -204.18, \quad a_3 = 3406.15, \quad a_4 = 11001.69, \quad a_5 = -236.54$$

$$R^2 = 0.9916.$$

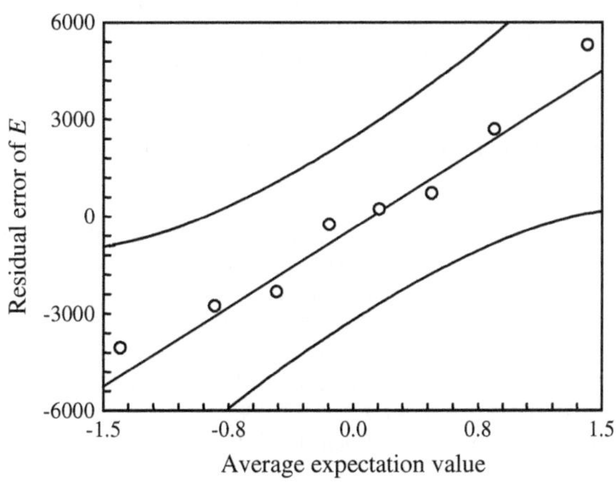

Fig. 4.10 Comparison of predicted and experiment data for activation energy of different coal types: scattered points are experimental data, straight line is the prediction, and the curves are prediction distribution range

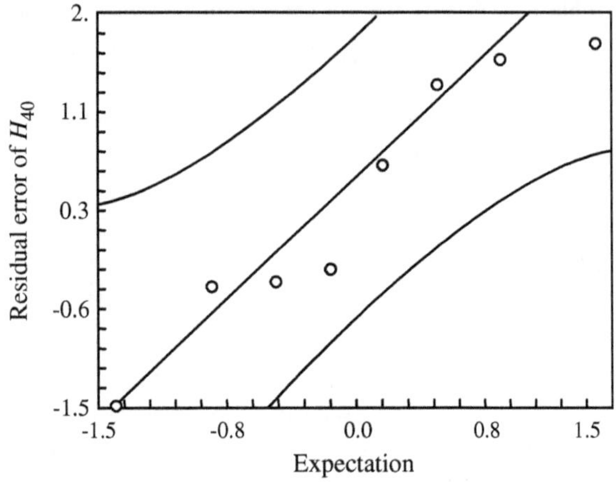

Fig. 4.11 Comparison of predicted and experiment data for hydration of different coal types: scattered points are experimental data, straight line is the prediction, and the curves are prediction distribution range

Coal hydration reactivity model:

$$H_{40} = a_1 O + a_2 S_{\mathrm{td}} + a_3 V_{\mathrm{d}} + a_4 \mathrm{CH}_2 + a_5 \mathrm{OH},$$
$$a_1 = 0.56, \quad a_2 = 0.78, \quad a_3 = 0.00322, \quad a_4 = -1.43, \quad a_5 = -0.10$$
$$R^2 = 0.8624.$$

References

1. Howard JB et al (1967) Ind Eng Chem Process Des Dev 6:74
2. Suuberg EM et al (1985) Fuel 64:1668
3. Suuberg EM et al (1985) Fuel 64:956
4. Gavalas GR et al (1981) Ind Eng Chem Fundam 20(2):113
5. Gavalas GR et al (1981) Ind Eng Chem Fundam 20(2):122
6. Shi L et al (2013) Fuel Process Technol 108:125
7. Li G et al (2014) Energy Fuels 29(2):980
8. van Krevelen DW (1981) Coal. Elsevier Scientific Publishing Company, Amsterdam p263
9. Nusselt WZ (1924) VDI 68
10. Badzioch S (1967) Bcura Month Bull 31(4):193
11. Kabayashi H (1977) In: 6th symposium (international) on combustion 411
12. Pitt GJ et al (1962) Fuel 41:267
13. Anthong D B et al (1975) In: 15th symposium on combustion, 1303
14. Suuberg EM (1977) Sc D thesis, University of Cambridge, 50
15. Given PH et al (1960) Fuel 39:147
16. Wiser WH et al (1967) Ind Eng Chem Process Des Dev 6:133
17. Suuberg EM et al (1978) Ind Eng Chem Process Des Dev 17:34
18. Gavalas GR et al (1981) Ind Eng Chem Found 20:113
19. Badzioch S et al (1970) Ind Eng Chem Process Des Dev 9:521
20. Wiser WH et al (1967) Ind Eng Chem Process Des Dev 6:133
21. Koch V et al (1969) Brennstoff-Chemie 50:369
22. Maria M et al (1983) Fuel 62:1393
23. Qiu JH (1994) Fuel Chem J 22:316

24. Johnston PR et al (1993) J Colloid Interface Sci 155:146
25. He XM (2010) Coal chemistry. Metallurgical Industry Press, Beijing, p 224
26. van Krevelen DW (1981) Coal. Elsevier Scientific Publishing Company, Amsterdam p287
27. Coats AW et al (1964) Nature 201:68
28. Marzec A et al (1994) Fuel 73(8):1294
29. van Heek KH et al (1994) Fuel 73(6):886
30. Vulava VM et al (2007) Chemosphere 68(3):554
31. Solomon PR (1987) Coal Sci Technol 11:601
32. Leveut B et al (1995) Fuel 74(11):1618
33. Liu ZH (1991) Introduction to thermoanalysis. Chemical Industry Press, Beijing
34. Solomon PR et al (1992) Prog Energy Combust Sci 18:133
35. Roberts J (1924) Fuel Sci Pract 3:301
36. Gillet A (1946) Rew Universelle Mines 89:145
37. Dulhunty JA (1953) Fuel 32:441
38. Saxby JD et al (1985) Proc Int Conf Coal Sci 15
39. Rohrback BG et al (1984) AAPG Bull 68:961
40. Connan J et al (1980) Geochim Cosmochim Acta 44:1
41. Karweil J (1956) Dtsch Geol Gesell 107(2):132
42. Lu S et al (1994) Exp Pet Geol 16(3):290
43. Lewan MD (1979) Science 203:897
44. Michels R et al (1994) Fuel 73(11):1691
45. Behar F et al (2003) Org Geochem 34(4):575
46. Tannenbaum E et al (1985) Geochim Cosmochim Acta 49(12):2589
47. Yang T et al (1983) Pet Explor Dev (6):29
48. Wiktorsson LP et al (2000) Fuel 79(6):701
49. Lu SF et al (1994) Pet Explor Dev 21(3):46
50. Lo HB (1991) Org Geochem 17(4):415
51. Huang DF et al (1987) Chinese Sci Bull 32(11):1266
52. Ganz HHJ (1991) Southeast Asian Earth Sci 5:19
53. Marzec A et al (1994) Fuel 73(8):129
54. Idris SS et al (2010) Bioresour Technol 101(12):4584
55. Dong J et al (2012) J Hazard Mater 243:80
56. Solomon PR et al (1992) Prog Energ Comb Sci 18:133
57. Marzec A et al (1992) Energy Fuel 6:97
58. Solomon PR et al (1994) Fuel 73(8):1371
59. Schulten HR et al (1992) Energy Fuel 6:103
60. Marzec A et al (1991) ACS Div Fuel Chem Prep 36:454
61. Chauvin R et al (1969) Bull Soc Chem Er 11:3916
62. Derbyshire F et al (1989) Fuel 68:1091
63. Nishioka M (1988) Energy Fuel 2:351
64. Marzec A (1982) Fuel 62:977
65. Bodzek D et al (1981) Fuel 60:47
66. Szeliga J (1983) Fuel 62:1229
67. Pajak J et al (1985) Fuel 64:64
68. Tian D et al (2010) Mining Sci Technol 20(4):562
69. Jurkiewicz A et al (1989) Fuel 68:1097
70. Jurkiewicz A et al (1990) Fuel 69:805
71. van Heek KH (1994) Fuel 73(6):894
72. Buxhanan AC et al (1994) Prep Pap ACS Fuel Chem 39(11):22
73. Son JB (1990) Coal science and chemistry. Elsevier, Bredenberg p332
74. Genetti DB et al (1995) In: 8th international conference on coal science, Spain

Chapter 5
Coal Gasification

Abstract Coal gasification is an important industrial process for converting raw coal into more useful and cleaner carbon feedstocks for use in power generation and as precursors for other transformations. Coal gasification involves complex heterogeneous physical and chemical changes that occur between coal and gaseous reactants. These are extremely complex processes owing to structural differences within coal composites. Nevertheless a clearer understanding of such processes is fundamental to developing clean coal technologies and there is a global renewed interest in coal gasification, motivated by the need to reduce reliance on declining oil reserves. In this chapter, we discuss models that can be used to describe these processes and the experimental techniques that are used in their investigation. The efficiency of the coal gasification is shown to be affected by a variety of factors including the coal type, maceral, additional of catalytic agents and other pretreatments. We explore the relationships between coal composition structure and reactivity through a presentation of our studies on various Chinese coal types.

5.1 Introduction to Coal Gasification

Coal gasification involves reactions between coal and a gas (air, oxygen, or water vapor). These reactions produce combustible gases and ash waste residues. After purification, the gas generated can be used directly, or in the synthesis of various chemical products. The removal of sulfides and nitrides from coal gasification greatly improves coal use, and this clean coal transformation provides useful carbon feedstocks for fuel and the chemical industry. Coal gasification technology has been developing since the 1960s. Research has focused on converting coal to low-calorific-value gases suitable for power generation and other industries, and high-calorific-value gases suitable for transportation and cheap substitutes for natural gas. Basic research on the gasification reaction has also gradually developed and expanded. However, coal gasification and pyrolysis reaction mechanisms remain poorly understood. No dynamic mathematical models have been established to

K.-C. Xie, *Structure and Reactivity of Coal*, DOI 10.1007/978-3-662-47337-5_5

describe the gasification of different types of coal or different macerals of the same type of coal. A better understanding of the theoretical basis of coal gasification is necessary to improve cleaner and more efficient coal transformations. Achieving complete gasification of the complex composite in the coal in a single process requires harsh conditions. This increases the capital investment in gasification and makes it difficult to reduce the operating costs. A better understanding of the reactivity of various coal types at different thermal conversion stages may allow the gasification process to be simplified to reduce the operating costs for direct removal of pollutants economically from coal.

5.1.1 Gasification Reactivity Research

One currently used technique to investigate gasification is analysis of the solid residue reduction rate with CO_2. Devolatilization at high temperature produces solid residues such as coke, semicoke and char. In this chapter coke is used as a general term for these solid residues. Many factors affect the coal gasification reactivity, such as coal rank and type, coal devolatilization and pretreatment conditions, coal mineral type and content, internal surface area, and reaction conditions. The coal gasification reactivity can be regarded as the coke gasification reactivity. This is because coke gasification is much slower than coke formation, so this approximation is acceptable in industry. There has also been more research on coke gasification dynamics than on coal gasification dynamics as the experiments are easier to perform under differential reaction conversion rate conditions. Under these conditions, the composition of the gas in contact with the coke surface remains constant, facilitating dynamic analysis. Fixed [1], fluidized [2], spouted [3], and entrained beds [4] have been used to study coke gasification reactivity. Thermogravimetric (TG) methods are also widely used [5, 6], because they provide a full picture of the differential conversion characteristics of coke gasification. Non-isothermal differential thermal analysis (DTA) has also been applied to study coke gasification power systems.

Dynamic studies show that the reactivity of coal is similar to that of coke at the start of coke formation processes. In the formation of coke from coal, the process of hydrocarbon gas generation during devolatilization typically finishes rapidly because of high coal devolatilization reactivity. The following gasification of the so formed coke is slow. The volume of the reaction vessels in the design reactors and industrial gasification furnaces depends mainly on the coke reactivity.

Coke gasification experiments mainly focus on three reaction systems: the reaction of coke with H_2 under pressure to produce methane; the reaction of coke with CO_2 to produce CO; and the reaction of coke with water vapor to produce CO and H_2. The reaction of coke with CO_2 can easily be performed in the laboratory with few side reactions. The research results can be compared with those of other gasification systems, and have attracted wide attention. These studies have mainly focused on the dynamics of gasification, the gasification reaction mechanism, catalysts and catalytic processes, and other factors that affect the coke reactivity.

5.1.2 Gasification Reaction Mechanisms

Studies have been performed on the gasification reactions of C–CO_2, C–H_2O, and C–O_2–H_2O [7, 8]. The formation of C–O complexes on the carbon surface by adsorption of oxygen has been proposed, based on analysis of the conversion rate constants and reaction rate constants for CO_2 and water vapor on the carbon surface. The presence of C–O surface complexes has been proved experimentally. It is generally believed that in the gasification process, the formation of C–O surface complexes and the reactions of carbon with air, oxygen, water vapor, and CO_2 occur in parallel. The reaction rate determined for one reactant can therefore be compared with the results for other reactants. An oxygen exchange mechanism has been widely used to explain the gasification mechanism. For different gasification agents, the common starting point of gasification is the formation of C–O complexes:

$$2C_f + O_2 \rightarrow 2C(O)$$
$$C_f + CO_2 \rightarrow C(O) + CO$$
$$C_f + H_2O \rightarrow C(O) + H_2$$
$$C(O) \rightarrow C_f + CO$$

In the reaction equations, C_f represents a vacancy, a potential active site that can bind oxygen-containing gases. There is no reason to believe that the structures of C–O complexes formed by dissociative adsorption from different sources of oxygen atoms will differ from each other. The C–O complexes can therefore all be denoted by C(O).

Gases adsorbed on solid surfaces form adsorption layers, which form reaction products after decomposition. This simple and basic concept has been developed in the past few decades. For gasification reactions, there are only two methods of adsorption on the carbon surface: adsorption of molecules or of dissociated molecules. Based on studies over the past 10 years, it is now generally agreed that molecular chemisorption does not occur. In the dynamic adsorbed state, chemical adsorption is dissociative, and gases can be adsorbed and dissociated at the same time or dissociated before adsorption in an independent intermediate step.

5.1.2.1 Concept of Active Sites

Studies of gasification reactivity have shown that coke gasification is affected by the coal rank, maceral content, and coke-forming conditions [9]. Changes in the coke-forming conditions can produce changes in the coke gasification rate of up to 60-fold [10]. The steam vapor gasification rate is nearly 100 times the gasification rate [11, 12]. Studies of the effect of the coke-forming conditions on the gasification reactivity show that when the coke-forming time is lengthened by 1 h, there is little difference in the coke mass: only a 2.2 % change. However, the reaction rate was three times higher [13]. We also found in gasification dynamic studies that the reactivity varies significantly with the maceral content. Many researchers have performed in-depth studies of the properties of C–O complexes in the gasification process, to introduce active point parameters to explain the reactivity differences among different cokes.

In 1986, Freund [14, 15] first used the instantaneous dynamic method (the TK method) to determine the active sites on coke, and estimated the number of points (*n* values) and the released amounts of CO after switching the sites. In subsequent research, the same method was used to determine the active sites on nine cokes. The research results show that the number of active sites explain the differences among coke gasification reactivities well. Freund also obtained a uniform gasification rate equation containing the active points. This means that as long as coke activity data are available, the gasification rate can be estimated. This equation has universal application. Active sites can be used to explain the coke reactivity; however, the exact effect of the active sites is not well understood. The active sites only represent experimental results; they do not have any specific physical meaning. This makes it difficult to understand and apply this concept. To explain the nature of the active sites, the idea of C–O complexes was introduced into studies of gasification dynamics. Huttinger and Fritz [16, 17] used TPD to study the amounts of CO released from gasified cokes, and suggested that the C–O complexes can be classified as stable or unstable. Only unstable complexes have active gasification sites. Zhang et al. [18, 19] suggested that in gasification the oxygen exchange process is key to the formation of C–O complexes. The generation of oxygen exchange is therefore the active point. They performed extensive research on the oxygen exchange process using TPD. Table 5.1 summarizes the research on coke gasification activity.

5.1.2.2 Determination of Active Sites

Various methods have been used to study active sites. Laine et al. [24] used the amount of O_2 adsorbed at low temperature to evaluate the number of active sites in the reaction. Freund et al. [14, 15] used the instantaneous switching reaction gas approach to interrupt the reaction at a given conversion rate, and used the amount of CO released as the basis for evaluating the active sites. Other researchers have suggested that the CO release observed in an instantaneous switching experiment is the result of the decomposition of surface C–O complexes. The number of surface C–O complexes is therefore directly used to denote the reactivity. It is interesting that Huttinger [16] and Lizzio et al. [13] believe that there are stable and unstable C–O complexes on the surface. However, the aims of their studies and their

Table 5.1 Summary of research on coke gasification active sites

Physical parameter	Author and publication year	Research method and main conclusions
Specific surface area	Smith [20], 1978	Correlation between N_2 and CO_2 surface area and reaction rate
	Adschiri [21–23], 1986	Unit surface reaction rate can be used to explain the differences among gasification rates of some coke samples
Reactivity surface area Activity surface area	Laine [24], 1963	Relationship between low-temperature O_2 chemical adsorption and oxidation of carbon
	Peter [25], 1985	Relationship between results of TPD study of ASA and reaction rate
	Lazzio [13], 1990	Relationship between results of study of RSA with TK and TPD and reaction rate
	Adschiri [22], 1991	Use of TK to demonstrate the association of different reactivities with ASA
Active sites	Freund [14, 15], 1986	TK method is used for the first time to determine activity
	Zhu [11, 12], 1991	TK; uniform gasification rate equation containing *n*
	Nozaki [26], 1990	TK; difference in reactivity originates from difference in *n*
Surface C–O complexes	Huttinger [16, 17], 1990	TPD is used to demonstrate that C(O) is divided into stable and unstable states
	Suzuki [27, 28], 1992	TPD is used to study oxide catalysis
	Zhang [18, 19], 1996	$^{18}O_2$-TPD is used to study oxygen transfer in gasification

experimental results were different. Both groups related the unstable C–O complexes to the CO release process after interruption of the reaction in the gasification experiments, and related the released amount of CO in the subsequent TPD process to the stable C–O complexes. Huttinger proposed that decomposition of the stable C–O complexes is the gasification control step, whereas Lizzio et al. proposed that the unstable C–O complexes are only gasification activity centers.

5.1.3 Catalytic Gasification Mechanisms

Catalytic gasification of coal has been studied worldwide for nearly 20 years [29]. Coal catalytic gasification has advantages such as accelerating the coal gasification rate, improving the carbon conversion rate, reducing the reaction temperature at a given gasification rate, decreasing energy consumption, and achieving gasification product selectivity, so research and development on this gasification technology have attracted much attention.

5.1.3.1 Catalytic Mechanism of Alkali Metals

Various compounds such as Na_2CO_3, Na_2O, and $(NaO)_2$ have been used. Catalysis by alkali metal salts is often described as an oxidization–reduction cycle [30]. In this oxidation–reduction cycle, Na is directly involved in the reaction.

$$2Na + CO_2 \rightarrow Na_2O + CO$$
$$Na_2O + C \rightarrow 2Na + CO$$

Similarly, the catalytic mechanism of K can be described by

$$2K + CO_2 \rightarrow K_2O + CO$$
$$K_2O + C \rightarrow 2K + CO.$$

Usually, it is difficult for aromatic hydrocarbons to react with oxygen or hydroxide ions. The presence of alkali metal salts may, through partial bonding with carbon change the electron cloud distribution of the carbon atoms on the coke surface, making the coke surface more reactive. As Fig. 5.1 illustrates, when O^-Na^+ combines with carbon 1 on the edge of the coke surface it attracts electron density from the carbon atoms at the edges, and the adjacent carbon atoms 2 and 3 lose some of their electron density [31, 32]. This weakens the bonds between carbon atoms 1 and 2, and 1 and 3, and simultaneously induces partial positive charges on carbon atoms 2 and 3. These positively charged carbon atoms can combine more easily with the negatively charged oxygen atoms. Alkali metals on the coke surface therefore increase reactivity by weakening bonds between carbons and facilitating gasification.

Quantum chemical calculations support this mechanism that the catalytic effect of the alkali metal is caused by changes in the electron cloud density distribution on the coke surface, as a result of the formation of C–O–M clusters at crystal defects on the carbon surface [31, 32]. CO_2 and H_2O molecules are preferentially adsorbed to the partially polarized carbon atoms easily form oxycarbides. The presence of C–O–M clusters facilitates the combination of edge carbon atoms with oxygen to form C–O complexes, which plays a catalytic role in gasification.

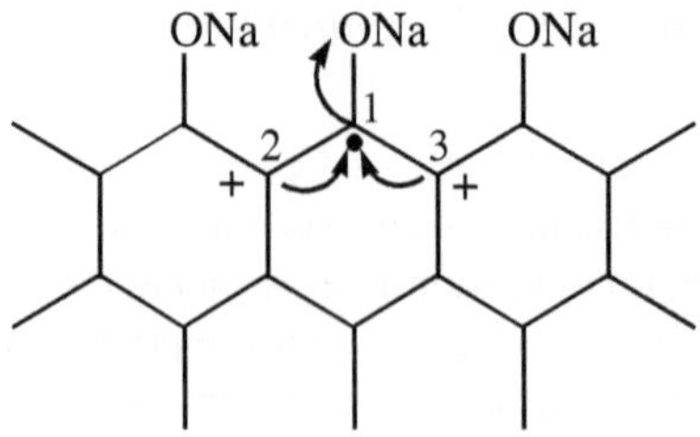

Fig. 5.1 Diagram of binding of alkali metal salts with carbons on coke surface edge

5.1.3.2 Catalytic Mechanism of Fe and Ca

Suzuki et al. [27, 28] used TPD to study the desorption properties of the surface of a coke sample impregnated with $Fe(NO_3)_3$. CO and CO_2 desorption peaks are observed in the TPD diagram. These results show that the CO desorption process at high temperature is closely related to the number of metal particles in the coke. This implies reduction of the metal oxide in the coke.

TPD shows that if ^{13}C is used as a tracer in the experiments, a ^{12}CO desorption peak is observed, but there is no ^{13}CO desorption peak when $^{13}CO_2$ is included in the atmosphere for high temperature gasification of coke samples. Suzuki's research suggests that the following oxygen exchange process occurs in the Fe catalytic reaction:

$$Fe_nO_m + CO_2 \rightarrow Fe_nO_{m+1} + CO,$$
$$Fe_nO_{m+1} + C \rightarrow Fe_nO_m + CO.$$

Various mechanisms for the catalysis of gasification by Ca compounds have been proposed. Ohtsuka et al. [33, 34] which suggest that when $CaCO_3$ is added as a catalyst, exchange of Ca and surface carboxylic ions occurs, forming $-(COO)_2Ca$, and this structure improves the reaction rate:

$$Ca^{2+} + 2(-COOH) \rightarrow -(COO)_2Ca + 2H^+.$$

The catalytic process with CaO is similar to that with $Fe(NO_3)_3$, and can be expressed as follows:

$$Ca_nO_m + CO_2 \rightarrow Ca_nO_{m+1} + CO,$$
$$Ca_nO_{m+1} + C \rightarrow Ca_nO_m + CO.$$

5.2 Thermogravimetric Behavior of Coal and Its Macerals Gasification

Thermogravimetric methods are the classic methods used for the study of gasification reactions, and much experience and results have been accumulated. The theoretical basis is well established. Isothermal operations can be performed and data processing is simple. The activation energy of coke reactivity is often calculated using thermogravimetric data.

5.2.1 Non-catalytic Gasification of Coal

5.2.1.1 Gas–Solid Reaction Model

The reaction model is expressed as $f(x)$, where x is the reaction conversion rate. The reaction rate can therefore be expressed as,

$$\frac{dx}{dt} = Ae^{-\frac{E}{RT}}f(x),$$

where A is the pre-exponential factor in the Arrhenius equation; E is the active energy (KJ mol^{-1}); t is the time (s); T is reaction temperature (K); R is gas constant (8.314 J mol^{-1} K^{-1}). In a temperature-programmed TG experiment, the temperature T and time t have a linear relationship expressed as $T = T_0 + \lambda t$, where λ is the heating rate, a constant (K s^{-1}). Substituting λ and setting $\alpha = E/(RT)$, the integral of the above equation can be given as:

$$\int \frac{dx}{f(x)} = -\frac{AE}{\lambda R}\int \frac{e^{-\alpha}d\alpha}{\alpha^2}.$$

Let the integral of the reaction model equal $F(x)$:

$$F(x) = \int \frac{dx}{f(x)},$$

and

$$P(\alpha) = -\int \frac{e^{-\alpha}d\alpha}{\alpha^2}.$$

The following equation can be derived:

$$\ln F(x) - \ln P(\alpha) = \ln\frac{AE}{\lambda R},$$

$$P(\alpha) = \frac{e^{-\alpha}}{\alpha} + \int \frac{e^{-\alpha}}{\alpha}d\alpha,$$

$$\int \frac{e^{-\alpha}}{\alpha}d\alpha = \ln\alpha - \alpha + \frac{\alpha^2}{2 \cdot 2!} - \frac{\alpha^3}{3 \cdot 3!}\cdots,$$

where α is a first approximation used for the expression. It follows that $\ln P(\alpha)$ and $\ln F(x)$ are linear functions of $1/T$. For a reaction following this mechanism, $\ln F(x)$ and $1/T$ must have a linear relationship, and the reaction model can be determined based from this relationship.

If a model is selected, the reaction equation can be transformed as follows:

Table 5.2 Common differential and integral expressions of gas–solid reaction models

No.	Reaction model	Differential form $f(x)$	Integral form $F(x)$
D_1	One-dimensional diffusion	$1/2x$	x^2
D_2	Two-dimensional diffusion	$1/[-\ln(1-x)]$	$x+(1-x)\ln(1-x)$
D_3	Three-dimensional diffusion (cylindrical symmetry)	$\frac{3}{2}[(1-x)^{\frac{2}{3}}-1]$	$(1-2x/3)-(1-x)^{\frac{2}{3}}$
D_4	Three-dimensional diffusion (spherical symmetry)	$\frac{3}{2}(1-x)^{\frac{2}{3}}[1-(1-x)^{\frac{1}{3}}]$	$[1-(1-x)^{\frac{1}{3}}]^2$
A_1	Random nucleation model ($n = 1$)	$1-x$	$-\ln(1-x)$
A_2	Random nucleation model ($n = 2$)	$2(1-x)[-\ln(1-x)]^{\frac{1}{2}}$	$[-\ln(1-x)]^{\frac{1}{2}}$
A_3	Random nucleation model ($n = 3$)	$3(1-x)[-\ln(1-x)]^{\frac{2}{3}}$	$[-\ln(1-x)]^{\frac{1}{3}}$
R_2	Shrinking core model (cylindrical symmetry)	$2(1-x)^{\frac{1}{2}}$	$1-(1-x)^{\frac{1}{2}}$
R_3	Shrinking core model (cylindrical symmetry)	$3(1-x)^{\frac{2}{3}}$	$1-(1-x)^{\frac{1}{3}}$

$$\ln\left(\frac{\lambda\frac{\mathrm{d}x}{\mathrm{d}T}}{f(x)}\right)=\ln A-\frac{E}{RT}.$$

The variables x and T in the above equations can easily be obtained by thermogravimetric methods. So, as long as the form of the reaction model $f(x)$ is appropriately selected, the value of the expression on the left-hand side of the equation can be calculated. This value can be used to produce $1/T$ plots to calculate E and A directly from the slope and the intercept.

Coal gasification is usually described as an irreversible gas–solid reaction. The study of coal gasification dynamics needs to account for both effects of material transfer and the reaction simultaneously. Many gas–solid reaction models have been reported in the literature. Several differential and integral forms of reaction models are listed in Table 5.2 [35–37].

5.2.1.2 Coal Gasification Reaction Model

Based on the theory discussed above, the CO_2 gasification reactivities of coals with different metamorphic degrees and various macerals were studied by thermogravimetric techniques. The data for the ultimate and proximate analyses of the coals and macerals are listed in Table 5.3. The preparation conditions and the ultimate analysis of the coke samples are listed in Table 5.4. We selected nine reaction models listed in Table 5.2 to examine coal gasification experimental data.

When the conversion rate of the coke sample is less than 15 %, the reaction rate is slow, so there is a large error in the calculation results for data taken from this segment. When the conversion rate is over 80 %, the diameter of the coke particles

Table 5.3 Proximate and ultimate analyses of coal samples (%)

Coal sample	Proximate analysis			Ultimate analysis				
	M_{ad}	A_d	V_d	C_{daf}	H_{daf}	O_{daf}	N_{daf}	$S_{t,daf}$
French bituminite	0.79	6.82	11.05	83.60	5.35	8.34	1.60	1.11
Fuxin long-flame coal	2.36	14.24	32.27	79.38	4.98	13.54	1.06	1.04
Pingshuo gas	0.98	20.53	28.46	79.76	5.33	12.78	1.46	0.67
Datong bituminite	1.57	8.69	30.05	84.57	4.41	9.12	0.89	1.01
Fenxi fat coal	–	4.42	31.21	87.24	5.14	4.33	1.48	1.81
Fengfeng meager coal	0.87	17.54	12.22	88.20	4.07	6.11	1.27	0.35
Dongshan lean coal	1.21	12.84	17.01	88.88	4.50	3.13	1.07	2.42
Jincheng anthracite	0.72	14.52	7.93	94.17	3.56	0.70	1.14	0.44
French vitrinite	–	–	–	81.72	5.49	10.09	1.44	1.26
French inertinite	–	–	–	84.34	4.46	8.61	1.41	1.19
French exinite	–	–	–	82.67	6.26	8.30	1.29	1.49
Pingshuo vitrinite	–	–	–	79.25	5.37	13.57	1.23	0.58
Pingshuo inertinite	–	–	–	84.73	3.96	9.89	0.66	0.76

shrinks so that the coke gasification mode may change. We therefore used thermogravimetric data for conversion rates ranging from 15 to 75 % to screen the model and calculate the activation energy and pre-exponential factor from Arrhenius plots.

Many methods of evaluating gasification reactivity are available, such as the CO_2 reduction rate and the gasification rate. When the conversion rate of fixed carbon reaches 50 %, the time taken for carbon conversion can also be used as the basis for the evaluation. These evaluation methods are commonly used in isothermal dynamic research. The half-life reactivity, the initial reactivity, and the maximum reactivity can be used to describe the programmed temperature-ramping method. In this section, the temperature when the conversion rate reaches 50 % is used to measure the gasification reactivity of the coke, i.e., the higher the temperature is when the conversion reaches 50 %, the lower the reactivity of the coke sample.

We appraised and selected the reaction models listed in Table 5.2, and obtained reaction models suitable for describing thermogravimetric gasification experiments. Figures 5.2 and 5.3 show the results for catalytic and non-catalytic gasification reaction models.

The relationship between $\ln F(x)$ and $1/T$ for a typical gasification reaction model should be linear. Figures 5.2 and 5.3 show the CO_2 gasification TG curves of French bituminous coal coke samples with and without KOH addition, respectively. Each line corresponds to a reaction mechanism. It can be clearly seen from the figures that the calculation results for the diffusion ($D_{1–4}$) and random core ($A_{1–3}$)

Table 5.4 Coke sample preparation conditions and ultimate analysis (%)

Coal sample code	Relevant coal sample	Preparation conditions		Ultimate analysis				
		coke-forming temperature (K)	Residence time (min)	C_{daf}	H_{daf}	N_{daf}	S_{daf}	O_{daf}
Fu	Fuxin long-flame coal	1073	5	94.23	1.25	0.83	0.97	2.72
FX	Fenxi fat coal	1073	5	91.51	1.86	0.78	2.01	3.84
FF	Fengfeng meager coal	1073	5	94.26	1.36	1.03	0.80	2.55
DS	Dongshan lean coal	1073	5	92.32	1.25	1.02	1.67	3.74
JC	Jincheng anthracite	1073	5	93.27	2.11	0.44	0.83	3.35
PS	Pingshuo gas coal	1073	5	92.85	1.55	1.87	0.87	2.86
PSV	Pingshuo vitrinite	1073	5	90.60	1.70	1.56	1.78	4.36
PSI	Pingshuo inertinite	1073	5	91.83	2.83	0.57	0.78	3.99
Fran1	French bituminous	1073	5	91.81	2.29	0.76	1.14	4.00
Fran2	French bituminous	1073	60	93.24	1.11	1.84	1.08	2.73
FranV	French vitrinite	1073	5	91.57	2.08	1.64	1.05	3.66
FranI	French vitrinite	1073	5	93.14	1.68	0.74	1.10	3.34
FranE	French exinite	1073	5	91.21	2.61	0.76	1.12	4.30
DT1	Datong bituminite	873	30	90.27	2.78	0.86	1.99	4.10
DT2	Datong bituminite	973	30	–	–	–	–	–
DT3	Datong bituminite	1073	30	–	–	–	–	–
DT4	Datong bituminite	1173	30	–	–	–	–	–
DT5	Datong bituminite	1273	30	93.39	1.18	0.58	0.79	4.06

models do not give straight lines. For the catalytic and non-catalytic processes, the best linear relationships are obtained using the spherically symmetric unreacted shrinking core model (R_3). It is therefore appropriate to use the unreacted shrinking core model to describe the reaction dynamic data obtained under the experimental conditions used.

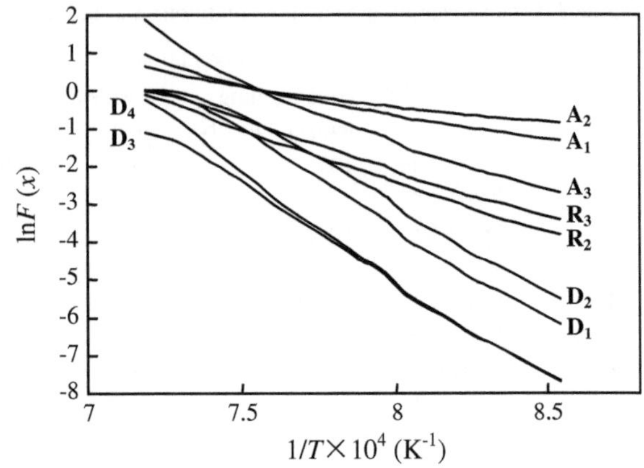

Fig. 5.2 Results for non-catalytic gasification reaction model using coke from French bituminous coal

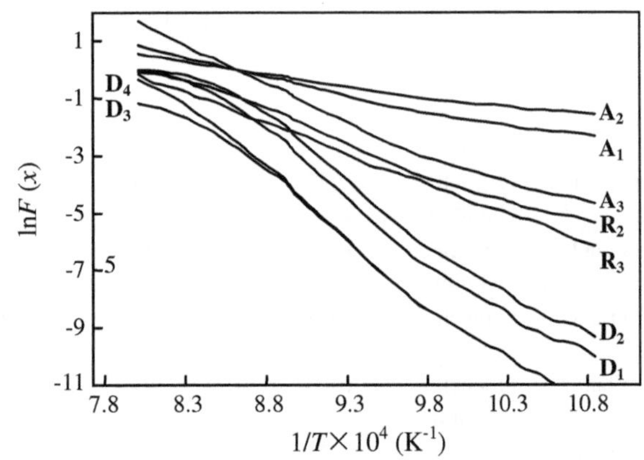

Fig. 5.3 Results for catalytic gasification reaction model using coke from French bituminous coal

The unreacted shrinking core model is often used to describe a non-catalytic gas–solid reaction process. Some research results have indicated that the unreacted shrinking core model is also suitable for processing coke CO_2 gasification data [38].

In the unreacted shrinking core model, it is assumed that coke gasification occurs on the particle surfaces, where the aperture effect is smallest. In this way, the specific surface area of the coke is related to the outer surface of the coke particles, and the effect of the specific inner surface area is ignored. It is also assumed that there is an ash layer in the reaction process. The ash layer expands as the reaction proceeds, but the particle size is unchanged. Third, it is assumed that the reaction rate is much lower than the diffusion rate of the reaction gas in the ash layer. The entire reaction is under chemical reaction control, and the effect of diffusion is negligible.

Some researchers have used imaging technology to study coke morphologies at different gasification conversion rates, and obtained images of longitudinal sections of the particles of the coke samples at several stages for conversion rates of 0.2, 0.7, and 0.95. Unreacted areas, atmospheric reaction zones, and the ash layer left after the reaction can be clearly observed in a series of photographs. When the reaction conversion rate reaches 95 %, the outer diameter of the ash layer is the same as the particle diameter. It can be clearly seen that a reaction belt moves continuously to the particle center as the reaction proceeds. The results of the photography experiments fit the assumptions of the unreacted shrinking core model. This illustrates that the unreacted shrinking core model is suitable for studying gasification reactions.

5.2.1.3 Reactivity in Coal Gasification

According to above results, both the catalytic and non-catalytic CO_2 gasification of coke can be expressed mathematically as:

$$\frac{dx}{dt} = 3Ae^{-\frac{E}{RT}}(1-x)^{\frac{2}{3}}.$$

This expression can also be written as:

$$\ln k = \ln\left(\frac{\lambda}{3(1-x)^{\frac{2}{3}}}\frac{dx}{dT}\right) = -\frac{E}{RT} + \ln A,$$

where λ is 20 K min^{-1} under the experimental conditions used in this section.

The part in parentheses can be easily obtained from the thermogravimetric curve. This allows the activation energy E and the pre-exponential factor A to be obtained from Arrhenius plots. Table 5.5 lists the gasification dynamic parameters of the coke samples, obtained from the thermogravimetric data. $T_{x=0.05}$ and $T_{x=0.5}$ are the temperatures at which the weight losses of the coke sample reached 5 and 50 %, respectively.

When the isothermal method and the programmed temperature-ramping method are used to study the gasification dynamic parameters, the results are similar. We used the isothermal method and the programmed temperature-ramping method in a thermogravimetric analyzer to study the CO_2 gasification dynamics of vitrinite coke from French bituminous coal. The experimental results were obtained using the unreacted shrinking core model. Table 5.6 lists the k values for the gasification reaction at different temperatures, obtained using the isothermal method. The activation energy and pre-exponential factor can be obtained from the slope and intercept of a plot of $\ln k$ versus $1/T$ (Fig. 5.4). The calculated activation energy is 258.6 kJ mol^{-1}, which is close to 268.4 kJ mol^{-1}, the result calculated using the programmed temperature-ramping method.

Table 5.5 Dynamic calculation results for non-catalytic gasification reaction

Coke sample code	E (kJ mol^{-1})	A	$T_{x=0.05}$ (K)	$T_{x=0.5}$ (K)
Fran1	256.6	20.63	1148	1306
Fran2	273.9	21.96	1169	1318
Fu	239.9	18.60	1196	1356
PS	262.9	20.83	1226	1355
FX	169.7	10.06	1334	1553
FF	195.4	14.49	1253	1371
DS	158.9	10.67	1212	1396
JC	229.6	16.90	1225	1397
PSV	194.5	14.55	1203	1353
PSI	225.1	17.55	1153	1336
FranV	268.4	21.59	1147	1302
FranE	298.9	23.87	1129	1293
FranI	226.0	17.48	1154	1312
DT1	250.7	18.18	1225	1398
DT2	234.2	16.58	1224	1401
DT3	226.2	15.80	1250	1417
DT4	198.2	12.62	1238	1469
DT5	144.3	7.37	1313	1499

Table 5.6 Isothermal gasification reaction rate constants of coke sample

Temperature (K)	1273	1283	1302	1330	1340
k (s^{-1})	0.000392	0.000560	0.000858	0.001164	0.001473

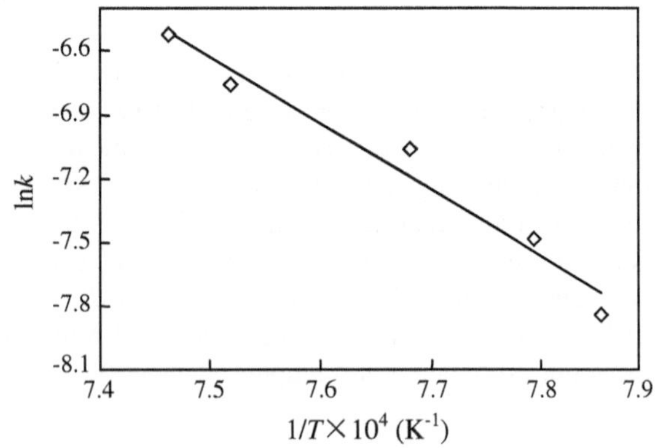

Fig. 5.4 Isothermal plot of relationship between ln$k$ and $1/T$ for vitrinite coke sample from French bituminous coal

5.2.2 Catalytic Gasification of Coal

5.2.2.1 Effect of Catalyst Amount

Table 5.5 shows the thermogravimetric curve for graphite powder with different amounts of catalyst. The figure shows that the test sample without any catalyst starts to gasify at a high temperature, above 1273 K. After addition of a catalyst, the reaction temperature decreases significantly, and as more catalyst is added, the reaction temperature decreases, and the reaction rate increases. The reaction temperature for a conversion rate of 50 % without the catalyst is 1513 K, and the reaction temperature decreases to 1453, 1373, and 1223 K, respectively, after the addition of 2, 4, and 7 wt% KOH. When the catalyst content is greater than 7 wt%, further increases in the amount of catalyst do not affect the gasification reactivity. This demonstrates that the thermogravimetric curves almost entirely coincide with one another. The results of this experiment show that there is a limit to the amount of alkali metal that can be effective (Fig. 5.5).

Figure 5.6 shows Arrhenius plots for the samples. The dynamic parameters calculated from the plots are listed in Table 5.7. The gasification activation energy of graphite powder without addition of a catalyst is 196.2 kJ mol^{-1}. There are obvious inflection points at 1360 K in the Arrhenius plots of the samples with 2 and 4 wt% of added catalyst. We calculated E and ln$A$ for the stages before and after the inflection point. The results for E and ln$A$ in the first stage were 161.2 and 169.7 kJ mol^{-1}, respectively, and those for the second stage were 265.1 and 282.9 kJ mol^{-1}, respectively. When the amounts added were more than 7 wt%, the gasification activation energies were all almost equal, about 157 kJ mol^{-1}. After adding the catalyst, the gasification activation energy decreased significantly and then the activation energy decreased gradually with increasing amount of added catalyst.

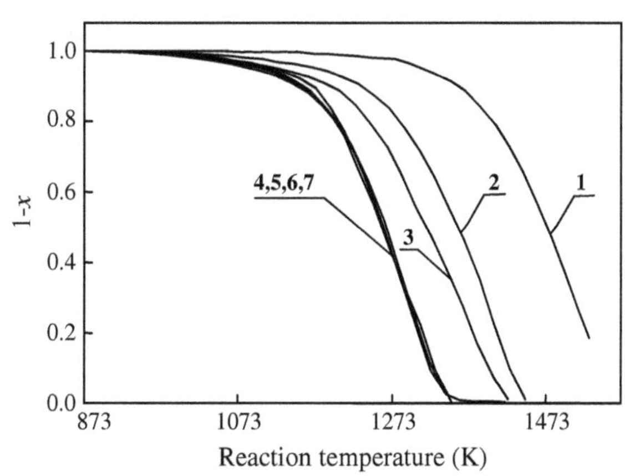

Fig. 5.5 Effect of addition of different amounts of KOH on graphite CO_2 reactivity: *1* 0 wt%; *2* 2 wt%; *3* 4 wt%; *4* 7 wt%; *5* 10 wt%; *6* 12 wt %; and *7* 16 wt%

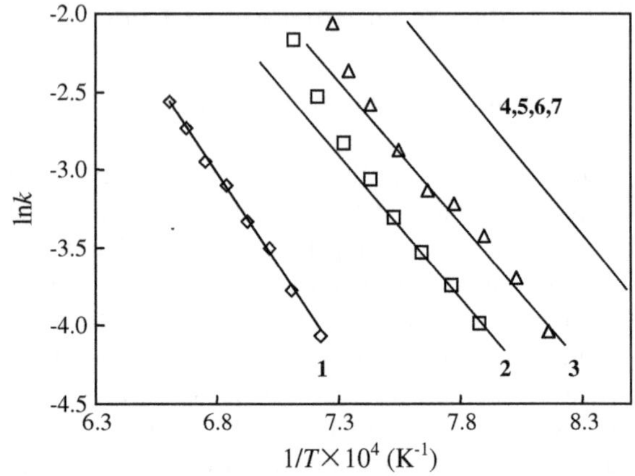

Fig. 5.6 Arrhenius plots for graphite with addition of different amounts of KOH: *1* 0 wt%; *2* 2 wt%; *3* 4 wt%; *4* 7 wt%; *5* 10 wt%; *6* 12 wt%; and *7* 16 wt%

Table 5.7 Dynamic calculation results for graphite gasification with addition of different amounts of catalyst

Added amount (wt%)	First stage		Second stage	
	Activation energy ($kJ\ mol^{-1}$)	ln*A*	Activation energy ($kJ\ mol^{-1}$)	ln*A*
0	196.2	13.02	–	–
2	161.2	11.32	265.1	20.50
4	169.7	10.55	282.9	22.71
7	160.6	12.62	–	–
10	155.2	12.07	–	–
12	158.8	12.44	–	–
16	154.9	12.09	–	–

The results show that when the amount of KOH added is more than 7 wt%, the thermogravimetric curve coincides almost completely with the Arrhenius plot. There are Arrhenius points at about 1360 K for the coke samples with 2 and 4 wt% of catalyst. It can be clearly seen from the thermogravimetric diagram that when the amount of catalyst added is more than 7 wt%, weight loss begins at 1090 K, and the conversion rate of the coke sample is 5 %. When the temperature reaches 1360 K, the conversion rate of the coke sample is 98 %, i.e., the catalytic gasification reaction is almost complete, whereas the carbon conversion rate for non-catalytic gasification at that point is still lower than 5 %. The temperature range for graphite gasification after addition of a catalyst is different from that in the absence of a catalyst, and the Arrhenius inflection point temperatures for coke samples containing 2 and 4 wt% of catalyst are 1360 K, which is the temperature at which non-catalytic gasification starts. The calculated data in Table 5.7 also show that the values of the dynamic parameters in the first stage are close to the catalytic gasification parameter values. The dynamic data for the second stage do not correspond to the dynamic values of either catalytic gasification or non-catalytic gasification. In this temperature range, the catalytic and non-catalytic gasification processes coexist. The dynamic data therefore reflect the coexistence of catalytic and non-catalytic processes. The dynamic calculation results only give the apparent activation energies. It is difficult to associate these results above with the intrinsic properties of the reaction. Based on the analytical results, large catalyst loadings are generally used in catalytic gasification studies, so that the calculation results accurately reflect the catalytic process.

5.2.2.2 Effect of $Ca(OH)_2$

Figure 5.7 shows the gasification TG curves of various coke samples (see Table 5.4 for their codes) with added $Ca(OH)_2$. Figure 5.8 shows the Arrhenius plots. The dynamic parameters calculated from the figure are listed in Table 5.8. In the

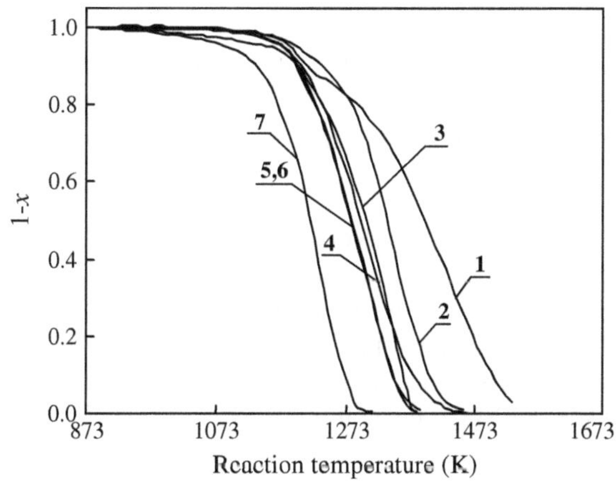

Fig. 5.7 Gasification thermogravimetric *curves* of coke samples with addition of $Ca(OH)_2$: *1* DS; *2* PS; *3* Fran; *4* FF; *5* Fu; *6* PSV; and *7* PSI

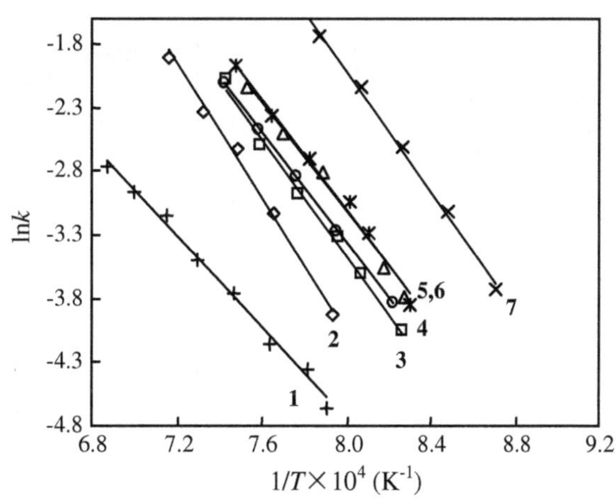

Fig. 5.8 Arrhenius diagrams of coke samples with addition of $Ca(OH)_2$: *1* DS; *2* PS; *3* Fran; *4* FF; *5* Fu; *6* PSV; and *7* PSI

Table 5.8 Dynamic results for $Ca(OH)_2$ catalytic gasification reaction

Code	E (kJ mol^{-1})	A	$T_{x=0.05}$ (K)	$T_{x=0.5}$ (K)
Ca-DS	138.2	9.30	1189	1398
Ca-Fran	189.3	14.74	1151	1304
Ca-FF	138.7	10.05	1176	1296
Ca-Fu	170.2	13.29	1175	1282
Ca-PS	230.7	18.07	1200	1339
Ca-PSV	174.8	13.71	1174	1282
Ca-PSI	209.6	18.22	1092	1220

experimental range, the gasification activation energies of the coke samples with added $Ca(OH)_2$ range from 140 to 230 kJ mol^{-1}; these are lower than the non-catalytic gasification activation energies (160–300 kJ mol^{-1}), listed in Table 5.5. A comparison of the reactivities of the coke samples, based on the temperatures at which the conversion rates reach 50 %, gives the following coke reactivity order: Pingshuo gas coal inertinite > Pingshuo gas coal vitrinite > Fuxin long-flame coal > Fengfeng meager coal > French bituminous coal > Pingshuo gas coal > Dongshan lean coal. The methods used here were also used to study the catalytic roles of $Fe(NO_3)_3$, NaOH, and KOH.

5.2.2.3 Effect of $Fe(NO_3)_3$

Figure 5.9 shows the gasification TG curve of coke samples with added $Fe(NO_3)_3$. Figure 5.10 shows the Arrhenius plots. The calculated dynamic parameters are listed in Table 5.9. The gasification activation energies of the coke samples with added $Fe(NO_3)_3$ range from 95 to 225 kJ mol^{-1}. A comparison of the reactivities of the coke samples, based on the temperatures at which the conversion rates reach

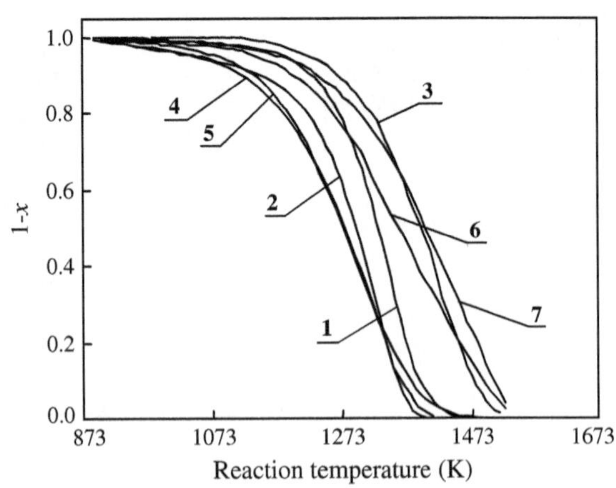

Fig. 5.9 Gasification thermogravimetric *curves* of coke samples with addition of $Fe(NO_3)_3$: *1* PS; 2 Fran; *3* DSF; *4* Fu; *5* FF; *6* DS; and *7* FX

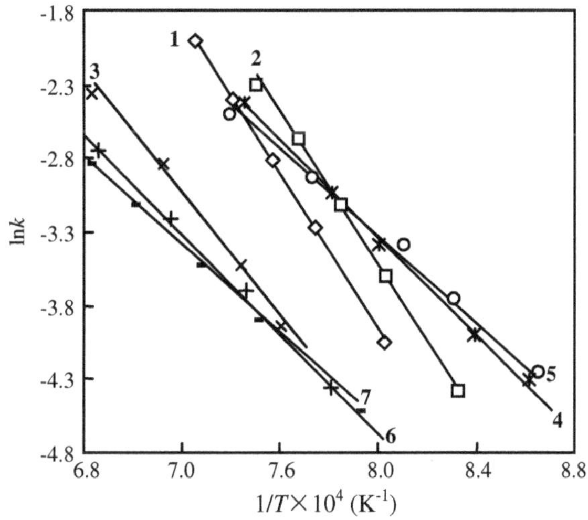

Fig. 5.10 Arrhenius diagrams of coke samples with addition of $Fe(NO_3)_3$: *1* PS; *2* Fran; *3* DSF; *4* Fu; *5* FF; *6* DS; and *7* FX

Table 5.9 Dynamic results for $Fe(NO_3)_3$ catalytic gasification reaction

Code	E (kJ mol^{-1})	A	$T_{x = 0.05}$ (K)	$T_{x = 0.5}$ (K)
Fe-DS	96.4	4.47	1144	1362
Fe-Fran	217.2	17.40	1048	1295
Fe-FX	115.6	5.79	1188	1414
Fe-FF	115.8	7.83	1084	1274
Fe-Fu	139.7	10.11	1052	1279
Fe-PS	222.3	17.43	1186	1331
Fe-PSI	179.5	12.54	1223	1393

50 %, gives the following coke reactivity order: Fengfeng meager coal > Fuxin long-flame coal > French bituminous coal > Pingshuo gas coal > Dongshan lean coal > Pingshuo gas coal inertinite > Fenxi coal.

5.2.2.4 Effect of NaOH

Figure 5.11 shows the gasification TG curves of coke samples with added NaOH. Figure 5.12 shows the Arrhenius plots. The calculated dynamic parameters are listed in Table 5.10. The gasification activation energies of the coke samples with added NaOH range from 75 to 180 kJ mol^{-1}. A comparison of the reactivities of the coke samples, based on the temperatures at which the conversion rates reach 50 %, gives the following coke reactivity order: Pingshuo gas coal inertinite > French bituminous coal > Fuxin long-flame coal > Pingshuo gas coal vitrinite > Fengfeng meager coal > Pingshuo gas coal > Dongshan lean coal.

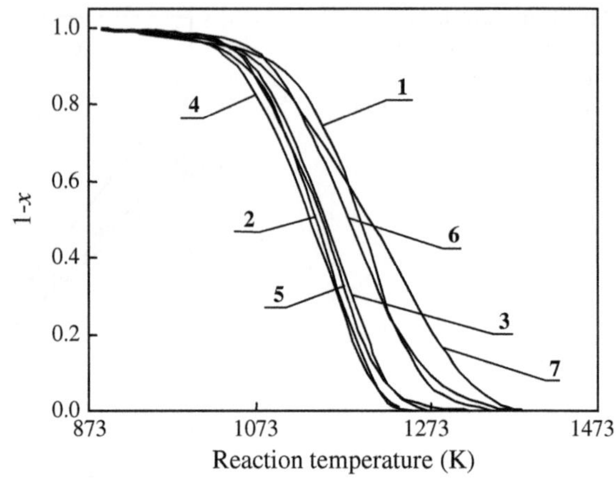

Fig. 5.11 Gasification thermogravimetric *curves* of coke samples with addition of NaOH: *1* PS; *2* Fran; *3* PSV; *4* PSI; *5* Fu; *6* FF; and *7* DS

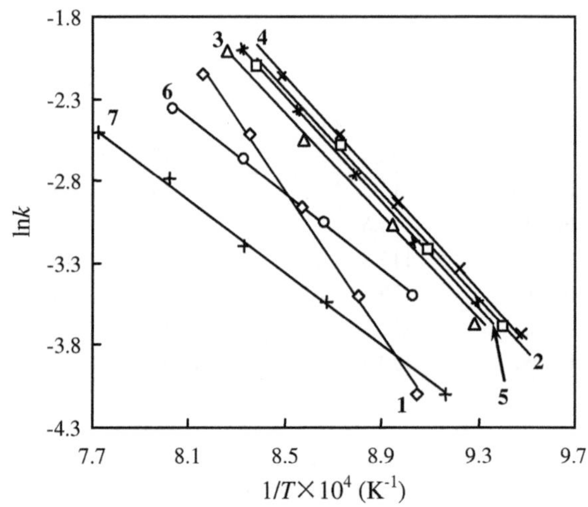

Fig. 5.12 Arrhenius diagrams of coke samples with addition of NaOH: *1* PS; *2* Fran; *3* PSV; *4* PSI; *5* Fu; *6* FF; *7* DS

Table 5.10 Dynamic results for NaOH catalytic gasification reaction

Code	E (kJ mol^{-1})	A	$T_{x=0.05}$ (K)	$T_{x=0.5}$ (K)
Na-DS	75.3	4.39	1045	1204
Na-Fran	146.1	12.75	1026	1143
Na-FF	101.5	7.51	1066	1180
Na-Fu	130.7	11.04	1040	1147
Na-PS	177.7	15.34	1040	1193
Na-PSV	126.7	10.56	1045	1151
Na-PSI	132.4	11.34	1021	1137

5.2.2.5 Effect of KOH

Figure 5.13 shows the gasification TG curves of coke samples with added KOH. Figure 5.14 shows the Arrhenius plots. The calculated dynamic parameters are listed in Table 5.11. The gasification activation energies of the coke samples with added KOH range from 65 to 195 kJ mol^{-1}. The coke reactivity order is as follows: Pingshuo gas coal inertinite > French bituminous coal > Fuxin long-flame coal > Pingshuo gas coal vitrinite > Pingshuo gas coal > Fengfeng meager coal > Dongshan lean coal. This order is the same as that for the addition of NaOH.

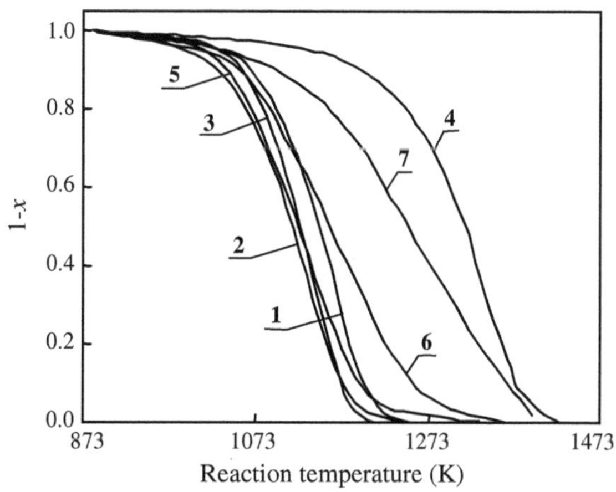

Fig. 5.13 Gasification TG curves of coke samples with addition of KOH: *1* PS; *2* Fran; *3* PSV; *4* PSI; *5* Fu; *6* FF; and *7* DS

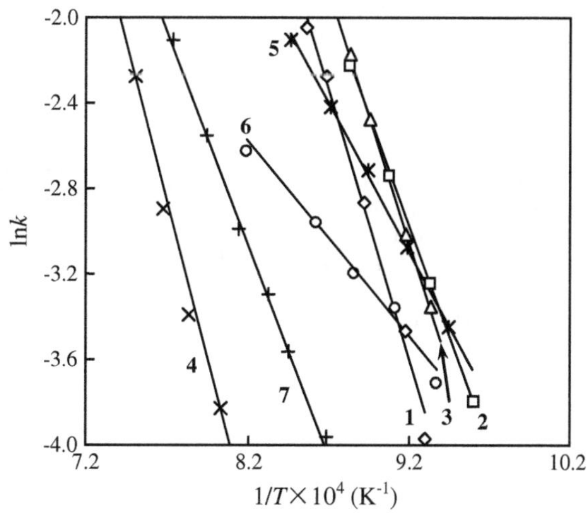

Fig. 5.14 Arrhenius diagrams of coke samples with addition of KOH: *1* PS; *2* Fran; *3* PSV; *4* PSI; *5* Fu; *6* FF; and *7* DS

Table 5.11 Dynamic results for KOH catalytic gasification reaction

Code	E (kJ mol^{-1})	A	$T_{x=0.05}$ (K)	$T_{x=0.5}$ (K)
K-DS	65.5	2.77	1047	1275
K-Fran	171.5	15.99	992	1116
K-FF	76.2	4.95	1007	1158
K-Fu	117.2	9.87	991	1122
K-PS	184.3	16.92	1033	1146
K-PSV	194.5	18.48	1031	1127
K-PSI	184.1	17.38	999	1116

5.2.2.6 Comparison of Effects of Catalysts on Gasification Reactivities of Coke Samples

Figure 5.15 shows the gasification thermogravimetric curves of coke samples from Fengfeng meager coal. The order of the catalytic effects of the four catalysts studied is KOH ≈ NaOH > $Fe(NO_3)_3$ > $Ca(OH)_2$. The dynamic data for these experiments and those for the other coke samples, listed in Tables 5.8, 5.9, 5.10 and 5.11, show that for a given coke sample, NaOH and KOH have the greatest catalytic effects, followed by $Fe(NO_3)_3$ and $Ca(OH)_2$.

The catalysts have different effects with different cokes. The data in Table 5.5 show that the order of the non-catalytic gasification reactivities of the cokes are as follows: French bituminous coal > Pingshuo gas coal inertinite > Pingshuo gas coal vitrinite > Pingshuo gas coal > Fuxin long-flame coal > Fengfeng meager coal > Dongshan lean coal. The reactivity order clearly changes after catalyst addition. For example, the reactivity of the coke sample from French bituminous coal is obviously higher than the reactivities of the coke samples from other coals. However, after addition of $Ca(OH)_2$, its reactivity is lower than that of the coke sample from macerals of Pingshuo coals, and of coke samples from Fuxin long-flame coal and Fengfeng meager coal. The reactivity orders of the coke samples

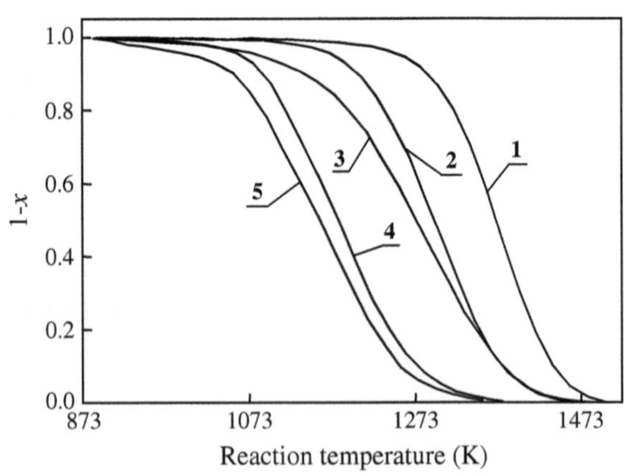

Fig. 5.15 TG *curves* of coke samples of Fengfeng meager coal with addition of catalysts: *1* FF; *2* FF + CaO; *3* FF + $Fe(NO_3)_3$; *4* FF + NaOH; and *5* FF + KOH

after addition of NaOH and KOH are basically the same. This suggests that there are similarities between these catalytic processes. These experimental results reflect differences among the coke structures. In later sections, the structures of the C–O compounds formed in the gasification process will be discussed, and this will help to explain these experimental observations.

The experimental results and calculations discussed in this section show that after addition of a catalyst, the gasification reaction activation energies, E, of the coke samples are lower than the values for non-catalytic gasification; additionally, the pre-exponential factor, A, of the catalytic process is relatively small. The $\ln A$ values for the non-catalytic gasification of coke samples range from 13 to 24, whereas the $\ln A$ values after catalyst addition range from 4 to 18. The gasification reaction rate is a combination of the activation energy and the pre-exponential factor. The lower the activation energy is, the more favorable the reaction. When the pre-exponential factor is low, the reaction is difficult. It is therefore not sufficient to evaluate the role of catalysts in reducing the activation energy merely from the catalytic process. This will be further discussed in terms of specific problems in subsequent sections on compensation effects.

5.3 Compensation Effects in Gasification Reactions

When calculating the CO_2 gasification kinetic parameters from thermogravimetric analysis curves, we find that there is a compensation effect between the activation energy and the pre-exponential factor of the CO_2 gasification reaction. In other words, the higher the activation energy is, the larger the pre-exponential factor is. This pair of parameters is mutually and linearly compensated. This can be described mathematically as,

$$\ln A = aE + b,$$

where a and b, known as the compensation coefficients, can be obtained experimentally. In some studies, it has been assumed that a is determined by the chemical-bond-breaking energy, whereas parameter b is related to certain characteristics of the reaction [39].

A pair of mutually independent parameters such as the activation energy and the pre-exponential factor of the reaction is usually used in describing the reaction rate. However, the presence of the compensation effect makes the problem complex because the higher the activation energy is, the more difficult it is for the reaction to occur. If there is a compensation effect, when the activation energy is large, the pre-exponential factor is also large, which is unfavorable to the reaction. The complementary relationship between the activation energy and pre-exponential factor makes their effect on the reaction complex. As a result of the compensation effect, therefore, it is impossible to determine the reactivity simply from the sizes of the activation energy and the pre-exponential factor. The discussion in Sect. 5.2.2.6 of this chapter shows that coke reactions are not always related to the size of the activation energy. The gasification reaction activation energy of a coke is reduced by adding a catalyst. There is a reasonable correlation between coke reactivity and the size of the activation energy. However, this does not apply to non-catalytic processes. Experimental results show that the gasification reactivity of enriched samples of French coke exinite is the highest, but its gasification reaction activation energy is also the highest. This is because of the compensation effect in the gasification process.

Compensation effects can often be seen in heterogeneous reactions such as some chemical reactions in solvents and the thermal decomposition of solid organic compounds. In this section, we discuss the compensation effect in the coal gasification reaction.

5.3.1 Compensation Effects and Coke Gasification Dynamic Parameters

5.3.1.1 Compensation Effects Among Non-catalytic Gasification Dynamic Parameters

Our study of gasification kinetics involved an adequate number of coal samples (including three types of coal maceral). These coal samples were subjected to different coke-forming conditions. The coke-forming temperature ranged from 600 to 1000 °C, and the time ranged from 5 min to 1 h. The coke samples selected cover almost all the factors affecting coke reactivity, such as type of coal, coke-forming conditions, and maceral content. Although coke samples differ in many ways, the results are surprisingly simple. All the coke gasification kinetic parameters are consistent with the compensation effect, which seems to be one of the essential characteristics of coke gasification. The effect has nothing to do with factors such as the coke type, maceral content, or coke-forming conditions, which affect the coke gasification reactivity. This is shown in Fig. 5.16.

5.3.1.2 Compensation Effects Among Catalytic Gasification Dynamic Parameters

The kinetic parameters of coal gasification also show a compensation effect after catalyst addition. However, this compensation effect and the non-catalytic compensation effect do not give the same straight line. Also, the compensation effects differ among the catalysts (Fig. 5.17).

Some research reports have shown that the kinetic parameters of different samples are at different positions; a curve was selected to describe the relationship between the kinetic parameters in the oxidation of a series of carbon materials with

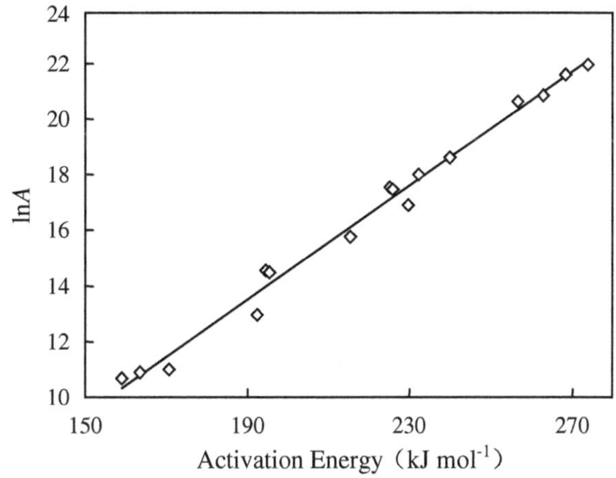

Fig. 5.16 Compensation effect among non-catalytic gasification parameters: activation energy and pre-exponential factor

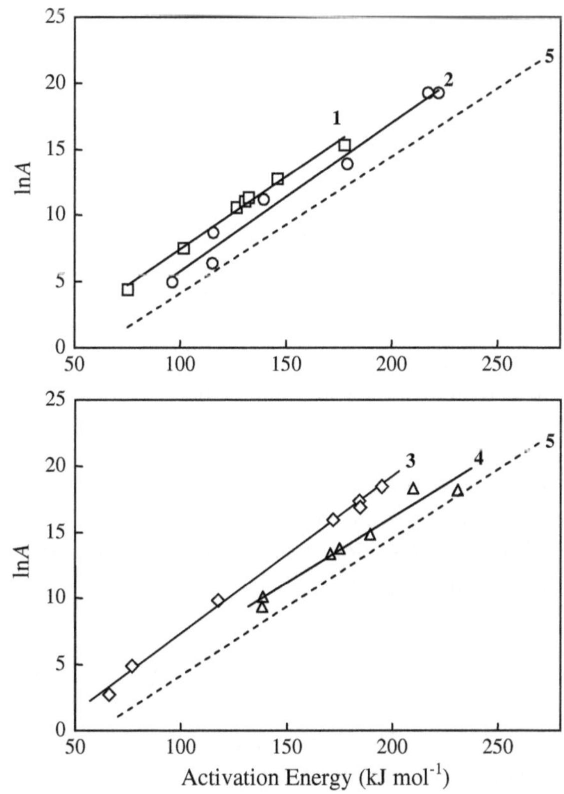

Fig. 5.17 Compensation effects between activation energy and pre-exponential factor in NaOH catalytic gasification: *1* NaOH; *2* $Fe(NO_3)_3$; *3* KOH; *4* $Ca(OH)_2$; and *5* No catalyst

different structures. We first examine the addition of different catalysts and then discuss their linear compensation relationships. This is different from the method in which the kinetic parameters are treated indiscriminately [40]. The experimental results show that with addition of a catalyst, the kinetic parameter data lie above the non-catalytic kinetic compensation line in the diagram. In terms of their relative positions, the compensation lines after adding KOH and NaOH are the highest among the five lines. The compensation lines after adding $Fe(NO_3)_3$ and $Ca(OH)_2$ lie in the middle, and that for non-catalytic conditions is at the bottom. The experimental results in Sect. 5.2.2 of this chapter show that the reaction order (from difficult to easy) of cokes on addition of a catalyst is KOH $\approx$ NaOH > $Fe(NO_3)_3$ > $Ca(OH)_2$ > non-catalytic process. This order is consistent with the relative positions of the compensation lines discussed in this section. When the gasification activation energies are the same, it can be seen from the diagram that the coke samples with added catalysts have higher pre-exponential factors than that without a catalyst, so the reaction rates are higher. It is therefore reasonable for the compensation lines for catalytic gasification kinetic parameters to have higher positions than that of the non-catalytic compensation line.

5.3.2 *Theoretical Analysis of Compensation Effect*

5.3.2.1 Transition-State Theoretical Analysis of Compensation Effect

Transition-state theory has been used to analyze the physical significance of the gasification kinetic parameters. The transition-state theory states that the first step in the reaction is formation of an unstable activated complex. In this complex, the interatomic distance is greater than the length of a normal chemical bond, and the activated complex bonds are weaker than normal bonds, but, as in normal molecules, translation, rotation, and vibration are possible to limited degrees. An unstable transition-state intermediate is also formed in coke gasification. The unstable C–O complex formed during gasification has a direct effect on the gasification of coke samples. The values of the kinetic parameters obtained from TG experiments, described in Sect. 5.2.1.3 of this chapter, reflect the nature of the unstable C–O complexes.

Based on transition-state theory, the reaction rate constants are calculated using the following equation:

$$k = \frac{k_B T}{h} \exp\left(\frac{\Delta S^{0\neq}}{R} - \frac{\Delta H^{0\neq}}{RT}\right),$$

where k_B is the Boltzmann constant; h is Plank's constant; $\Delta H^{0\neq}$ and $\Delta S^{0\neq}$ are the enthalpy and entropy of activation of the unstable C–O complex formed in the gasification process.

The relationship between the activation energy and the activation enthalpy of a chemical reaction can be expressed as

$$E = \Delta H^{0\neq} + nRT,$$
$$E \approx \Delta H^{0\neq}.$$

By definition, the activation enthalpy is the energy difference between the activated complex and the reactants, i.e., the reactant potential energy required to form the activated complex state. This reactant energy originates from the energy of collisions between pairs of reaction molecules. The greater the activation enthalpy of the activated complex is, the greater the energy barrier to formation of the activated complex. It is therefore not easy to form an activated complex. If the activation enthalpy is small, the energy barrier to formation of the activated complex is small. It is then easy to form an activated complex and the C–O complex is weakly bonded. In gasification, $E \gg RT$; therefore, when the activation energy is equal to the activation enthalpy, the error will be small.

The relationship between the activation entropy and the pre-exponential factor can be expressed as:

$$\ln A = \ln\left(\frac{k_B T}{h}\right) + \frac{\Delta S^{0\neq}}{R}.$$

It can be seen that there is a linear relationship between A and the activation entropy. When the activation entropy is positive, the activated complex formed in the reaction has increased entropy, i.e., the C–O complex has more degrees of freedom of motion. This is an unstable activated complex, i.e., the activated complex is not strongly bonded and decomposes easily. If the activation entropy is negative, the activated complex has fewer degrees of freedom of motion, it is strongly bonded, and difficult to decompose. When a bimolecular reaction forms an activated complex, it loses the translational and rotational degrees of freedom, so the activation entropy is generally negative. Under such circumstances, the more negative the activation entropy is, the more strongly the unstable activated complex is bonded and the more difficult it is to decompose. Conversely, the less negative the activation entropy is, the more readily the activated complex decomposes.

The above analysis shows that there is a relationship between the gasification reaction kinetic parameters obtained from TG experiments and the nature of the C–O complexes formed in the reaction. The compensation effect among the kinetic parameters is a compensation relationship between the activation energy and the pre-exponential factor. This can be attributed to the linear relationship between the activation entropy and the activation enthalpy of the formed unstable C–O complexes. According to the Eyring equation, it can be seen from the change in the activation energy and pre-exponential factor that the compensation effect between the activation energy and the pre-exponential factor results in a linear relationship between the activation entropy and the activation enthalpy of the C–O complexes:

$$\Delta H^{0\neq} = T_i \Delta S^{0\neq} + \text{Constant},$$

where T_i is the kinetic equilibrium temperature, at which the reaction has the reaction rate constant k_i.

When C–O complexes are formed during gasification, if the active sites on the surface of the coke sample combine easily with CO_2 to form C–O complexes, $\Delta H^{0\neq}$ (E) for formation of the C–O complexes is small. The CO_2 molecules and the coke surface are also strongly bonded. The degrees of freedom of the C–O complexes formed are small. As a result, $\Delta S^{0\neq}$ (ln$A$) is also small. It is not difficult to understand from this analysis why a reduction in the activation energy accompanies a decrease in the pre-exponential factor. Establishment of the compensation effect is determined by the nature of the C–O complexes.

5.3.2.2 Physical Meaning and Calculation of Compensation Coefficient

The following equation can be obtained by logarithmic processing of the Arrhenius equation:

$$\ln A = \frac{E}{RT_i} + \ln k_i.$$

Based on the linear equation for the compensation effect, the compensation coefficients can be expressed as

$$a = \frac{1}{RT_i},$$
$$b = \ln k_i.$$

Table 5.12 lists the compensation coefficients and kinetic equilibrium temperatures for gasification with different catalysts. The final two columns of data in the table are the average values of those listed in Tables 5.5 and 5.8. These represent the average initial reaction temperatures of the gasification processes and the reaction temperature at which the average conversion rate reaches 50 %. Essenhigh et al. [41] proposed that the kinetic equilibrium temperature is close to the average

Table 5.12 Calculated compensation effect coefficients

Catalyst	a	b	T_i (K)	$T_{x=0.05}$ (K)	$T_{x=0.5}$ (K)
NaOH	0.1089	−3.43	1104	1040	1165
KOH	0.1177	−4.36	1022	1014	1151
$Fe(NO_3)_3$	0.1012	−4.91	1188	1117	1325
$Ca(OH)_2$	0.0994	−3.86	1210	1165	1303
Non-catalytic process	0.0989	−5.02	1216	1183	1338

Table 5.13 Reaction rate constants and pre-exponential factors at kinetic equilibrium temperature

Sample	ln*A*	ln*k*	Sample	ln*A*	ln*k*	Sample	ln*A*	ln*k*
Na-DS	4.39	−3.82	Ca-DS	9.30	−4.44	DS	10.67	−6.14
Na-FF	7.51	−3.57	Ca-Fran	14.74	−4.08	FX	10.06	−7.02
Na-PSV	10.56	−3.28	Ca-FF	10.05	−3.74	FF	14.49	−6.18
Na-Fu	11.04	−3.23	Ca-Fu	13.29	−3.63	FranE	23.87	−7.94
Na-PSI	11.34	−3.12	Ca-PS	18.07	−4.86	PSI	17.55	−6.26
Na-Fran	12.75	−3.20	Ca-PSV	13.71	−3.67	JC	16.90	−7.38
Na-PS	15.34	−4.07	Ca-PSI	18.22	−2.62	FranV	21.59	−6.55
K-DS	2.77	−4.95	Fe-DS	4.47	−5.29	Fu	18.60	−6.76
K-FF	4.95	−4.03	Fe-Fran	17.40	−4.59	Fran1	20.63	−6.51
K-Fu	9.87	−3.94	Fe-FX	5.79	−5.91	PSV	14.55	−6.88
K-Fran	15.99	−4.22	Fe-FF	7.83	−3.89	PS	20.83	−6.97
K-PSI	17.38	−4.31	Fe-Fu	10.11	−4.03	FranI	17.48	−8.55
K-PS	16.92	−4.80	Fe-PS	17.43	−5.08	Fran2	21.96	−7.01
K-PSV	18.48	−4.40	Fe-PSI	12.54	−5.63			

value of the midpoint in the reaction temperature range. According to this view, the kinetic equilibrium temperature is a mathematical average of the temperatures. However, the computational results listed in Table 5.12 show that the kinetic equilibrium temperature is closer to the initial reaction temperature of the coke sample, which differs from the results reported in the literature.

The pre-exponential factors and the reaction rate constants at the kinetic equilibrium temperatures, listed in Table 5.13, show that the differences among the pre-exponential factors of the samples vary greatly, but at the kinetic equilibrium temperatures, the reaction rate constants are almost equal. For example, ln*A* of the coke samples are in the range 10.6–23.9. The ln*k* values at the kinetic equilibrium temperature range from 6.2 to 7.9. These results demonstrate that at the kinetic equilibrium temperature, the reaction rate constants of various reactions in the same series are equal.

5.3.2.3 Relationship Between Catalytic Mechanism and Compensation Effect

The formation of C–O complexes in gasification can be described as:

$$M-O+C-C_f \rightarrow M\cdots O\cdots C-C_f,$$

$$M\cdots O\cdots C-C_f \rightarrow [M\cdots O\cdots C]^{\neq}\cdots C_f,$$

$$[M\cdots O\cdots C]^{\neq}\cdots C_f \rightarrow [M\cdots O-C]+C_f,$$

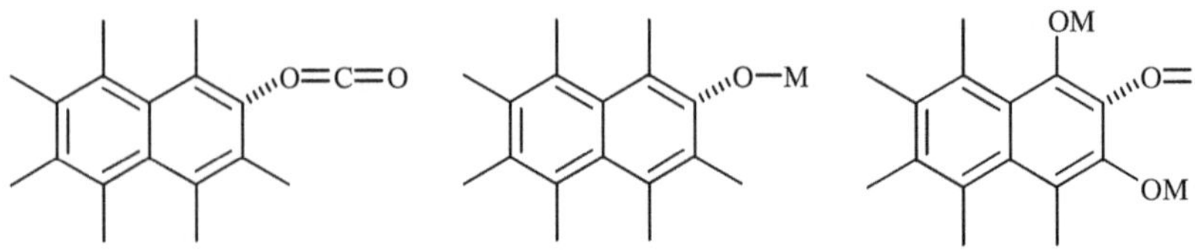

1 2 3

Fig. 5.18 structures of surface C–O complexes: *1* Non-catalytic gasification; *2* Ca- and Fe-catalyzed gasification, M=Ca, Fe; and *3*, K- and Na-catalyzed gasification, M=K, Na

where M–O represents an oxygen-containing reactant (e.g., CO_2, H_2O, O_2, CaO, γ-Fe_2O_3). First, the oxygen-containing reactants move to, and collide with, the coke surface. In effective collisions, the carbon atoms on the coke surface can form partial bonds with the oxygen in the reactants. At this point, carbon and oxygen only attract each other and the chemical bonds between adjacent carbon atoms lengthens, weakening the bond, but not destroying it. As the reaction proceeds, the distance between the carbon and oxygen decreases and the carbon–carbon bond lengthens further prior to formation of a C–O bond. However, at this time in the bonding process, the carbon–carbon bond is not yet completely broken. An unstable intermediate C–O complex is formed. The carbon–carbon bond then breaks, and CO molecules separate from the complex, completing the gasification process.

The unstable C–O complexes formed by non-catalytic and catalytic gasification of coke are different in terms of structure and stability, and there is a compensation effect among the kinetic parameters. Three diagrams of the structures of C–O complexes, based on the results of studies of C–O complexes in non-catalytic and catalytic processes, are shown in Fig. 5.18.

The C–O complex formed in non-catalytic gasification has the simplest structure and is produced from a CO_2 molecule that is directly chemically adsorbed on an activated carbon atom on the coke surface. The C–O complexes formed when Ca and Fe salts are used as catalysts are formed by bonding between carbon atoms and CaO and γ-Fe_2O_3, respectively. The alkali metal is not directly involved in the reaction with carbon during gasification, so the C–O complexes formed in K- and Na-catalyzed reactions are actually CO_2 molecules chemically adsorbed on the coke surface. The carbon atom that forms the complex combines with a basic O^-Na^+ group; this group weakens and lengthens the carbon–carbon bond, and induces a positive charge on the adjacent carbon atom, enabling formation of the C–O complex.

The structural differences among the C–O complexes result in different compensation effects in different gasification processes. The electronegativities of Ca and Fe are smaller than that of carbon. They are more likely to lose electrons, so the electron cloud densities around the oxygen atoms in CaO and γ-Fe_2O_3 are higher than that in a CO_2 molecule. Therefore, when CaO and γ-Fe_2O_3 combine with positively charged active carbon atoms on the coke surface, the energy barrier to C–O complex formation is relatively small. The action of an alkali metal compound is different from those of CaO and γ-Fe_2O_3. The presence of an alkali metal results in an increase in the positive charges on carbon atoms on the coke surface, so CO_2 molecules are less likely to combine with active carbon atoms on the surface; therefore the energy barrier to C–O complex formation and the activation energy in the gasification reaction are reduced.

To summarize, the presence of a catalyst changes the mechanism of C–O complex formation. The structural differences among the C–O complexes lead to different compensation effects among the kinetic parameters. The coke surface edge structure is actually far more complex than that shown in Fig. 5.18. In the coking process, the presence of cross-linking bonds distorts the lattices of different cokes to different degrees, and these distortions probably prevent the bonding of catalyst particles to the coke surface. The ease with which different cokes form C–O complexes in the presence of catalyst particles therefore differs. This leads to changes in the order of the reactivities of cokes on addition of catalysts.

5.3.2.4 Theoretical Meanings of Compensation Effect Results

There is still some controversy regarding the authenticity of the compensation effect. Based on analysis of the results of nearly 1000 thermogravimetric experiments, Zsako [39] believes that the compensation effect observed for kinetic parameters calculated from thermogravimetric curves are not compensation effects in the real sense. He used the thermogravimetric method to study the thermal decomposition of nearly 400 compounds and analyzed the kinetic data. The $\ln k$ values obtained were similar, in the range -1 to -0.93. In X–Y plots of the kinetic data, the change range is therefore relatively narrow, and the variation is misinterpreted as being on cause by the compensation effect. In fact, the decomposition temperatures of these compounds vary from 310 to 1100 K. Their reaction mechanisms are different. The structures of the transition complexes during decomposition are also different, so these cannot be located on the same line as the compensation effect. In contrast, Essenhigh et al. [41] proposed that the reaction rate constants obtained in the experiments are a subset representing the real values of the rate constants, with errors. In general, for random experimental points, the experimental kinetic parameters obtained change randomly. Therefore, "It is very difficult to suggest the results from the compensation are caused merely by the results of the mathematical and statistical calculations." This proposition regarding the compensation effect has still to be verified.

Most researchers suggest that the appearance of a compensation effect indicates similar reaction mechanisms during reactions [42, 43]. Based on a comparison of the compensation effects of different reaction processes, Essenhigh suggested that the gasification compensation phenomena of C–O_2, C–H_2O, and C–CO_2 gasification are similar, and that this is because the reaction processes all involve the decomposition of oxygen surface complexes with similar structures. Compensation effects and reaction mechanisms are therefore different, and should be studied separately. The compensation effect identified in Zsako's study [39] of the decomposition of nearly 400 compounds, was not a true compensation effect because each reaction featured different mechanisms. In our research, the methods for calculating the gasification kinetic parameters are different for catalytic and non-catalytic gasification processes. The directly obtained kinetic parameters reflect the essential characteristics of the gasification process. We therefore believe that these kinetic parameters reflect the properties of the C–O complexes in the gasification process. The compensation effects illustrated here are real.

Discussion of the compensation effect enables us to understand catalytic and non-catalytic coal gasification processes. It can be seen from the definition of the compensation effect made by the International Union of Pure and Applied Chemistry that the conditions for establishing a compensation effect are that when a set of homologs go through the same process, if the reaction changes are systematic, then the activation enthalpies and the activation entropies in the reaction process show a linear compensation. Establishment of a compensation effect is related to the properties of the activated complexes formed during the reaction process. When the activated complexes formed have similar structures and stabilities, the activation enthalpy and entropy can compensate each other. The experimental results clearly show that after adding various catalysts, the compensation relationship lines for the gasification kinetic data of coke samples are different from that for the non-catalytic process. This means that the transition complexes formed during the catalytic and non-catalytic processes have different structures and stabilities. The activation energy and pre-exponential factor data in Sect. 5.2.2 of this chapter were used to compare the kinetic parameters of catalytic and non-catalytic gasification processes. The results were as follows. The activation energy of the catalyzed gasification reaction is less than that of the non-catalytic process, indicating that the presence of the catalyst decreases the energy barrier to formation of the unstable activated complex, and C–O complexes are easily formed. The pre-exponential factor when a catalyst is added is also smaller than that of the non-catalytic process, indicating that the activation entropy of the C–O complex formed in the catalytic process is smaller than that of the complex formed in the non-catalytic process. The vibrational degrees of freedom of the activated complex in the catalytic reaction are restricted, and this affects the degrees of freedom of motion. These results suggest that metal ions reduce the energy barrier to C–O complex formation in the gasification process, decreasing the activation energy of the reaction. At the same time, their presence also limits the degrees of freedom of motion of the C–O complex, so the reaction pre-exponential factor also decreases.

To summarize, establishment of a compensation effect is related to the nature of the activated complexes formed in the reaction process. Easy formation of an activated complex requires a small $\Delta H^{0\neq}$ of formation. However, combination limits the degrees of freedom of motion of the activated complex, with the result that $\Delta S^{0\neq}$ is small. This is why a kinetic compensation effect is established. The existence of a compensation effect is an essential characteristic of CO_2 gasification of coke. Its establishment is independent of the coke type, maceral content, coke-forming conditions, and other factors that affect the coke gasification reactivity. However, the effects of the coke type, maceral content, and coke-forming conditions on the coke reactivity are comprehensively embodied in the compensation effect. As a result of the presence of a compensation effect, it is impossible to use only the value of the activation energy or the pre-exponential factor to evaluate coke reactivity. A compensation effect among the coke CO_2 gasification kinetic parameters is also seen for catalytic reactions. The compensation law is different from that for the non-catalytic process. In kinetic data plots, the compensation relationship line for the catalytic process is located above that for the non-catalytic reaction. These results can be used to judge whether a catalyst is involved in the coke gasification process. The different compensation laws for catalytic and non-catalytic reactions reflect the different structures and stabilities of the activated complexes.

5.4 Studies of C–O Complexes Over Coal Surface with TPD

In gasification, the unstable transition-intermediate complexes that exist over the coal's surface play an important role in the gasification. The active intermediates are generally surface C–O complexes. Gasification consists of decomposition of the surface C–O complexes, which are the major active sites of gasification.

Over the past 10 years, various techniques have been used to measure and characterize the surface C–O complexes of coals or cokes. The results show that the formation of surface complexes is related to oxygen transfer in the gasification process. The key to understand the gasification mechanism is an in-depth study of the formation of surface C–O complexes. Different researchers have produced various descriptions of C–O complexes. It is generally thought that cokes contain many C–O complexes of different stabilities, which can release CO during TPD. Lizzio et al. [13] used temperature-programmed desorption to define the reaction surface area (RSA) in different ways, and suggested that it is appropriate to use the RSA to explain differences among gasification reactivities. An examination of the role that the surface plays during gasification shows that C–O compounds on the coke surface are the gasification activity centers. Lizzio proposed that the ratio of the RSA to the total surface area (TSA) should be equal to the ratio of the amount of unstable C–O compounds to the total amount of C–O compounds.

However, there are few reports in the literature regarding the effects of complexes on gasification reactivity. Based on previous studies, we used TPD to study surface complex formation on Pingshuo coke and acid-pickled samples in an H_2 atmosphere, and the relationship between the complexes and the gasification reaction.

5.4.1 Pretreatment of Samples

Pingshuo coal and acid-pickled deashed coal were used to produce Pingshuo ash-containing and acid-pickled coke samples, respectively. The coke-making conditions were the same as those described in Table 5.4. The inorganic composition data for the Pingshuo coal and coke samples are listed in Table 5.14.

To study the coke surface C–O complexes, special treatment of the coke sample surface is required. The treatment process is performed using a TPD apparatus. The sample (150 mg) was placed in the TPD apparatus, and heated to 1273 K at a rate of 10 K min^{-1} in an H_2 atmosphere (45 mL min^{-1}), thereby eliminating the original C–O complexes on the coke sample and producing coke samples with clean surfaces. The original C–O complexes were completely decomposed during the treatment. The obtained samples are referred to as clean-surface coke samples. The clean-surface samples were placed in a furnace and heated to a given temperature. The furnace atmosphere was then switched to CO_2 (60 mL min^{-1}) and maintained for a certain period of time, to enable formation of C–O complexes on the sample surface. Then the CO_2-pretreated coke samples were treated at different temperatures for different times. These samples are referred to as CO_2-treated/processed coke samples.

The TG curves in Fig. 5.19 show the weight changes of the Pingshuo coke samples during TPD in N_2 and CO_2 atmospheres. The figure that shows the Pingshuo acid-pickled coke does not show an obvious weight loss during heating in N_2, whereas the original/raw coke sample shows a 20 % weight loss in N_2 in the range 973–1073 K. The surface-cleaned original/raw Pingshuo coke sample, like the acid-pickled coke sample, shows no weight loss during heating in N_2. After treatment in CO_2 for 30 min at 873 K, the clean-surface coke sample shows a weight loss in the range 973–1073 K when subjected to TPD.

Table 5.14 Analytical data for Pingshuo raw coal and original coke ash (%)

Raw coal petrographic analysis				Raw coal inorganic composition		
Vitrinite	Semivitrinite	Inertinite	Exinite	Clay	Pyrite	Carbonate
59.9	4.4	29.9	5.8	7.9	0.5	0.5
59.9	4.4	29.9	5.8	7.9	0.5	0.5
Main inorganic components in ash of raw coal coke sample						
SiO_2	Al_2O_3	Fe_2O_3	CaO	MgO	TiO_2	SO_3
55.50	32.83	2.81	3.63	1.03	1.02	1.48
55.50	32.83	2.81	3.63	1.03	1.02	1.48

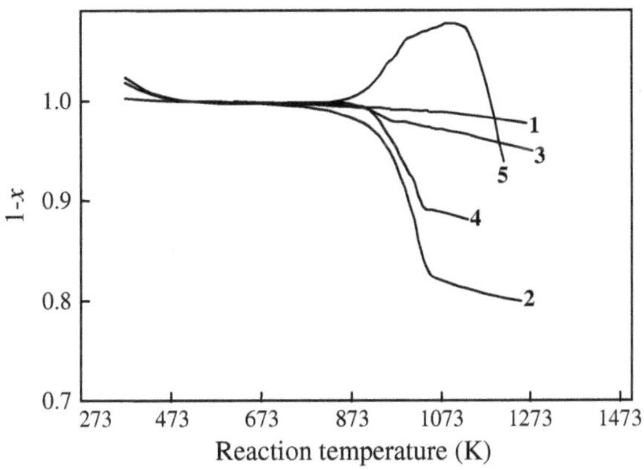

Fig. 5.19 Thermogravimetric *curves* of various Pingshuo coke samples under different conditions: *1* thermal decomposition of pickled clean coke sample in N_2; *2* thermal decomposition of original/raw coke sample in N_2; *3* thermal decomposition of clean-surface coke sample in N_2; *4* thermal decomposition of clean-surface coke sample in N_2 pretreated at 873 K In CO_2; *5* CO_2 gasification of clean-surface coke sample

5.4.2 TPD Study of C–O Complexes

5.4.2.1 TPD Results of Pingshuo Cokes

Figure 5.20 shows the gas releases from Pingshuo and French Cokes samples. There are three release periods in the TPD process. When the temperature is below 423 K, the coke sample is dehydrated; this will not be discussed in detail. There is a release period in the temperature range 823–1093 K. Gas chromatography shows that main product is CO. There is another release period when the temperature is above 1073 K, and gas chromatography shows that the main product is still CO. The figure shows that CO release from Pingshuo acid-pickled coke samples only occurs above 1073 K. There is no release stage in the range 823–1093 K. The figure also shows the TPD curves of coke samples Fran1 and Fran2 (see Table 5.4 for definitions). The figure shows that there is no release period in the temperature range 823–1093 K. When the temperature is above 1073 K, the amounts of CO released are not the same: the amount of CO from Fran 2 is smaller than that from Fran1.

5.4.2.2 TPD Study of CO_2-Treated Pingshuo Coke

In terms of gasification reactivity, it is important that the coke surface is able to form C–O complexes in the gasification process. We used CO_2 treatment to study the formation of C–O complexes on the coke surface. Figure 5.21 shows the experimental results for the Pingshuo original/raw coke samples and acid-pickled

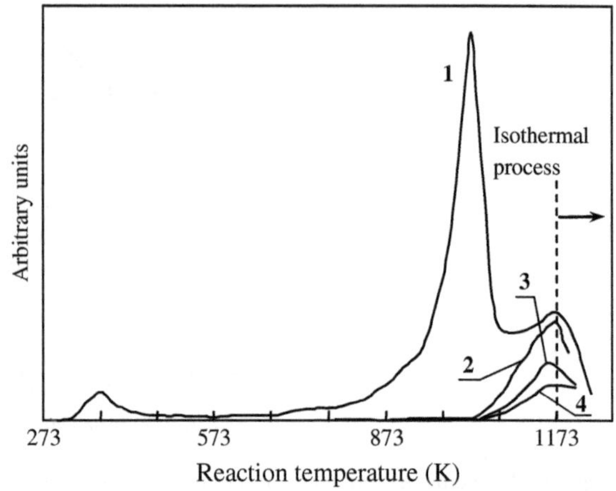

Fig. 5.20 TPD *curves* of Pingshuo and French cokes: *1* Pingshuo raw coal coke; *2* Pingshuo acid-Pickled coal coke; *3* Fran1; and *4* Fran2

coke samples. Figure 5.21 shows that after the surface-cleaning treatment, the Pingshuo original/raw coke sample no longer showed CO release at 1000 K in TPD. However, after a CO_2 treatment at 873 K, the clean-coke samples of Pingshuo coal recovered the CO release peak at 1000 K, indicating that the CO_2 treatment helps the formation of new C–O complexes on the surfaces of the coke samples. Figure 5.22 shows that in the entire TPD process, the Pingshuo acid-pickled coke only shows CO release above 1073 K. Even after CO_2 treatment, unlike the original/raw coke, the acid-pickled coke does not show CO release at 1000 K. These results indicate that CO release by Pingshuo coke at 1000 K is probably related to the ash content of the coke.

XRD patterns show that the Pingshuo original/raw coke contains CaO. The acid-pickled coke sample contains no CaO, but contains CaF_2. It was difficult to distinguish clear lattice peaks from the clean-surface coke XRD pattern, and the conversion process of CaO in the coke sample is unclear. After CO_2 treatment at

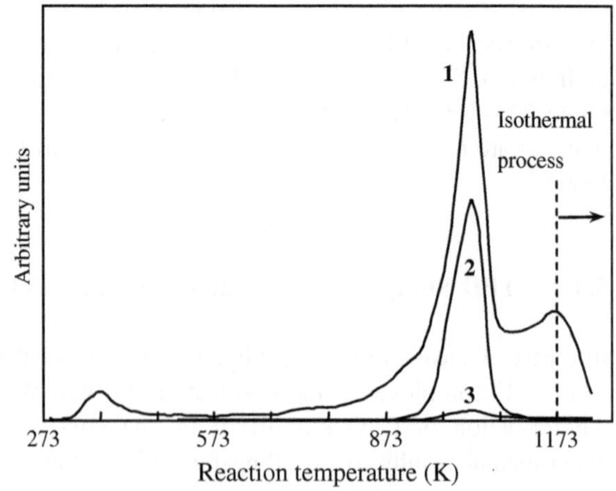

Fig. 5.21 TPD *curves* for Pingshuo original/raw cokes: *1* Pingshuo original/raw coke sample; *2* after CO_2 treatment at 873 K; *3* after surface-cleaning treatment

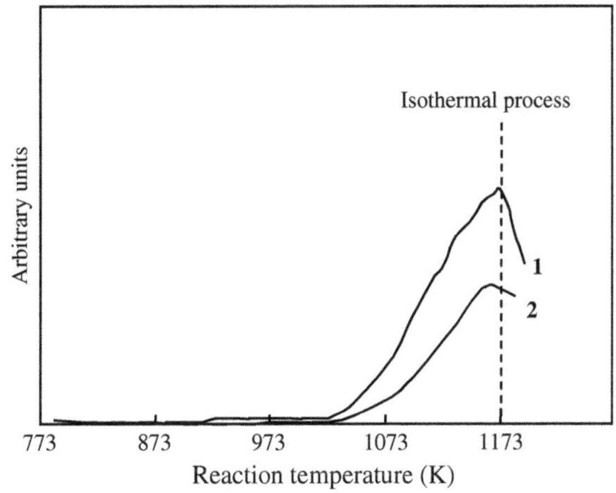

Fig. 5.22 TPD *curves* for Pingshuo acid-pickled coke samples: *1* Pingshuo acid-pickled coke sample; and *2* after CO_2 treatment at 873 K

873 K, the CaO lattice peak was recovered for the clean-surface sample. These results, in combination with the experimental results shown in Fig. 5.22, illustrate that the CO release peak at 1000 K is closely related to the CaO in the coal ash. The CO release at 1000 K is the result of a reaction between CaO and carbon atoms on the coke surface. The CO release period during TPD of Pingshuo original/raw coke samples at 1000 K occurs not only because of the C–O complexes but also due to the reaction between the coal ashes and the cokes.

Figure 5.19 shows the CO_2 gasification curve of the clean-surface coke sample from Pingshuo original/raw coke. The thermogravimetric curve shows a weight increase from 823 K; this is generated by CaO on the coke surface. With increasing temperature, the mass of the coke sample increases gradually. When the temperature is above 1000 K, the rate of weight increase slows down. This is because although CaO is being generated, the reaction between CaO and carbon releases CO. At temperatures above 1110 K, the TG curve shows a weight loss process. It can therefore be seen that when the temperature is above 1100 K, CO release from the reaction between CaO and carbon is faster than CaO formation. This illustrates that the ash contained in coal has pronounced effect on the TPD result.

5.4.2.3 TPD Study of Metal Compound Addition

Addition of $Ca(OH)_2$

Figure 5.23 shows the TPD curves of Dongshan coke samples after adding Ca $(OH)_2$. In the surface treatment process, $Ca(OH)_2$ decomposes to CaO, which remains in the coke samples. After the surface treatment, TPD of the sample did not show significant CO release. After treatment of the sample at 873 K in CO_2 for 30 min, the sample showed a sharp release peak at 970 K during TPD. Gas chromatography shows that the product was CO. The experimental results for

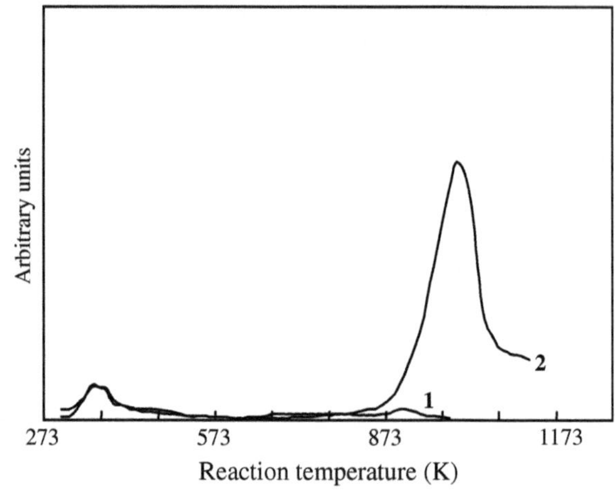

Fig. 5.23 TPD *curves* of Dongshan cokes with added $Ca(OH)_2$: *1* after surface-cleaning treatment and *2* after CO_2 treatment at 873 K

Dongshan coke were similar to those for Pingshuo ash-containing coke described above. This clearly indicates that in the CO_2 gasification process, CaO reacts directly with the carbon atoms of the coke sample. There is an obvious difference between the crystalline states of the Ca salts in the two samples; the *d* values do not correspond to the values for the standard lines of $CaCO_3$, CaC_2, and elemental Ca. It is therefore difficult to identify the crystalline form of the Ca salt in the clean-surface samples.

Addition of $Fe(NO_3)_3$

Figure 5.24 shows the TPD curves of Dongshan coke samples after adding Fe $(NO_3)_3$. The sample does not show any significant release peak after surface-cleaning treatment. However, the cleaned sample shows a CO release peak at 930 K

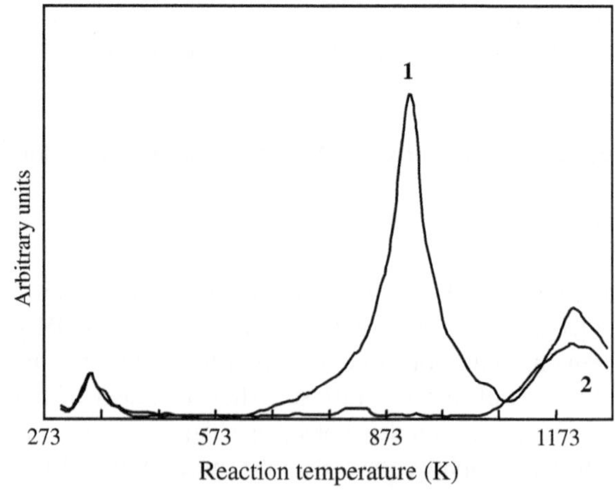

Fig. 5.24 TPD *curves* of Dongshan cokes with added $Fe(NO_3)_3$: *1* after surface-cleaning treatment and *2* after CO_2 treatment at 873 K

after treatment with CO_2 at 873 K. Comparison with standard cards shows that Fe is present as γ-Fe_2O_3 after treatment of the coke sample in a CO_2 atmosphere, and as Fe_3C in the clean-surface sample. These experimental results show that γ-Fe_2O_3 is the catalytically active form of $Fe(NO_3)_3$ in gasification. At 930 K, γ-Fe_2O_3 reacts with carbon atoms on the Dongshan coke surface to form CO, with simultaneous reduction of the γ-Fe_2O_3 to Fe_3C in the coke. Fe_3C is oxidized to γ-Fe_2O_3 at 873 K in a CO_2 atmosphere. These results are similar to those for Ca salts.

Addition of Alkali Metals

Figure 5.25 shows the TPD curves of Dongshan coke samples after addition of NaOH. Neither the clean-surface sample nor the sample treated with CO_2 showed an obvious CO release peak similar to those in Figs. 5.23 and 5.24. The Na salt formed in the gasification could not be identified by XRD process.

The TPD and XRD experimental results for Dongshan cokes after adding KOH are similar to those for the addition of NaOH. These results show that the catalytic reaction mechanisms of compounds of K and Na, which are alkali metals, in the gasification process are different from those of CaO and γ-Fe_2O_3. In gasification, CaO and γ-Fe_2O_3 react directly with the carbon atoms on the coke surface, whereas K and Na cannot react directly with the carbon atoms on the coke surface to release CO molecules.

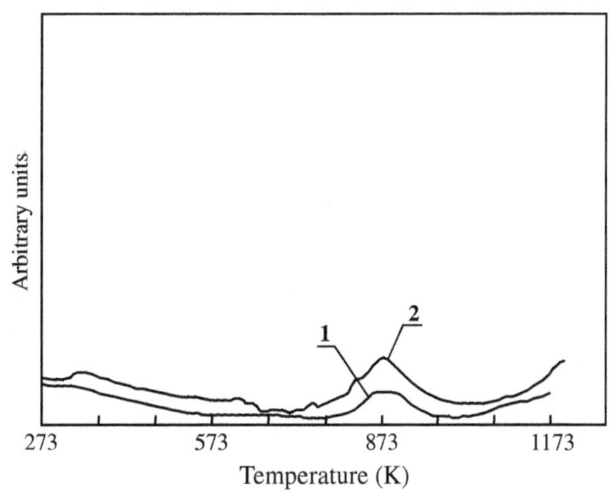

Fig. 5.25 TPD *curves* of Dongshan cokes with added NaOH: *1* after surface-cleaning treatment and *2* after CO_2 treatment at 873 K

5.4.3 *Analysis of CO_2 Gasification of Cokes Based on TPD Results*

The C–O complexes formed during coke gasification are unstable. It is impossible to detect them using TPD. The TPD method can only detect CO release as a result of decomposition of the original oxygen functional groups on the coke surface. This process generally occurs when the temperature is above 1073 K. The release process above 1073 K, is related to the decomposition of oxygen functional groups on the coke surface. Different coke surfaces contain different numbers of oxygen functional groups. When the temperature is above 1073 K, the volume of CO released by decomposition of the oxygen functional groups on the coke surfaces are different for different cokes. It takes longer to produce coke from Fran2 than from Fran1. When coke formation takes a long time, the number of oxygen functional groups on the surface is reduced. The CO release volume in the TPD process is therefore smaller for Fran2 than for Fran1. This reflects the characteristics of the oxygen functional groups on the surface.

Suzuki et al. [27] successfully detected the presence of ^{13}CO in isotope-labeling experiments. This proves that the CO_2 molecule passed an oxygen atom to the coke surface during gasification. There must therefore be C–O complexes present during gasification. The C–O complexes are unstable, so they cannot be studied using TPD. The present knowledge of C–O complexes is speculative. However, the use of TPD in combination with XRD is an effective way of studying the catalytic roles of metal ions.

5.5 Structure–Reactivity Relationships in Coal Gasification

5.5.1 *Factors Affecting Gasification Reactivity*

5.5.1.1 Effects of Maceral Components

Coal macerals originate from different structural components of plants; therefore the coke gasification reactivity is related to maceral composition. Cokes with different coal macerals have different specific surface areas and active center densities; therefore the reactivities of different macerals in coke samples differ greatly.

The data in Table 5.5 show that the gasification reactivities of coke samples with different maceral components differ. Figure 5.26 shows the Arrhenius plots of three different coke maceral samples. The figure shows that at gasification temperatures above 973 K, the order of the temperatures at which 50 % conversion is reached is as follows: that for exinite coke is the lowest, 1293 K; those for the vitrinite and original samples are close to each other, 1302 and 1306 K; and that of inertinite is the highest, 1312 K. This implies that exinite has the highest reaction rate, and therefore the highest reactivity. In these experiments, exinite from French

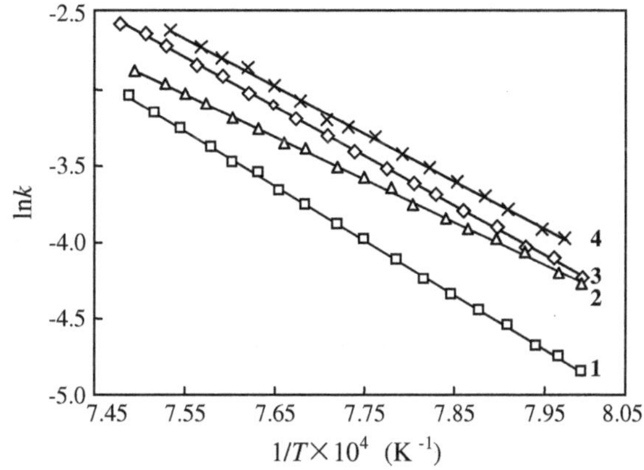

Fig. 5.26 Arrhenius plots of French bituminous coal and coke macerals: *1* FranE; *2* Fran1; *3* FranV; and *4* FranI

bituminous coal has the highest activation energy, 298.9 kJ mol^{-1}, that for vitrinite is 268.4 kJ mol^{-1}, and inertinite has the lowest activation energy, 226.0 kJ mol^{-1}. The activation energy of the raw coal is 256.6 kJ mol^{-1}, close to that of vitrinite. The order of the reactivities of the three main maceral components is exinite coke > raw coal ash-free sample ≈ vitrinite coke > inertinite coke.

Figure 5.27 shows the Arrhenius plots for Pingshuo coal. The vitrinite activation energy is lower than that of inertinite. The temperatures for 50 % conversion are 1336 K for inertinite coke and 1353 K for vitrinite; the former has a higher gasification reactivity than the latter. The coke maceral reactivity results for Pingshuo coal are inconsistent with those for the French coal. These results show that the coke maceral components of coke samples from different coals have different gasification reactivities, because of differences among the maceral molecular structures. Furthermore, the orders of the gasification reactivities of the coke maceral samples are not necessarily the same.

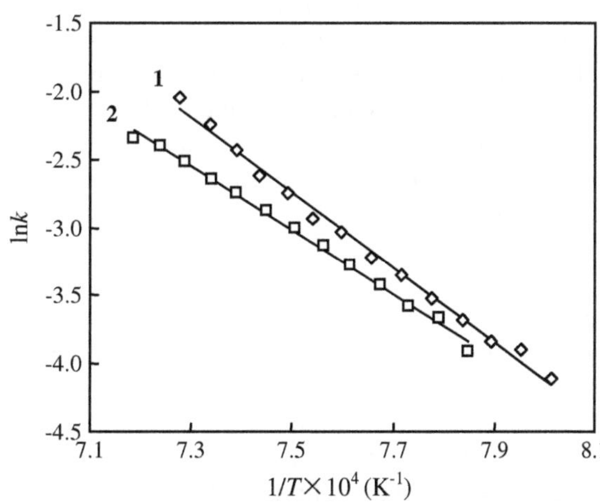

Fig. 5.27 Arrhenius plots of Pingshuo coal and coke macerals: *1* PSF and *2* PSV

The findings of Franciszek et al. [44] indicated that vitrinite had the highest reactivity and gasification under the experimental conditions used. They proposed that water vapor and CO_2 reactivities of the macerals are consistent. The order was determined to be vitrinite > raw coal > exinite > inertinite. Many of the results are inconsistent, showing that the maceral content affects coke reactivity, but the effect is complicated. Macerals of coal and coke have different influences on the gasification reactivity because coal types and coking processes differ.

5.5.1.2 Effects of Coal Devolatilization Processes

Coal gasification can be approximately divided into two stages: coal pyrolysis and gasification of the generated coke. The gasification reactivity of the generated coke depends on the conditions used in the pyrolysis stage. The coke-forming conditions therefore affect the gasification reactivity. It is generally believed that the coke reactivity is related to the coke formation temperature. The higher the temperature and pressure the longer the coking time, and the greater the difference in the reactivity is, although the coke yield may show little change. Research by Shang et al. [45] showed that with increasing coking time the reactivity is reduced by 63 % although the amount of coke decreases by only 2 %. Regarding the effect of the pyrolysis temperature on the gasification reactivity, van Muhlen et al. [46] found that the longer the coke stays at the final coke-forming temperature, the lower the gasification reactivity of the coke is. It has been suggested that harsh coke-forming conditions decrease the number of active centers on the coke surface. The coke-forming conditions significantly affect the development of the coal inner surface and important factors such as the number of active centers.

Figure 5.28 shows the relationships between temperature and concentrations of CO produced from the reactions between CO_2 and coke samples produced at final temperatures of 1100, 750, and 450 °C, at a heating rate of 20 °C min^{-1}. These relationships are used to express the gasification reactivities of the coke samples. There is little difference among the reactivities of the coke samples produced at 1100 °C. However, there are significant differences among the reactivities of the coke samples produced at 750 °C. The order of the reactivities is long-flame coal > anthracite > lignite > coking coal > fat coal > meager coal.

Comparison this sequence with that of the pyrolysis activation energies at a heating rate of 20 K min^{-1} (given in Chap. 4, Table 4.4, 4.5, 4.6 and 4.7) shows that coals with high pyrolysis activation energies have low gasification reactivities during pyrolysis. The gasification reactivities of the coke samples are correspondingly high. When the coke-forming temperature is 450 °C, the order of the gasification reactivities of the coke samples is basically the same as that of the pyrolysis activation energies. The order is coking coal > lignite > fat coal > meager coal > long flame coal > anthracite. At the same time, the temperature at which CO is generated is delayed from 400–450 °C to 580–600 °C. This shows that 450 °C is not a suitable coke-forming temperature, and that unreleased coal tar has a significant effect on the gasification characteristics of the coke. Tests on the pyrolysis

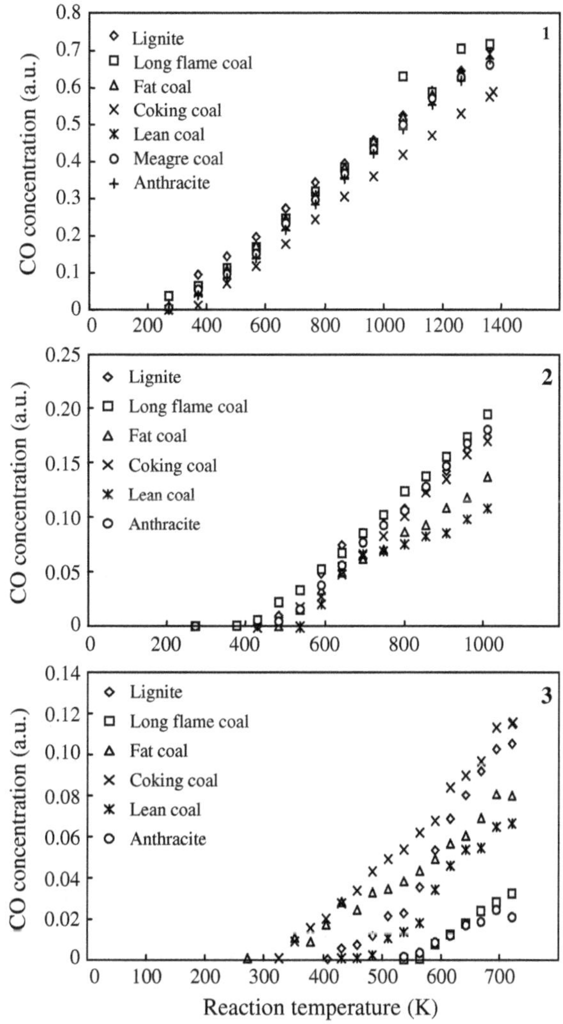

Fig. 5.28 Relationship between CO concentration and temperature in gasification of coke samples (Coke-forming heating rate 20 °C min^{-1}) at coke-forming temperatures of: *1* 1100 °C; *2* 750 °C; and *3* 450 °C

products produced from coal at final temperatures of 450 and 750 °C (see Chap. 4, Sect. 4.2) show that the activation energy order is similar to that of the amount of C_{20} and higher compounds released from the coal at 750 °C. In other words, the gasification reactivity of low-temperature coke is proportional to the heavy hydrocarbon content of the coal. As the coke-forming temperature increases and the amounts of volatile components in the coke residues decrease, the coke reactivities of different types of coal gradually converge.

Figure 5.29 shows that when the temperature is increased to 1100 °C at a rate of 1200 °C min^{-1}, the coke gasification reactivity order is meager coal > lean coal > fat coal > coking coal > long-flame coal > anthracite > lignite. This order is significantly different from the reactivities of cokes produced at a heating rate of

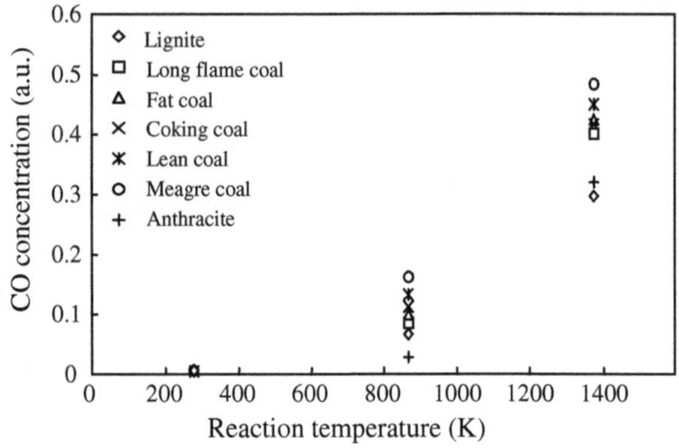

Fig. 5.29 Relationship between CO concentration and temperature in gasification of coke samples (coke-forming heating rate 1200 °C min^{-1}, final temperature 1100 °C)

20 °C min^{-1}. The last three types of coal in the sequence, i.e., those with the lowest reactivities, produce the cokes with the highest reactivities. This order is basically the same as that of the pyrolysis activation energies of coal at 20 °C min^{-1}. Fast pyrolysis shortens the pyrolysis process, weakening the general rule that the high-temperature section is favorable for reactions with high activation energies and the low-temperature section is favorable for reactions with low activation energies; the condensation degree of low-rank coal in the low-temperature section is therefore reduced, and depolymerization is enhanced, leading to a decrease in differences among the activation energies and the mean value of the activation energy distribution. The coke-forming temperature is also an important parameter affecting coke reactivity.

Figure 5.30 shows the order of the coke gasification reactivities for different coke-forming temperatures. Under fast pyrolysis conditions, at 1200 °C min^{-1}, the order of the coke gasification reactivities for different final coke-forming temperatures is always 750 °C > 450 °C > 1100 °C, regardless of the coal type. The reactivity of the coke sample is highest at 750 °C, because the rapid rise in temperature creates large amounts of highly active tar-forming compounds on the coke surface. This is basically in agreement with the results obtained by Echterhoff et al. [47].

It is clear from Fig. 5.30 that for coke-forming temperatures at two heating rates, there is a boundary point P at about 830 °C (750 °C < P < 1100 °C) in the coke reactivity curve. If the coke-forming temperature is above P, the coke gasification reactivity under slow coke-forming conditions is higher than that under fast coke-forming conditions. If the coke-forming temperature is below P, the opposite holds. Above P, i.e., when the coke-forming temperature is above 830 °C, because the release of CH_2, aromatic hydrocarbons, and some functional groups accelerates under slow pyrolysis conditions, a large number of tar components escape from the coal particle surface, so the coke surface layer contains large amounts of tar precursor compounds. These compounds provide many active sites for fast reactions. At this point, the second-order derivative of the fast pyrolysis product concentration versus temperature change curve is less than zero. The devolatilization process

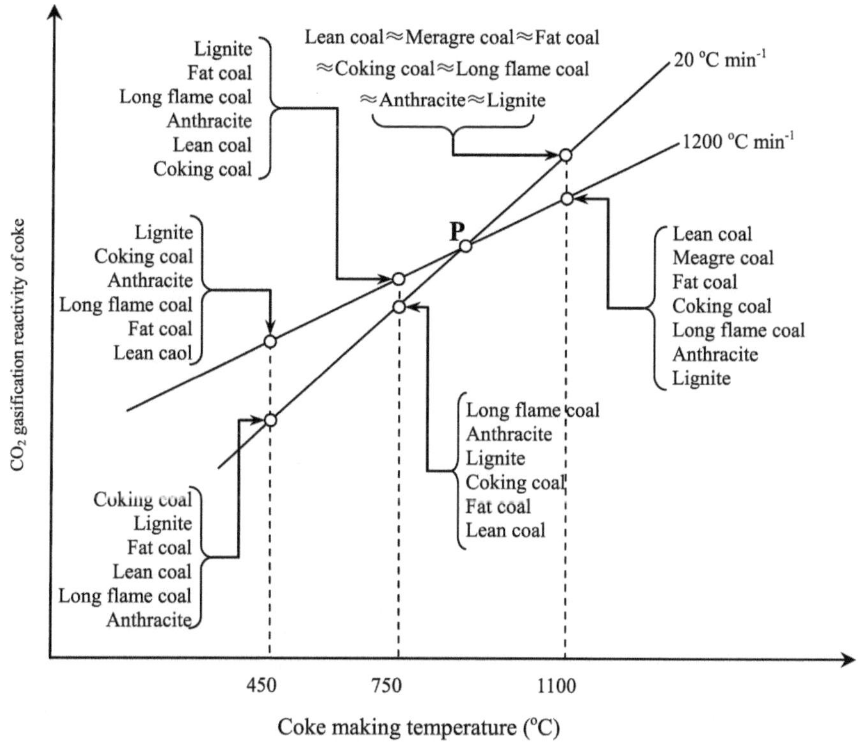

Fig. 5.30 Relationship between order of CO_2 gasification reactivities of coke and coke-forming conditions

enters a late phase and the active coke components are depleted, so the coke gasification reaction rate is lower than that under slow coke-forming conditions.

To summarize, coke CO_2 gasification has a negative correlation with the coal pyrolysis reactivity under slow coke-forming conditions. The higher the pyrolysis activity is, the lower the gasification reactivity. When the coke-forming temperature is above 830 °C, the gasification reactivity of the coke when the temperature is slowly increased is higher than that for a rapid temperature increase. However, when the coke-forming temperature is below 830 °C, the opposite holds. Under fast pyrolysis, the coke reactivities of different types of coal are similar. The coke gasification reactivity has a positive correlation with the amount of unreleased tar precursors in the residues.

5.5.1.3 Effects of Ash Content

The catalytic effect of ash on gasification was observed at an early stage in gasification research [48]. There is a linear relationship between the metal oxide content

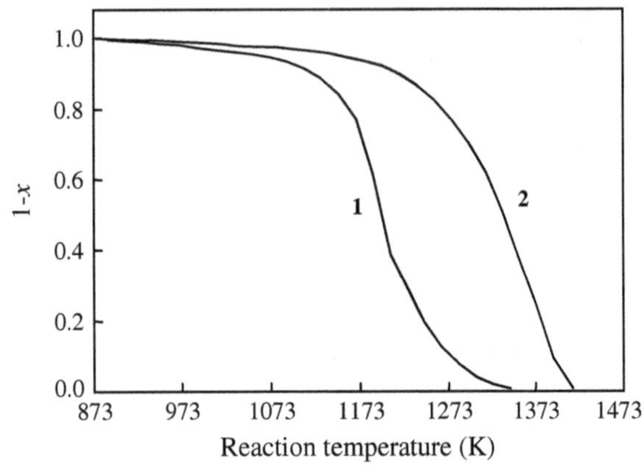

Fig. 5.31 Effects of ash content on gasification reactivities of Pingshuo coke samples: *1* ash-containing and *2* ash-free

of coal and coal reactivity. In general, the alkali, alkaline-earth, and transition metals in coal have catalytic effects. Alkali and alkaline-earth metals are the main catalysts in coal.

Figure 5.31 shows the CO_2 gasification TGA plots for ash-containing and ash-free Pingshuo cokes. The gasification reactivity of the ash-containing coke sample is higher than that of the ash-free sample because the ash catalyzes the gasification process. Acid-pickling deashing decreases the alkali and alkaline-earth metal contents of coal, so the gasification activity declines.

Sulfur in coal has the most adverse effect on gasification because it can form stable compounds with transition metals, e.g., Fe–S, which inhibit catalysis. Research by Matsumoto et al. [49] shows that a gas phase content of H_2S as low as 500 ppm can inhibits catalysis. In particular, in water vapor gasification using Fe as the catalyst, catalyst poisoning occurs and catalyst regeneration is slow. Usually, sulfur inhibition can be compensated for by increasing the reaction temperature and pressure. Coal also contains a large amount of aluminosilicates. At high temperatures, reactions of these aluminosilicates and alkali metals generate water-insoluble compounds, which have no catalytic effect, thereby reducing the catalytic effect. Although the gasification conditions used in the studies differ, and the conclusions are not completely consistent, it is generally believed that K salts have the best catalytic effect, followed by Na salts.

5.5.1.4 Effect of Surface

Changes in the pore structures of gasified coal and coke have a significant effect on mass transfer in coal gasification. Cokes have complex porous structures with wide pore distribution ranges, resulting from the non-uniform arrangement of aromatic structures in coal. Different pore sizes have different inhibiting effects on internal mass transfer. It is generally believed that diffusion of the reaction gas is activated in pores of size less than 0.5 nm. To make full use of microporous active sites, a large number of inlet pores need to be connected to the micropores, i.e., macro- and meso-pores to reduce the diffusion distance. Reaction gas diffusion from macropores to the micropore openings is fast. In the micropores, the gas concentration is close to the surface concentration. Gan et al. [50] showed that the percentage of macropores in low-rank coals is greater than that in high-rank coals. More air inlets are therefore available in the gasification of low-rank coal, so the mass transfer limitations are low.

In coke gasification, the specific surface areas of various cokes also change. Johnson's [51] thermogravimetric analysis study of the gasification of various cokes shows that the specific surface area decreases slightly during gasification, and the changes in the surface areas of various cokes are similar. It was also found that acid-pickling deashing before gasification had little effect on the specific surface area, but after acid pickling, the reactivity decreased significantly. Superficially, the change in the specific surface area is related to the change in reactivity. However, the reactivity of the coke of low-rank coal is higher than would be expected based on the additional surface area. This means that other chemical factors are involved. The results of our earlier research on the structural changes of micropores in different metamorphic cokes in CO_2 gasification [52] show that the average pore size of coke of low metamorphic degree (such as Pinglu coke) increases with increasing conversion rate when the conversion rate is higher than 0.4. The average pore diameters of coals with higher degrees of metamorphism (such as Datong bituminite coal, Dongshan coke coal, and Jincheng anthracite coal) increase with increasing conversion rate, and are not limited by the conversion rate being greater than 0.4. Our studies also show that the specific surface area of raw coal coke decreases linearly with increasing conversion rate, whereas the specific surface area of deashed coal coke first increases and then decreases, giving a saddle-shaped curve [53].

5.5.1.5 Relationship Between Coal Rank and Reactivity

Coke reactivity decreases with increasing coal rank. This is accepted by the majority of researchers [54, 55]. Studies of the gasification reactivities of different cokes and water vapor, air, CO_2, and H_2 show that the gasification reactivity order is lignite/brown coal > bituminous coal and bituminous coke > coke and pitch coke. Our long-term studies of various coal samples have also shown that the higher the coal metamorphic degree, the lower the reactivity, with both CO_2 and water vapor. However, some researchers have suggested that the effect of coal rank on reactivity is more completed.

Takarada et al. [56] analyzed the gasification reactivities of 34 different types of coal, ranging from peat to anthracite, and suggested that the gasification reactivities of low-rank coals are not always higher than those of high-rank coals. They used the concept of the reactivity index, R, to characterize the coke, defined as:

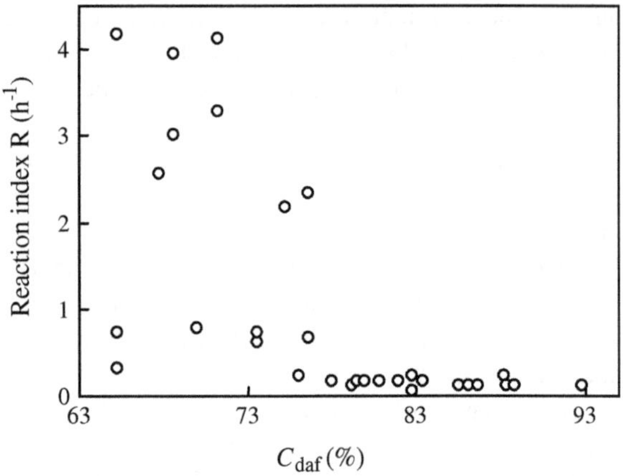

Fig. 5.32 Reactivity index changes with changes in coal content. Reprinted from Ref. [56], Copyright 1985, with permission from Elsevier

$$R = \frac{2}{\tau_{0.5}},$$

where $\tau_{0.5}$ is the time (h) it takes for the fixed carbon conversion rate to reach over 50 %.

The results of this research (Fig. 5.32) show that when the carbon content is greater than 78 %, R is small (less than 0.1 h^{-1}). When the carbon content is less than 78 %, the reactivity is much higher than that of high-rank coal, but the volatility is also significant. At this stage, the correlation between reactivity and carbon content is poor. Takarada et al. therefore suggested that the coke reactivity is related not only to the coal rank, but also to the coal oxygen functional groups and inorganic compound content of the coke.

Figure 5.33 shows the relationship between coal rank and reactivity. The figure shows that regardless of the coke preparation effects, the coke sample has the highest reactivity when the carbon content of the raw coal is 84 %. Raw coal with a

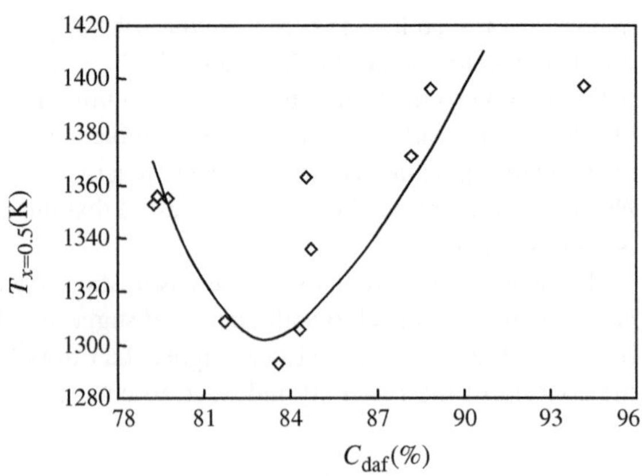

Fig. 5.33 Relationship between coke reactivity and raw coal carbon content

carbon content of 84 % has many special properties such as higher swelling and caking. Such raw coal has good polymerization of aromatics, more crosslinking bonds, and higher CO_2 gasification reactivity of its pyrolysis products, indicating that the coke reactivity is significantly affected by the raw coal structure.

5.5.2 Relationship Between Coke Crystal Structure and Gasification Reactivity

The crystal structure parameters of 14 different cokes were determined with the corresponding coal types, and preparation conditions of the cokes are listed in Table 5.4. In the XRD pattern, the (002) facet peak represents the degree of parallel orientation of the aromatic nuclei surface network, and the 100 facet peak represents the size of the surface network. The diffraction peak angle and half-peak width were determined according to the literature [57]. The *d* spacing of the coke crystal structural unit, the size of each layer, L_a, and the stacking height, L_c are listed in Table 5.15.

For French bituminous coal, the *d* and L_a values of the coke macerals and raw coal are similar. The L_c of inertinite is higher than those of the other coke samples, and that of the raw coal is similar to that of vitrinite. For Pingshuo gas coal, the L_c of inertinite is higher than that of vitrinite; this is similar to the case for French bituminous. The XRD patterns indicate that the crystal structures of different coke maceral samples generated under the same preparation conditions from the same coal are generally different from each other. The crystal size of inertinite is larger than that of vitrinite; the vitrinite crystal size is similar to that of raw coal.

Table 5.15 Crystal structure parameters

Code	Diffraction angle (rad)		Crystal structure parameters (Å)		
	(002) peak	(100) peak	*d*	L_c	L_a
DS	24.4	42.7	3.65	10.63	43.75
FF	22.4	44.2	3.97	10.62	29.24
FX	25.2	43.3	3.53	14.19	43.75
Fu	23.4	43.3	3.80	9.77	34.96
JC	24.9	42.9	3.58	12.15	38.78
PS	24.7	44.7	3.60	10.63	43.90
PSI	23.0	–	3.87	11.30	–
PSV	23.9	43.2	3.72	10.61	38.87
DT1	23.5	43.3	3.79	9.98	–
DT5	24.0	43.8	3.71	13.06	43.80
FranI	22.8	43.2	3.90	11.93	34.96
FranV	23.4	43.0	3.80	9.99	34.95
Fran1	23.4	43.5	3.80	9.98	35.15
Fran2	23.6	43.3	3.77	11.31	38.89

For the French bituminous coke samples Fran1 and Fran2, L_a of Fran2 is larger than that of Fran1, because of the longer coke preparation time for Fran2. For the Datong bituminite samples DT1 and DT5, the d and L_c values of DT1 are smaller than those of DT5, so the crystal structure depends on the coke preparation temperature and time. Generally, a higher preparation temperature and longer time generate larger crystals.

The gasification reaction starts from the edge of the coke surface, so only carbon atoms at the edge can act as active sites for gasification; those located in the bulk are inactive. The crystal structure of coke is similar to that of graphite; the crystal sizes obtained under different coke preparation conditions are different. The crystal structures of different maceral components from the same coal are also different. Generally, the crystal size of inertinite coke is larger than those of vitrinite and raw coal, which have similar sizes. The preparation process also affects the coke crystal structure: longer time and higher temperatures generate larger crystals. The crystal size affects the gasification reactivity; large crystals do not favor gasification.

If it is assumed that the coke crystal is a cylinder with three multi-aromatic rings arranged in parallel, the volume of the coke crystal, V_{mc}, can be estimated as

$$V_{mc} = \pi L_c \left(\frac{L_a}{2}\right)^2.$$

Figure 5.34 shows the relationship between crystal volume and gasification reactivity, based on the reactivity data in Table 5.5. The figure shows that the larger the crystal is, the higher the 50 % conversion temperature for gasification, and therefore the poorer the reactivity.

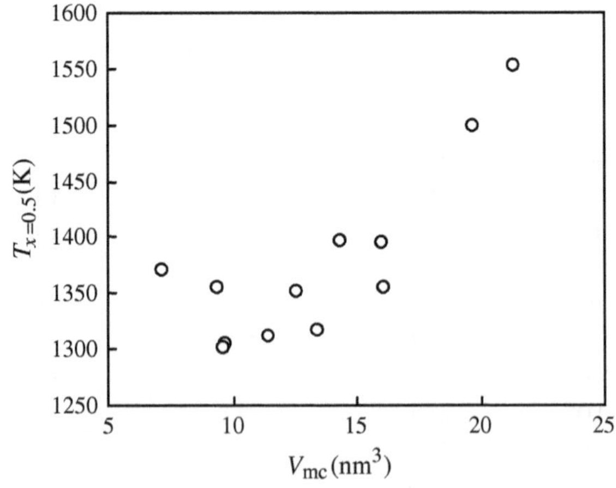

Fig. 5.34 Relationship between crystal volume and gasification reactivity

5.5.3 Relationship Between Coke Surface Properties and Gasification Reactivity

C–O complexes are formed by the combination of a negatively charged oxygen atom with a positively charged carbon atom. The active sites for the gasification are positively charged carbon atoms, which provide Lewis acid centers on the coke surface. As is well known, gasification of a complete six-carbon ring is difficult. The presence of oxygen results in the formation of oxygen-containing functional groups on the coke surface, causing defects in the crystal lattice, generating gasification sites. Oxygen therefore has a significant effect on active sites at the coke surface. In this section, the surface acid–base properties and crystal structures are investigated, using methods such as XRD and titration. Oxidation of the coke surface with strong nitric acid increases the number of oxygen-containing functional groups, enabling the effect of oxygen-containing groups on gasification reactivity to be studied. The relationship between coke surface properties and gasification reactivity is also discussed [6].

5.5.3.1 Acid–Base Properties of Coke Surfaces

Oxygen-containing functional groups make the coke surface acidic. Based on the principle of acid–base neutralization, the acidic properties of the coke surface can be measured by the adsorption of organic bases. The concentration of Lewis acid centers is indicated by benzidine adsorption. The adsorption of the base by the coke was calculated, to give the concentration of surface acid centers.

The benzidine adsorption results are listed in Table 5.16. The amounts adsorbed by the coke sample ranged from 0.6 to 0.11 mmol g^{-1}; the adsorption was affected by the coke preparation conditions. The adsorption by Fran1 was 1.5 times that by Fran2. Table 5.16 shows that the benzidine adsorption by the coke sample is closely

Table 5.16 Benzidine adsorption by coke samples

Coke code	Coke oxygen content (wt%)	Benzidine adsorption (mmol g^{-1})
DS	3.74	0.0772
FF	2.55	0.0619
FX	3.84	0.0751
Fu	2.72	0.0593
JC	3.35	0.0910
PS	2.86	0.0623
PSI	3.99	0.0800
PSV	4.36	0.1025
DT1	4.10	0.1097
Fran1	4.00	0.0914
Fran2	2.73	0.0604

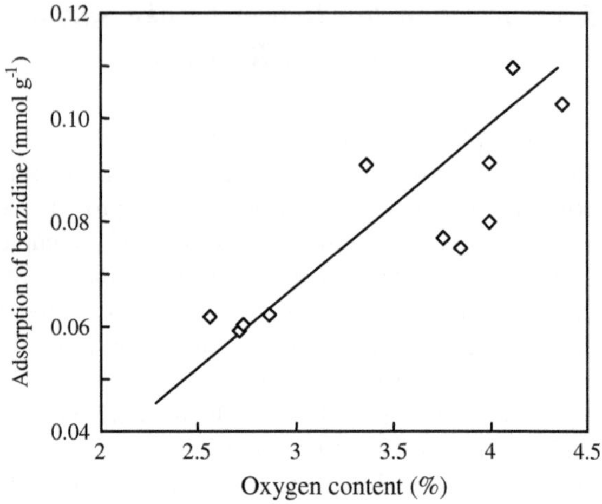

Fig. 5.35 Relationship between oxygen content and surface adsorption of benzidine

related to the oxygen content of the coke. Figure 5.35 shows that the higher the oxygen content of the coke sample (see Table 5.4), the more benzidine it adsorbs. This suggests that the presence of oxygen changes the charge distribution at the coke surface, and results in formation of Lewis acid centers.

5.5.3.2 Relationship Between Coke Surface Acid–Base Properties and Gasification Reactivity

The test sample was Fenxi fat coal coke (the basic data are shown in Table 5.3, and coke preparation conditions are listed in Table 5.4). An ash-free coke sample (<100 mesh) was prepared and mixed with strong nitric acid in a 1:10 volume ratio. The mixture was stored in the dark overnight. The sample was filtered and washed with distilled water until it was neutral. The sample was dried at 105 °C, giving the oxidized sample for the test.

The crystal structure parameters of Fenxi coke prepared using nitric acid are listed in Table 5.17. Preparation using nitric acid significantly increases the hydrogen and oxygen contents. After dipping in nitric acid for 70 h, the hydrogen content of coke sample FX-4 increased from 1.86 to 2.72 %, and the oxygen content increased from 3.84 to 5.47 %; the crystal size did not change significantly. This suggests that preparation using nitric acid mainly oxidizes the edge of the coke

Table 5.17 Ultimate analysis and crystal structure parameters of Fenxi coke samples

Coke sample	Ultimate analysis (daf, wt%)					Crystal structure parameters (Å)		
	C	*H*	*O*	*N*	*S*	*d*	L_c	L_a
FX	91.51	1.86	3.84	0.78	2.01	3.51	14.19	43.75
FX4	89.21	2.72	5.47	0.78	1.83	3.56	13.09	43.72

sample, and enriches the oxygen-containing groups at the surface, without changing the crystal size.

The data in Table 5.18 show the change in benzidine adsorption by the coke surface after nitric acid treatment. This represents the amount of surface Lewis acid centers after the treatment. The adsorption of benzidine by Fenxi raw coal is 0.0751 mmol g^{-1}; after treatment with nitric acid, the benzidine adsorption increases. After treatment for 5 h, the benzidine adsorption doubles, and it triples after treatment for 24 h. The increased base adsorption indicates that there are more Lewis acid centers on the surface.

A thermogravimetric study of the CO_2 gasification kinetics of treated Fenxi coke samples showed that the reactivities of the treated samples increased slightly (Table 5.19). The original coke sample had an activation energy of 169.7 kJ mol^{-1}; after treatment with nitric acid, the activation energy decreased by about 30–40 kJ mol^{-1}. The longer the nitric acid treatment time was, the more significant the decrease in the activation energy.

After nitric acid treatment, the surface properties of the Fenxi coke samples changed: the hydrogen and oxygen contents increased, but the crystal structure did not change significantly. Titration results showed that the ability to adsorb benzidine increased significantly after the treatment. These results indicate that a large amount of oxygen-containing functional groups are formed on the surface as a result of the treatment. The amount and strength of Lewis acid centers at the edge of the coke sample crystal increase.

CO_2 gasification tests on samples treated with nitric acid showed that the gasification activation energies were lower for the treated samples. This decrease in the activation energy is related to the increase in the number of oxygen-containing functional groups on the sample surface. The gasification activation energy is

Table 5.18 Benzidine adsorption by Fenxi coke samples before and after nitric acid treatment

Coke sample	Treatment	Benzidine adsorption (mmol g^{-1})
FX	Ash-free coke sample	0.0751
FX-1	Nitric acid, 5 h	0.1256
FX-2	Nitric acid, 24	0.2143
FX-3	Nitric acid, 50 h	0.2037
FX-4	Nitric acid, 70 h	0.1722

Table 5.19 CO_2 gasification kinetic parameters for samples treated with nitric acid

Coke sample	Activation energy (kJ mol^{-1})	Pre-exponential factor
FX	169.7	10.06
FX-1	136.0	7.31
FX-2	136.6	7.56
FX-3	127.4	6.35
FX-4	129.8	7.15

related to the energy required to form C–O complexes. Oxygen strongly attracts electrons, so the combination of oxygen-containing functional groups with the crystal surface can polarize carbon atoms. The positively charged carbon atoms can easily combine with negatively charged oxygen atoms in CO_2 molecules to form C–O complexes; this decreases the gasification activation energy, and is the reason for the decrease in the activation energy after nitric acid treatment. The change in the coke sample gasification kinetic properties caused by oxygen-containing functional groups is very similar to the effect of O^-Na^+ groups, but the change at the surface edge caused by O^-Na^+ groups is much more significant.

5.5.3.3 Relationship Between Gasification Reactivity and Coke Surface Structure in Coal Gasification

There are certain trends in the changes in coke reactivity in coal gasification. It has been reported that coke conversion affects the rate of gasification. Zhu et al. [11, 12] found that the changes in reactivity with respect to conversion give a convex curve; at around 50 % gasification conversion, the number of active sites on the coke and the specific surface area both reach maxima. Lizzio et al. [13] reported changes of 20 % in the gasification rate constant for 10–85 % conversion. After the conversion reaches 85 %, the reaction rate is much slower than the initial rate; this decrease in reactivity has a significant impact on the design and manufacture of gasification and combustion reactors. Sha et al. [58] proposed that this decrease in carbon reactivity is the result of loss of catalytic minerals. Sharma et al. [59] studied the changes in crystal structure during gasification, using high-resolution transmission electron microscopy; the results indicate that as the carbon conversion increases, the crystal structure of the residual coke changes, and the gasification reactivity for coke at 98 % conversion is not lower than the initial reactivity, suggesting that the coke crystal structure is not the only factor determining the coke reactivity.

Residual Coke Samples Preparation

Pinglu gas coal, Dongshan lean coal, and Xiqu coals were studied. The raw coal was first deashed to less than 1.5 % by acid washing, followed by coke preparation in a muffle furnace at 1173 K, to give a raw coke sample; the cokes were gasified with CO_2 in a fixed-bed reactor at 1123 K, to obtain residual coke. The effect of ash content was determined using Dongshan and Pinglu coals without acid washing.

Table 5.20 shows the ultimate analytical results for the gasification residual cokes. For the cokes, the carbon content is generally about 85–90 %, and the oxygen and sulfur contents are less than 5 %. The difference between the partially gasified and residual cokes is that the carbon content decreases; as the conversion increases, the carbon content tends to decrease, whereas the oxygen and sulfur contents increase significantly. The oxygen contents of Dongshan, Xiqu (XQ), and Pinglu (PL) cokes are about 5 %, whereas those of residual cokes at 33 %

Table 5.20 Ultimate analyses of residual coke samples (%, daf)

Sample code[a]	*C*	*H*	*N*	*O* + *S*
DS (2 %)	47.77	1.52	4.04	46.68
DS (21 %)	65.39	1.53	3.44	29.64
DS (30 %)	68.24	1.74	2.69	27.34
DS (40 %)	73.16	1.98	2.02	22.84
DS (67 %)	69.40	1.97	1.59	27.05
DS	91.36	1.51	1.42	5.71
PL (5 %)	71.78	2.59	4.80	20.82
PL (16 %)	69.57	2.24	3.63	24.56
PL (21 %)	65.45	2.32	3.33	28.90
PL (46 %)	71.96	2.87	2.57	22.60
PL	92.18	1.44	1.77	4.61
XQ (10 %)	63.32	1.75	3.05	31.88
XQ (30 %)	70.53	2.20	3.90	23.37
XQ (70 %)	73.08	1.93	2.54	22.45
XQ	94.01	2.05	2.30	1.64
DS[b] (15 %)	78.93	2.00	2.60	16.47
DS[b] (38 %)	84.82	1.78	1.55	11.86
DS[b] (85 %)	84.10	1.72	2.61	11.57
DS[b]	90.89	1.48	1.92	5.71
PL[b] (5 %)	67.89	2.12	5.13	24.86
PL[b] (63 %)	85.09	1.73	1.23	11.95
PL[b]	94.49	1.70	1.89	1.93

[a]Percentages in parentheses are residual weights during preparation. [b]Ash-containing samples

conversion are as high as 27 %. The increase in oxygen with increasing conversion implies that during gasification oxidation occurs at the surfaces of the coke samples, and the increases in the oxygen contents are probably the result of C–O complex formation during gasification.

Characterization of Partial Gasification Residual Coke

The XRD data for the cokes of the three ash-free coals are listed in Table 5.21. The data show that the distances between the crystal layers of the three residual cokes are roughly the same. As the conversion increases, *d* does not change significantly, but L_c tends to decrease. Xiqu coke has an L_c value of 18.3 Å; this changes to 12.1 Å when the conversion reaches 90 %. The L_c value of Dongshan coke is 16.7 Å, and changes to 15.4 Å when the conversion reaches 80 %; the L_c value of Pinglu coke is 14.2 Å, and changes to 13.5 Å when the conversion reaches 85 %. It can be concluded that the gasification reaction occurs at the outer edge of the coke crystal structure; therefore as the gasification conversion increases, the crystal structure is not damaged, with only a slight decrease in stacking height.

Table 5.21 XRD parameters of coke samples after partial gasification

Sample code[a]	(002) peak		(100) peak		Crystal structure (Å)		
	Angle (°)	Half-peak width (°)	Angle (°)	Half-peak width (°)	L_c	L_a	d
XQ (10 %)	25.1	7.1	–	–	12.1	–	3.6
XQ (30 %)	25.4	5.7	–	–	14.9	–	3.5
XQ (70 %)	25.4	5.1	42.6	5.1	16.7	34.2	3.5
XQ	25.3	4.6	43.5	4.5	18.3	38.9	3.5
DS (21 %)	25.5	5.5	–	–	15.4	–	3.5
DS (30 %)	25.3	5.5	42.7	4.9	15.3	35.3	3.5
DS (40 %)	25.5	5.7	42.9	5.2	14.9	33.6	3.5
DS (67 %)	25.4	5.1	43.1	5.3	16.7	33.0	3.5
DS	25.1	5.1	43.3	5.2	16.7	33.6	3.5
PL (5 %)	25.5	5.5	–	–	15.5	–	3.6
PL (16 %)	25.1	6.3	43.8	5.4	13.5	32.5	3.6
PL (21 %)	25.8	6.5	43.6	5.7	13.2	30.7	3.5
PL (46 %)	24.7	5.8	44.0	5.8	14.5	30.2	3.6
PL	24.8	6.0	44.1	5.3	14.2	33.1	3.6

[a]Percentages in parentheses are residual weights during preparation (wt%)

In the FTIR spectra peaks were observed at 3450, 2830–2950, 1580, 1720, and 1100 cm^{-1}. The peak at 3450 cm^{-1} arises from hydroxyl vibrations, that at 2830–2950 cm^{-1} from aliphatic vibrations, that at 1580 cm^{-1} from aromatic C–C vibrations, that at 1720 cm^{-1} from C=O vibrations, and that at 1100 cm^{-1} from C–O stretching vibrations. All the peak areas are listed in Table 5.22. The 1600 cm^{-1} peaks for Dongshan and Pinglu cokes are very weak, but are stronger for the gasification residual cokes. Convolution of the residual coke peaks at 1500–1750 cm^{-1} shows that during gasification, the 1700 cm^{-1} peak is strengthened, suggesting the formation of carbonyl oxygen-containing groups on the coke surface. It is the formation of such groups in the gasification residual coke that makes the oxygen content higher than that of coke. This result further shows that oxygen is adsorbed by the coke to form C–O complexes during gasification.

Table 5.22 Peak areas of raw coke and partial gasification residual cokes for three coals[a]

Group and wavenumber (cm^{-1})		PL (5 %)	PL	DS (21 %)	DS	XQ (10 %)	XQ
–OH	3450	523	383	237	333	469	773
C–H	2830–2950	39	27	–	–	6	–
C=O	1720	32	–	104	–	42	–
C–C	1580–1600	394	171	982	158	558	322
C=O	1000–1400	1681	1842	4337	1836	2269	921

[a]Percentages in parentheses are residual weights during preparation (wt%)

Gasification Reactivity of Residual Coke

The reactivity of gasification residual coke with CO_2 was measured and the activation energy and pre-exponential factor were calculated from the TG curve. Figures 5.36 and 5.37 show the thermogravimetric curves for the CO_2 gasification of Dongshan, Pinglu, and Xiqu ash-free coke samples. The curves show that the reaction rates of the gasification residual cokes prepared under different conversion conditions are different. Table 5.23 shows the calculation results for all the parameters of all the coke samples. The results show that as the conversion increases, the reactivity of the residual coke increases. It can also be seen by comparing the gasification kinetic parameters that the activation energies of these residual cokes are very similar to that of raw coke, but the pre-exponential factors are larger. For the ash-containing coke samples, the presence of ash makes calculation of the kinetic parameters and analysis of the crystal structures difficult, but the thermogravimetric analysis show that the reactivity increases as the conversion increases during gasification, regardless of whether the samples contain ash.

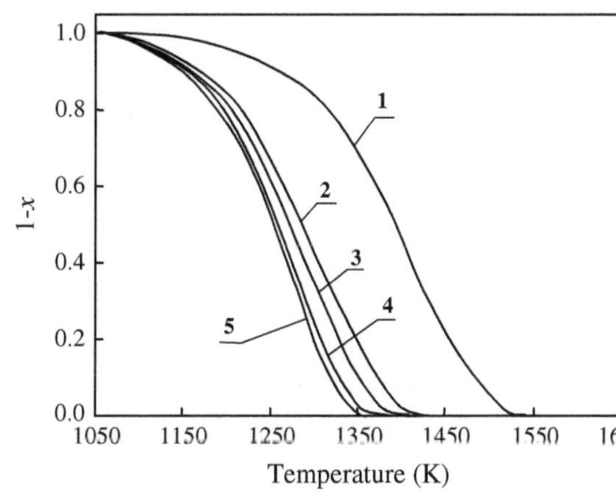

Fig. 5.36 Thermogravimetric analysis *curves* for CO_2 gasification of residual cokes of acid-washed dongshan raw coke: *1* DS; *2* DS (67 %); *3* DS (40 %); *4* DS (30 %); and *5* DS (21 %)

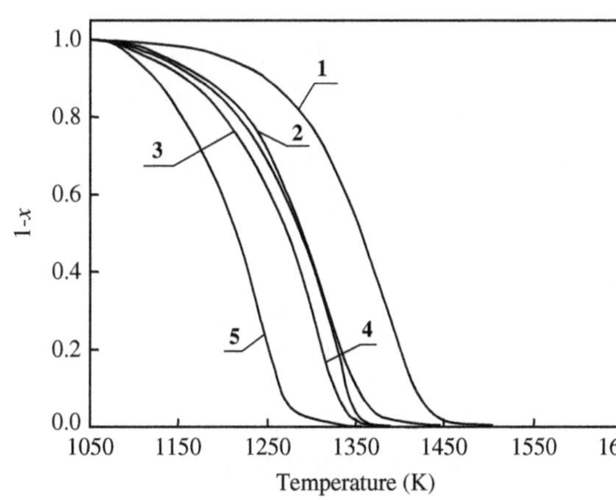

Fig. 5.37 Thermogravimetric analysis *curves* for CO_2 gasification of residual cokes of acid-washed Pinglu raw coke: *1* PL; *2* PL (46 %); *3* PL (21 %); *4* PL (16 %); and *5* PL (5 %)

Table 5.23 Kinetic parameters for CO_2 gasification of coke and residual cokes of acid-washed Dongshan, Pinglu, and Xiqu cokes

Sample code[a]	Activation energy (kJ mol^{-1})	Pre-exponential factor	$T_{x=0.5}$ (K)
DS (21 %)	166.8	13.1	1260
DS (30 %)	167.8	13.3	1255
DS (40 %)	158.7	12.0	1275
DS (67 %)	153.2	11.5	1286
DS	158.9	10.7	1396
PL (5 %)	219.6	19.2	1221
PL (10 %)	195.7	15.5	1290
PL (16 %)	217.5	17.8	1278
PL (21 %)	197.9	15.7	1291
PL (46 %)	220.3	17.8	1291
PL	215.5	16.4	1358
XQ (10 %)	157.0	11.5	1261
XQ (30 %)	144.1	10.2	1305
XQ (70 %)	138.7	9.2	1358
XQ	143.1	9.7	1456

[a]Percentages in parentheses are residual weights during preparation (wt%)

The XRD results show that as the conversion of the residual cokes increases, there are no significant changes in the crystal surface structure, and the structures of the raw coke and the coke at 80 % conversion are very similar. However, as the coke conversion increases, the (002) and (100) peaks tend to become weaker, suggesting a decrease in the number of crystalline particles as the conversion increases. Coke gasification occurs from the external surface to the internal area. Each individual coke particle consists of randomly arranged crystalline particles. During gasification, the reactions of the crystals located at the surfaces of the coke particles start first; those located inside have difficulty reacting, because of mass transfer effects. It is only after the outer crystalline particles have reacted that the inner structure is able to react. Therefore, although the coke particles shrink during gasification, the XRD results show that the crystal structure is retained.

Ultimate analysis shows that after partial gasification, the amount of oxygen in the coke sample is significantly higher. FTIR analysis shows that there is no peak at 2350 cm^{-1}, but the carbonyl group peak at 1700 cm^{-1} is strengthened significantly. This indicates that the increase in the oxygen content is not caused by chemical adsorption of CO_2 at the coke surface, but by the formation of C–O complexes during gasification. The formation of an oxide layer on the crystal surface during gasification changes the surface properties and increases the reactivity. Thermogravimetric analysis of gasification residual cokes shows that the higher the conversion of the residual coke, the higher the reactivity.

Figures 5.38 and 5.39 show the thermogravimetric analysis curves for acid-washed and ash-containing partial gasification residual cokes (5 %). The

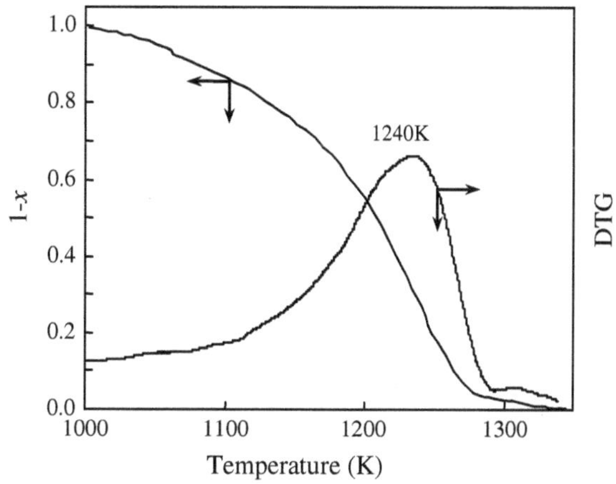

Fig. 5.38 Thermogravimetric analysis and DTG curves for CO_2 gasification of raw coke and partial gasification residual coke (5 %) of acid-washed Pinglu coal

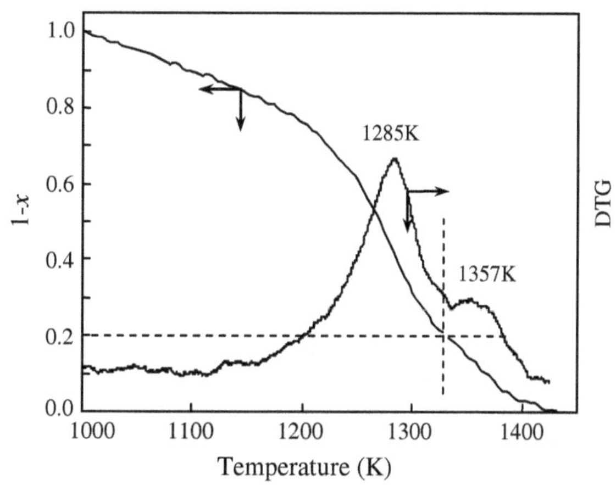

Fig. 5.39 Thermogravimetric analysis and DTG *curves* for CO_2 gasification of raw coke and partial gasification residual coke (5 %) of ash-containing Pinglu coal

gasification of the acid-washed coke occurs at 1000–1400 K, and a peak appears in this temperature range in its DTG curve, with the maximum at 1240 K. Two peaks appear in the DTG curve of the ash-containing coke, at 1285 and 1357 K respectively. The thermogravimetric analysis curves show that under the same conversion conditions, the ash-containing sample has a lower gasification rate than the acid-washed one, and at 1340 K, the ash-containing sample shows a significantly slower weight loss after the conversion reaches 80 %. This indicates that the reactivity of the residual coke decreases after the conversion reaches 80 %.

Catalysis by the ash content was discussed in previous sections, but when the ash content is high, it can also inhibit gasification, because of the effect of mass transfer. The ash content of Pinglu coke is about 20 %, and it reaches 83–85 % when the amount of coke remaining is 5 %. Under such conditions, a large quantity of ash is present outside the unreacted carbon; this affects the mass transfer in the gasification reaction, and decreases the reaction rate. Figures 5.38 and 5.39 demonstrate these points. For the same gasification residual coke prepared under the same conversion conditions from the same raw coke, the ash-containing coke has a lower reaction rate; this phenomenon is not observed for acid-washed samples. This indicates that the decrease in the gasification rate after 80 % conversion for the partial gasification residual cokes is caused by the ash content.

References

1. Furimsky E (1985) Fuel Process Technol 11(2):167
2. Zhao CS (2010) Fuel Process Technol 91(8):805
3. Thamavithya M et al (2008) Fuel Process Technol 89(10):949
4. Lee SH et al (2010) Energy 35(8):3239
5. Huo W et al (2014) Bioresour Technol 159:143
6. Duman G et al (2014) Fuel Process Technol 118:75
7. Nilsson S et al (2013) Fuel 105:764
8. Montoya A et al (2002) Fuel Process Technol 77–78:125
9. Oboirien BO et al (2011) Fuel Process Technol 92(4):735
10. Nozaki T et al (1990) Fuel Process Technol 24:277
11. Zhu Z et al (1994) Fuel Chem J 22(3):321
12. Zhu Z et al (1992) Chin J Chem Eng 43(4):401
13. Lizzio AA et al (1990) Carbon 28:7
14. Freund H (1986) Fuel 65:63
15. Freund H (1985) Fuel 64:657
16. Huttinger K (1990) Carbon 28(4):453
17. Fritz OW et al (1993) Carbon 31:923
18. Zhang ZG et al (1988) Energy Fuel 2:679
19. Zhang ZG et al (1996) Energy Fuel 10:169
20. Smith IW (1978) Fuel 57:409
21. Adschiri T et al (1986) Fuel 65:1688
22. Adschiri T et al (1991) AIChE J 37:897
23. Adschiri T et al (1986) Fuel 65:927
24. Laine NR et al (1963) J Phys Chem 67:2030
25. Peter C et al (1985) Fuel 64:1447
26. Nozaki T et al (1990) Fuel Process Technol 24:277
27. Suzuki T et al (1989) Fuel 68:626
28. Suzuki T et al (1992) Energy Fuel 6:343
29. Zhang F et al (2015) Fuel Process Technol 130:107–116
30. Xu SQ et al (2011) Fuel 90(5):1723
31. Chen SG et al (1997) Energy Fuel 11:421
32. Chen SG et al (1993) J Catal 141:102
33. Ohtsuka Y et al (1996) Energy Fuel 10:431
34. Ohtsuka Y et al (1986) Fuel 65:1653
35. Varhegyi C et al (1996) Energy Fuel 10:1208
36. Yoon H et al (1978) AIChE J 24:885
37. Leppalahti J et al (1995) Fuel Process Technol 43:1
38. Li S et al (1995) Fuel 74:456
39. Zsako JJ (1996) Therm Anal 47:1679
40. Cuesta A et al (1993) Energy Fuel 7:1141
41. Essenhigh RH et al (1990) Energy Fuel 4:171
42. Li SF et al (1994) Fuel 73:413

43. Lee WJ et al (1995) Fuel 74:1387
44. Franciszek C et al (1991) Fuel Process Technol 29:57
45. Shang JY et al (1984) Fuel 63:1604
46. Muhlen HJ et al (1985) Fuel 64:944
47. Echterhoff H et al (1961) Coke Oven Raw Mater Conf 20:403
48. Burgert B et al (2011) Fuel Process Technol 92(10):2048
49. Matsumoto S et al (1986) Carbon 24:277
50. Gan H et al (1972) Fuel 51(4):272
51. Elliott MA et al (1991) Chemistry of coal utilization. Chemical Industry Press, Beijing
52. Xie KC et al (1990) Gasification kinetics and minerals role. Shanxi Science & Education Press, Taiyuan, p 293
53. Xie K C et al (1990) Gasification Kinetics and Minerals Role. Shanxi Science & Education Press, Taiyuan, p 303
54. Radovic LR et al (1983) Fuel 62:849
55. van Heek KH et al (1985) Fuel 64:1405
56. Takarada T et al (1985) Fuel 64:1438
57. Fan X (1996) Metal X-ray analysis. China Machine Press, Beijing, p 55
58. Sha XZ et al (1990) Fuel 69:1564
59. Sharma A et al (1999) 10th ICCS 371

Chapter 6
Coal Depolymerization and Liquefaction

Abstract Coal is generally an insoluble material which is an obstacle to solution based chemical analysis techniques. It can be useful to consider coal as a complex polymeric substance, composed of a wide variety of monomeric units with a disorganized structure that resists dissolution. Depolymerization and liquefaction studies concern processes that break down these insoluble polymers into soluble constituents. This can be achieved indirectly via gasification, however direct methods involving solvent extraction liquefaction, and catalytic liquefaction can give a deep insight into the chemical structure and reactivity of the precursor coal that is used. In this chapter we discuss the various conditions such as temperature solvent and catalysts that are used for liquefaction of coal and their applications to the study of different Chinese coal types. We introduce the mechanisms that contribute to specific composition of soluble fractions, and pay particular attention to the use of infrared spectroscopy in the study of these components.

6.1 Introduction to Coal Depolymerization and Liquefaction

The composition of coal is complex and diverse, and depends on the metamorphic environment, and the degree of metamorphism and coalification. In particular, its high degrees of polymerization and condensation (it is difficult to dissolve coal without any treatment, even in strongly polar solvents) make it difficult to study coal structure and reactivity. However, coal can undergo catalytic reactions under certain conditions, and data on coal structure and reactivity can be obtained from these chemical reactions. Information on coal reactivity can be obtained by studying chemical reaction mechanisms, and structural information can be obtained by studying the types of groups and chemical bonds present, and the low-molecular-weight compounds dissolved in the organic solvent after coal depolymerization. Based on this theory, coal depolymerization and liquefaction are discussed in this chapter.

K.-C. Xie, *Structure and Reactivity of Coal*, DOI 10.1007/978-3-662-47337-5_6

6.1.1 Coal Depolymerization and Liquefaction Methods

Coal depolymerization and liquefaction are categorized as indirect liquefaction, solvent extraction liquefaction, and catalytic liquefaction. The latter two are direct liquefaction methods. Biocoal liquefaction is not considered as an independent system because of its low conversion rate.

6.1.1.1 Indirect Coal Liquefaction

Coal is gasified to syngas, which has CO and H_2 as its main components, during indirect liquefaction. Syngas with a certain CO/H_2 ratio is catalytically converted to liquid fuel after desulfurization and denitrogenation, e.g., by Fischer–Tropsch (F–T) synthesis:

$$nCO + 2nH_2 \rightarrow -(CH_2)_n - + nH_2O$$
$$nCO + nH_2O \rightarrow nH_2 + nCO_2$$
$$2nCO + nH_2 \rightarrow -(CH_2)_n - + nCO_2$$

Another method of indirect coal liquefaction is the methanol-to-gasoline (MTG) process. In MTG, coal is first gasified to syngas. The syngas is then converted to methanol, from which gasoline is synthesized.

In recent years, other processes have been in the R&D or demonstration stages, such as TIGAS (Topsoe integrated gasoline synthesis), AMSG (Asia-Mitsubishi syngas gasoline), SMDS (Shell middle distillate synthesis), and MFT (modified F–T). These processes are two-stage improved F–T syntheses.

6.1.1.2 Direct Liquefaction

Solvent extraction liquefaction involves coal liquefaction with hydrogen donor extraction by a solvent containing aromatic compounds or another solvent. Hydrogenated aromatic compounds easily donate hydrogen to the coal. Solvent extract liquefaction includes the solvent refinery process, extraction with a hydrogen-donating solvent, liquid solvent extraction, and supercritical gas extraction.

Catalytic liquefaction consists of coal liquefaction by catalytic hydrogen donation. This liquefaction process is usually performed in a fixed-bed reactor. Catalytic liquefaction includes methods such as the hydrogen coal process (H-Coal) etc.

Methods for hydro-coprocessing coal and heavy oil, bitumen, residual oil, and tar include the CANMET single-stage method, HRI two-stage method, ARC and CCLC two-stage methods, and PYROSOL three-stage method [1].

6.1.2 Low-Temperature Coal Liquefaction Depolymerization Reactions

Coal liquefaction and catalyst behavior are relatively easy to control at low reaction temperatures, i.e., below 300 °C. The principle of low-temperature coal liquefaction is the catalytic cleavage of C–C, C–O, C–N, and C–S crosslinking bonds in large coal molecules, to minimize secondary reactions and avoid changes in the structure of the macromolecular skeleton. Several low-temperature coal liquefaction processes have been studied in depth: coal reduction in 1,2-ethylenediamine lithium, electrochemical reduction liquefaction, reductive alkylation, depolymerization and liquefaction in BF_3–phenol solution, transfer of alkylbenzenesulfonic or benzenesulfonic acid as a coal liquefaction catalyst, Friedel–Crafts alkylation and acylation, hydrotreatment with a metal halide as a catalyst, coal depolymerization with a strong base such as NaOH or KOH, and reaction of coal in a CO–H_2O system.

6.1.2.1 Mechanism of Low-Temperature Depolymerization and Liquefaction

The amount of pyridine-soluble material in products increases and the sulfur content decreases significantly after extraction of coal with 1,2-ethylenediamine lithium. This is mainly the result of the breaking of aromatic rings, ether bond cleavage, and reduction of carbonyl to alcohol. When a high-volatile bituminous material is dissolved in an amine, after chemical or electrochemical hydrogenation, up to 30 % of the product is soluble in benzene and up to 78 % is soluble in pyridine. The reaction mechanism indicates that the aromatic compound is reduced to an alicyclic compound in an amine solution in two steps: solvated electron transfer to the substrate to form anionic groups, which is followed by proton addition to the anionic groups to form radicals. Electrons and protons are continuously generated to form the final products.

The alkylation products have high solubilities in benzene and pyridine. A model compound study showed that in alkali-metal-catalyzed alkylation in tetrahydrofuran (THF), electrons are transferred form the alkali metal, forming aromatic anions, and then the aromatic anions are halogenated. This process is accompanied by bond breaking, such as the fracture of aldehyde groups and ether bond cleavage. The anion alkylation reactions of halogenated alkanes and anthracene are similar, and can be described as [2, 3]:

$$(\mathrm{Coal})^{n-} + n\mathrm{RX} \rightarrow n\mathrm{R} - (\mathrm{Coal}) + n\mathrm{X}^-$$

where

n is the negative charge on the coal anion;
R is an alkyl group;
X is halogen.

In the depolymerization of high-volatile bituminous coal in a BF_3–phenol system, the product solubility in phenol increases from 19 % to 60 %. The main reason for coal degradation is bond breaking between aliphatic carbons and aromatic carbons (C_{al}–C_{ar}), and transformation of phenol. Phenol is also found embedded in coal derivatives [4, 5].

Recently, a new direct coal liquefaction process that uses no catalyst was shown to operate at temperatures below 360 °C [6–9]. This method, allowed bituminous coal to be sequentially separated into several fractions of compounds with different molecular weights without decomposing the coal. The process used a flowing steam of solvents, including tetralin, 1-methylnaphthalene, coal-derived oil, and water. These results showed that up to 80 % of the organic matter in coal could be separated into a soluble fraction that could be used for fuel, [10–14] fine chemicals, [15] or raw material for further production [6].

6.1.2.2 Metal Halide Catalytic Coal Depolymerization and Liquefaction

In coal liquefaction, the major chemical reactions in the presence of hydrogen are hydrogenolysis of C–C, C–O, C–N, and C–S bonds, and hydrogenation of aromatic rings; these reactions convert coal to smaller-molecule liquid products. In high-temperature liquefaction, free radicals are formed by bond homolysis. Hydrogen addition to these free radicals results in further cracking to small molecules.

Molten metal halides have been widely used in coal liquefaction. In the early 1970s, Shell developed a molten zinc iodide-catalyzed coal and coal extract hydrocracking process. This technique converts coal and coal extracts to low-sulfur fuel oil and gasoline at 300–500 °C and 3.447–10.34 MPa [16].

A mixture of $ZnCl_2$ and $SnCl_2$ is an effective catalyst for coal hydrotreatment with a short residence time. Bodily et al. [17, 18] studied impregnated metal halide pyrolysis of coal samples. The study shows that metal halides enhance the production of volatile compounds, leading to an increase in the number of micropores.

Tanner et al. [19] studied the effects of Lewis acid catalysts ($AlCl_3$, $FeCl_3$, $SbCl_3$, $HgCl_2$, $ZnCl_2$, and $SnCl_2$) on the liquid products. Experiments were performed below 400 °C to limit the extent of the pyrolysis reaction. This study also indicated that even in a non-hydrogen atmosphere, $ZnCl_2$ can increase the solubility of coal. It also shows the relationship between liquefaction efficiency and Lewis acid catalyst acidity. With a moderately acidic $ZnCl_2$ catalyst, the yield of soluble products is 46.8 %, whereas the yield is only 20.3 % with a strongly acid catalyst such as of $AlCl_3$, as a result of agglomeration of aromatic compounds. The effects of the acidities of various metal halides have been investigated, e.g., the Lewis acidity of Fe Cl_3 is greater than that of $ZnCl_2$ but the catalytic activity of $FeCl_3$ is lower than that of $ZnCl_2$.

Oblad et al. [20] studied the catalytic hydrogenation liquefaction of New Mexico coal with $ZnCl_2$ below 375 °C in an autoclave. They examined the effects of the $ZnCl_2$ content (0.5–50 wt%), hydrogenation treatment temperature (200–375 °C), hydrogenation treatment pressure (1.86–13.79 MPa), and reaction time on the product distributions of oil (cyclohexane soluble), asphaltene (benzene soluble, cyclohexane insoluble), and asphaltenes (THF soluble, benzene insoluble). The study showed that THF-soluble product yields of 82–88 % were obtained with 20 % $ZnCl_2$ as the catalyst under the conditions reaction temperature 350–375 °C, reaction pressure 13.79 MPa, and reaction time 1 h. In the range 200–375 °C, the product molecular weight decreased and the oil fraction yield increased with increasing temperature. The asphaltene fraction yield decreased from 92 % (320 °C) to 25 % (375 °C). The compositions of the oil, asphaltene, and asphaltenes fractions obtained at 315 °C were analyzed. The average molecular weights of the three fractions are functions of the hydrogenation reaction temperature: in the range 315–375 °C, the average molecular weight of the asphaltene fraction dropped from 1217 to 938, that of asphaltenes from 728 to 653, and that of the oil fraction was unchanged at approximately 430.

Mobley et al. [21] studied the effects of $ZnCl_2$ on model compounds with ether bonds. They found that in the range 136–305 °C, unsaturated cyclic ethers, furans, and diaryl ethers were stable, whereas saturated cyclic ethers such as THF aralkyl ethers were cracked.

Some reaction took place at 136 °C, and the reaction intensified with increasing reaction temperature. However, the product distribution was unchanged. Based on these results, it can be concluded that it is difficult to break Ar–O–Ar bonds in $ZnCl_2$-catalyzed coal liquefaction. The results also show that $ZnCl_2$ can be used as a catalyst for removing sulfur from dibenzyl thioethers and diphenyl sulfides. The effects of Lewis catalysts on diarylalkanes such as Ar–(CH_2)–Ar and Ar–Ar were studied using a high-pressure autoclave.

We found that in the range 400–350 °C, diarylalkanes with methylene groups between two aromatic rings, such as diphenylmethane and 1-benzenemethylnaphthalene, undergo catalytic hydrogenation dealkylation to form products containing aromatic and methyl groups. This shows that a proton acid species, $ZnCl_2H^+(ZnCl_2OH)^-$, participates in aromatic ring protonation, followed by dealkylation through a Friedel–Crafts reaction.

Jensen et al. [22] studied the removal of aromatic rings from alkanes by $ZnCl_2$ impregnated on silica gel. It has been found that the hydrocarbon bonds in benzene broke at 200–300 °C. However, the aromatic ring in the dealkylation reaction becames dominant above 300 °C. When the aromatic hydrocarbon was a triple-bonded alkyne (*sp*^{1} hybridized), the removal is difficult, but removal was easy when the substituted alkane was branched.

Li et al. [23] investigated the dissolution behavior of low-rank coal in *N*-methyl-2-pyrrolidinone (NMP) with additional LiCl at temperatures up to 300 °C. Their results showed that the $[Li{\cdot}n\text{NMP}]^+Cl^-$ complex that formed increased the dissolution yields by weakening hydrogen bonds between carboxyl groups and decreasing the aggregation of the coal, because of the high basicity of Cl^-.

Takahashi et al. [24, 25] also reported that addition of lithium halides (LiCl, LiBr, LiI) increased CS_2/NMP extraction yields for several bituminous coals at room temperature. They attributed these increased yields to acid–base interactions between acidic coal sites and halogen anions and/or breaking of π-cation interactions in coal.

6.1.2.3 Base-Catalyzed Coal Depolymerization and Liquefaction

In the study of low-temperature coal depolymerization and liquefaction, it has been found that a strong alkali can catalytically crack ether and resin bonds in the coal macromolecular structure at 200–300 °C. Makabe et al. [26, 27] studied the effects of the reaction conditions on the alkali-catalyzed depolymerization of different ranks of vitrinite. In the range 200–300 °C, the extraction rates of products into pyridine and ethanol increased with increasing reaction temperature. At a reaction temperature of 300 °C, for coal with a carbon content less than 83 %, the reaction product was almost completely dissolved in pyridine after 1 h. Molecular-weight loss is mainly caused by ether bond cleavage. Taiheyo coal (C 77.9 %) depolymerization in ethanol was studied at 380–450 °C. The results show that the yields of pyridine-soluble and benzene-soluble products were 95 and 85 %, respectively, after 1 h at 400 °C.

A comparison of the products of coal–ethanol and coal–ethanol–NaOH systems showed that the average molecular weight of the pyridine-soluble products (750) was larger than that of the benzene-soluble products (580). Analysis showed that the main gaseous products of coal–ethanol–NaOH were H_2, C_2H_4, and C_{1-4} alkanes.

Ouchi [28] studied the effects of different alcohols as solvents on coal depolymerization and liquefaction at 290 °C for a reaction time of 1 h with NaOH as the catalyst, and a coal:catalyst:solution ratio of 1:1:10. The solvents employed were methanol, ethanol, *n*-propanol, isopropanol, *n*-butanol, isobutanol, *tert*-butyl alcohol, *n*-amyl alcohol, and dimethyl alcohols.

The results showed that the yields of pyridine-soluble products were 95.1 and 96.4 %, with methanol and ethanol as the solvent respectively. With isopropyl alcohol and butanol as solvents, the yields of pyridine-soluble compounds were more than 95 %. Elemental analysis showed that the H/C values of all the extracts were larger than that of raw coal. Infrared (IR) spectroscopy indicated that ether bonds were cleaved and some aromatic compounds were hydrogenated in coal–alcohol–alkali systems. Based on these experiments, it was concluded that hydrogen transfer occurred from the solvent to the coal. In a similar experiment, Lu et al. [29] reported on the degradation of lignite through the dissociation of intermolecular interactions followed by esterification at low temperatures, and then alkanolysis at higher temperatures.

Base-catalyzed depolymerization is different for coals with different metamorphic degrees. The study by Ollad showed that young coal with a low carbon content (C < 83 %) dissolves in methanol– and ethanol–NaOH systems, and the pyridine and benzene extraction rates are higher than 95 and 65 %, respectively. The extraction rates are lower for coals with higher metamorphic degrees.

Anderson et al. [30] determined the statistical average molecular structures of the products of Clear Creek (C 79.3 %) coal in ethanol–NaOH. The reactions were performed at 300 and 320 °C for 100 min. The average molecular weight of the depolymerization products extracted with pyridine at 320 °C was 540, whereas at 300 °C, it was 908. The average numbers of rings were 2.1 and 2.6, the average numbers of aromatic rings were 1.5 and 2.3, and the molecular weights of the average structural units were 24.6 and 327, at 320 and 300 °C, respectively.

CaO-catalyzed methanolysis of lignite has been shown to be a feasible liquefaction method. This has advantages in terms of higher liquefied product yields and higher *n*-hexane soluble fraction and gas yields compared with those of liquefaction in a H_2 atmosphere [31, 32]. This is explained by methanol and CaO promoting the pyrolysis and producing active H which stabilizes free radical fragments from the pyrolysis of the lignite.

Sakbut et al. [33] studied the conversion of low metamorphic grade coal in alcohol–alkali systems. They found that the carboxyl group, COOH, which is an electron acceptor, accelerates the hydrogenation reaction via an electronic transmission mechanism.

6.1.2.4 Effects of Reaction Time on Coal Depolymerization and Liquefaction

Generally, coal is prepared as a slurry with hydrogenated aromatic compounds (hydrogen donor solvents) before coal depolymerization or liquefaction. The slurry is then heated to the reaction temperature under a hydrogen atmosphere. Short reaction times are used in studies of the coal liquefaction mechanism to obtain the initial depolymerization products and avoid mixing of the products with secondary reaction products. In these studies, the reaction time is generally no more than 5 min because further or secondary reactions take place at such high temperatures and pressures, which will cause problems in mechanistic studies.

The initial depolymerization products from coal can be obtained by rapid heating. However, the results show that the molecular weights of these initial products are very high because they contain a large amount of functional groups as well as aromatic nuclei. These functional groups are present in the multidimensional carbon network and they are not affected by fast pyrolysis. In some reactions, even if the reaction time is very short (<3 min), the yield of pyridine-soluble products is very high [34].

Neavel [35] studied short-term reactions and the results showed that the coal decomposition rate is very fast at 400 °C. However, at this temperature, a hydrogen-donating solvent cannot immediately donate hydrogen atoms to form free radicals, which can stabilize coal pyrolysis. The early stage of coal liquefaction is therefore similar to coal pyrolysis. Wiser [36, 37] has already identified this problem. He suggested that free radicals come from hydrogenated aromatics in the coal itself. It is difficult to mix solid coal particles and the hydrogen-donating solvent fully in the early stages of the reaction; they mix well when the coal is melted and degraded.

The reactions in the fast pyrolysis liquefaction of model compounds with different crosslinks have been studied. Benjamin et al. [38] found that alkylbenzenes, benzene, aromatic compounds, dibenzofurans, dibenzothiophenes, polyphenol ethers, and thioethers have high thermal stabilities. Short alkyl bridges, i.e., $-(CH_2)_n-$, and diaryl ethers, i.e., $C_{ar}–O–C_{ar}$, both occur as connections between aromatic nuclei. They are thermally and chemically stable under non-catalytic conditions. The thermal stabilities of these compounds decrease if there is a phenol substituent in the compound.

It has been found that alcohols act as solvents in alkali–alcohol systems during coal depolymerization and liquefaction. The alkali mainly catalyzes catalytic cracking of ethers and esters. Partial depolymerization can occur with alkali catalysts in various alcohols.

In summary, methanol acts as a weak hydrogen donor solvent under certain reaction conditions. The transferrable hydrogens in the aromatic rings in coal are the main hydrogen donor source in the initial stage of coal depolymerization and liquefaction. This stabilizes coal pyrolysis and limits free-radical polymerization reactions. When the reaction temperature reaches 400 °C, the coal decomposition rate is very fast. The initial cracking is not selective, and secondary reactions are accelerated, resulting in high yields of heavy oils and low yields of light oils. Hydrogen donor solvents cannot immediately donate hydrogen at this temperature. The early stage of liquefaction is therefore similar to coal pyrolysis. In the initial liquefaction stage at low reaction temperatures (<350 °C), the coal particles are inert. Depolymerization does not occur and it is difficult to mix the solid coal particles and hydrogen-donating solvent well. Coal depolymerization and liquefaction are improved if coal pyrolysis occurs with a catalyst at low temperature (<300 °C).

6.2 Studies of Depolymerization and Liquefaction of Shenfu Coal

The amounts of ether-, methanol-, and THF-soluble compounds increase significantly in coal depolymerization in alkali–methanol systems. With BC II (a basic catalyst) as the catalyst, methanol itself is involved in the reaction, providing $-CH_3$, –OH, and $-OCH_3$ for participation in the reaction. Coal depolymerization takes place in this system and new chemical reactions occur. In this section, a detailed study of the product distribution in Shenfu coal depolymerization and liquefaction is presented.

Analytical data for the Shenfu coal used in the experiments are shown in Table 6.1.

Pentane, diethyl ether, methanol, and THF were selected as extractants for Shenfu coal extraction. The extraction was performed at the boiling point of the extractant to avoid reactions. Depolymerization and liquefaction of Shenfu Coal

Table 6.1 Analytical data (%) for Shenfu coal

Coal	Proximate analysis				Ultimate analysis				
	V_{daf}	A_d	M_{ad}	FC_{daf}	C_{daf}	H_{daf}	N_{daf}	$S_{t,d}$	O_{daf}
Shenfu	40.64	2.21	2.39	59.35	80.14	5.52	1.83	0.22	12.29

were performed in a supercritical methanol solvent under non-catalytic and alkali-catalyzed conditions, respectively. The depolymerization products were extracted and separated using pentane, diethyl ether, methanol, and THF.

6.2.1 Base-Catalyzed Methanol–Coal Depolymerization

Base-catalyzed methanol–coal depolymerization was performed in a high-pressure autoclave. Methanol was used as the hydrogen-donating solvent; 10–48 % KOH, NaOH, or BC II was used as the depolymerization catalyst. The reaction process is shown in Fig. 6.1.

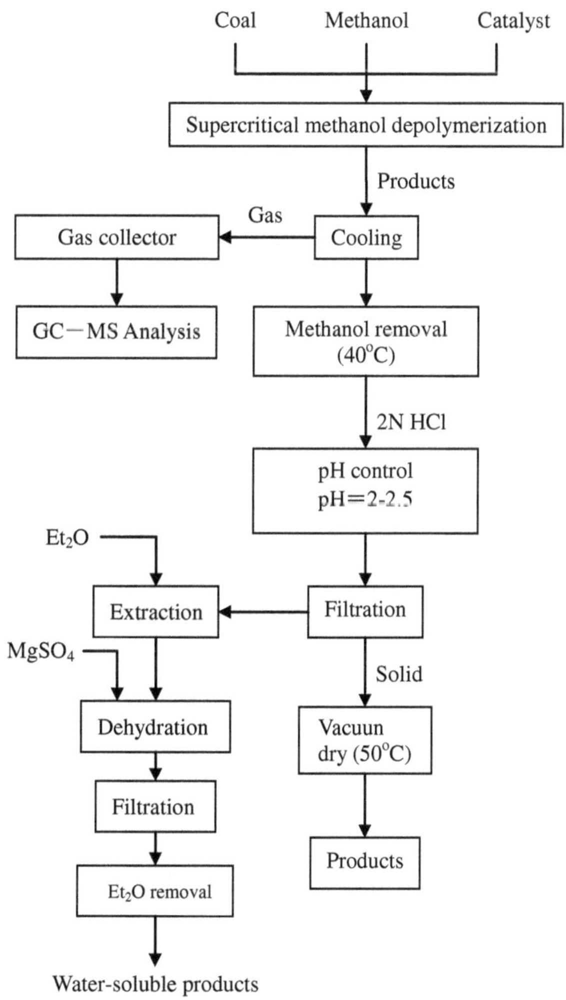

Fig. 6.1 Coal depolymerization in supercritical methanol system

The water-insoluble organic compounds (i.e., the depolymerized coal) were dried, giving the products of the base-catalyzed methanol–coal reaction.

The water-soluble organic compounds were recovered from the filtrate.

6.2.2 Extraction and Separation of Products in Base-Catalyzed Methanol–Coal Reaction

The mixture of alkali-catalyzed methanol–coal products was dried and extracted with pentane. The pentane-insoluble products were dried, weighed, and extracted with diethyl ether to separate the diethyl ether-soluble and -insoluble compounds. The dried diethyl ether-insoluble products were extracted with methanol, to obtain methanol-soluble and -insoluble products. The same method was applied to the methanol-insoluble products, using THF as the solvent, for 36–48 h, to separate the THF-soluble and -insoluble products. The extraction process is shown in Fig. 6.2.

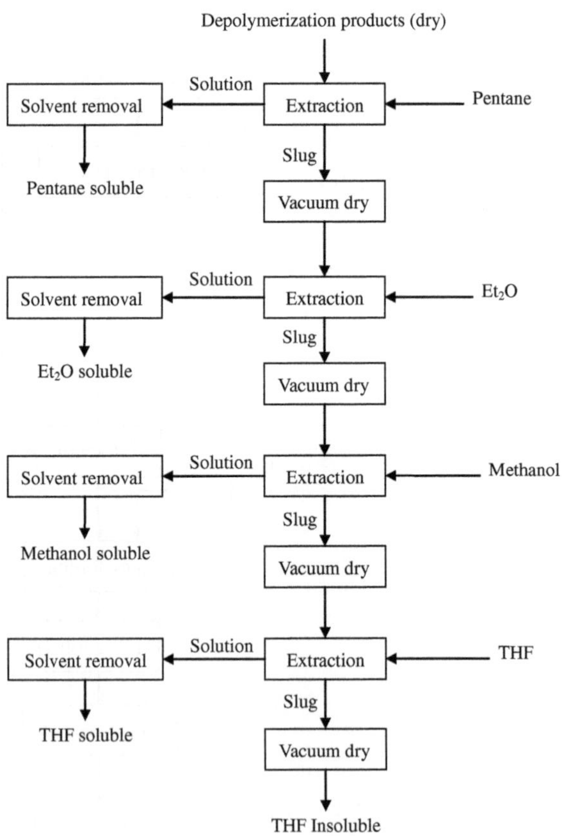

Fig. 6.2 Extraction of depolymerization products

6.2.3 Distribution of Extraction Products of Shenfu Coal Depolymerization Under Different Reaction Conditions

The extraction fractions of raw Shenfu coal, and the pyrolysis products of Shenfu coal in supercritical methanol and those from the BC II catalytic reaction were obtained. The extraction separation results are shown in Table 6.2.

6.2.4 Fourier-Transform (FT) IR Analysis of Extracted Fractions from Shenfu Coal Depolymerization

The pentane-soluble and ether-soluble compounds in the light extraction fraction were analyzed using gas chromatography-mass spectrometry (GC-MS). Phase structural information on small molecules was obtained. The surface functional groups and the heavy THF-soluble and -insoluble products in the alkali–methanol system were analyzed using FTIR spectroscopy. The IR spectra of Shenfu coal, and the THF-soluble and -insoluble products were similar; five peak ranges were observed: the OH stretching vibration at 3400 cm^{-1}, the aromatic C–H stretching vibration at 3000–3100 cm^{-1}, the aliphatic C–H stretching vibration at 2800–3000 cm^{-1}, absorptions from oxygen-containing groups and the C=C groups in the benzene ring at 1000–1800 cm^{-1}, and the CH structure fingerprint region at 300–700 cm^{-1}.

Table 6.2 Extraction products of Shenfu coal depolymerization under different reaction conditions

Extraction product		Raw coal	Non-catalytic methanol supercritical products	BC II catalytic depolymerization products
Total conversion (%)		6.7	8.33	75.21
Gas/pressure (MPa)		–	–	5.86/0.228
Water soluble (%)		–	–	1.49
Solid and liquid products (g)		–	18.89	17.82
Distribution of solid and liquid products (%)	Pentane soluble	–	1.19	16.04
	Diethyl ether soluble	1.7	3.03	29.06
	Methanol soluble	1.4	0.71	6.72
	THF soluble	3.5	3.09	17.50
	THF insoluble	93.3	91.67	24.79
	Total	99.9	99.28	94.05

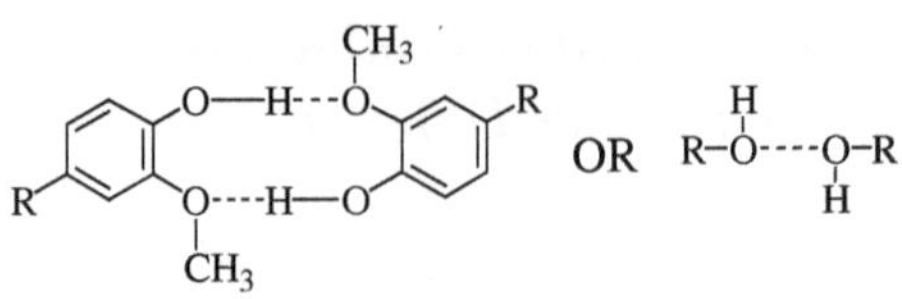

Fig. 6.3 Dimeric structure of Shenfu coal

6.2.4.1 OH Groups

Shenfu coal and its THF-soluble and -insoluble products show a strong absorption at 3400 cm^{-1}. The O–H or N–H bond stretching vibration absorption is at 3400 cm^{-1}. The free OH group vibration absorption is located at 3650–3590 cm^{-1}. The shoulder peak at 3620 cm^{-1} in Shenfu coal shows the presence of a small amount of free phenol or alcohols. The ether extraction of Shenfu coal proved the presence of 4-methylbenzyl alcohol and 2,3-dimethylphenol. Structure I in Shenfu coal is shown Fig. 6.3.

Effects of THF Extraction on OH Group Absorption

The intensities of the absorption peaks at 3400 cm^{-1} for the THF-soluble and -insoluble products are higher than those for Shenfu coal. This may be because the oxygen-containing structure reacts with methanol to form a hydrogen-bonded polymeric structure during methanol extraction of the coal macromolecular structure. This polymeric structure is relatively stable, and THF extraction cannot break the hydrogen bonds to release methanol. The formation of this polystructure leads to an increase in intensity at 3400 cm^{-1} for THF-soluble and -insoluble products.

Table 6.3 presents the proximate and ultimate analyses for Shenfu coal and some of its reaction products. The data show that the oxygen content of the THF-insoluble products is 15.25 %, which is 2.96 % higher than that of the raw coal. This confirms that the THF-insoluble products contain methanol. The IR peak at 3400 cm^{-1} indicates a high content of OH groups in the THF-soluble products. The FTIR spectra show that the absorption peaks of the free OH groups in the different fractions are very similar. However, the content of polymer I increases, whereas that of polymer II decreases. The reaction can be displayed as follows:

Polymer II → Monomer + Polymer I
Polymer I also depolymerizes:

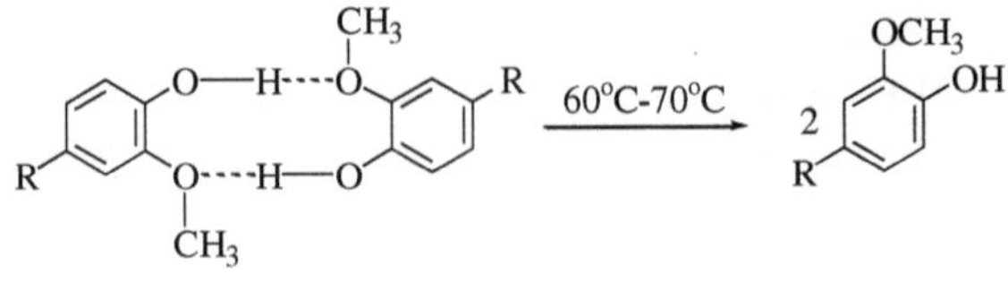

Table 6.3 Shenfu coal analysis (%)

Sample	Proximate analysis				Ultimate analysis				
	V_{daf}	A_d	M_{ad}	FC_{daf}	C_{daf}	H_{daf}	N_{daf}	$S_{t,d}$	O_{daf}
Coal	40.64	2.21	2.39	59.35	80.14	5.52	1.83	0.22	12.29
THF-insoluble products	37.36	4.05	5.49	62.64	80.28	4.25	1.53	0.18	15.25
Non-catalytic supercritical methanol extraction THF-insoluble products	30.57	4.42	4.19	69.42	80.28	4.69	1.47	0.17	13.36
BC II catalytic depolymerization THF-soluble products	51.21	1.27	2.93	48.78	81.46	6.57	0.96	0.16	10.85
BC II catalytic depolymerization THF-insoluble products	31.32	12.77	4.86	68.67	84.72	8.08	0.89	0.16	6.15

Characterization of OH Groups in THF-Soluble and -Insoluble Products of Supercritical Methanol Depolymerization

A comparison of the OH absorption peaks of the THF-soluble and-insoluble products before and after supercritical methanol depolymerization shows that the absorption of the THF-insoluble material was unchanged, but that of the THF-soluble fraction varied. The free alcohol O–H stretching vibration peak at 3634 cm^{-1} can be clearly observed in the supercritical methanol coal depolymerization products. The free OH groups might come from two sources: one is alcohols or phenols from coal depolymerization products, and the other is methanol adsorbed during pyrolysis or extraction.

Characterization of OH Groups in BC II Catalytic Products

The O–H vibration absorption peak at 3434 cm^{-1} of the THF-insoluble products decreased significantly after the BC II catalytic reaction. This shows that a higher proportion of bond breaking and depolymerization occurred, indicating that the oxygen atoms produced from bond breaking and depolymerization are present in small groups. These small groups can be dissolved in pentane, diethyl ether, methanol, and THF.

In summary, the OH group content of coal is high, and a small number of them are present as free phenols or alcohols; most of the OH groups are present in polymeric structures, mainly dimers and multimers. The results of THF extraction show that the oxygen content increases as a result of dimer formation by coal and methanol during extraction. This dimeric structure is relatively stable, and cannot be extracted with THF. In coal depolymerization in supercritical methanol, some of the OH structures are depolymerized, so free OH groups are formed in the THF-soluble fraction, and the intensity of the IR absorption at 3634 cm^{-1} is strong. The BC II catalytic reaction changes the structures of the OH groups significantly. After the reaction, OH groups show up on small molecules, and then are extracted by ether, methanol, and THF. This results in a decrease in the intensity of the absorption peak at 3434 cm^{-1}, and the oxygen content is greatly reduced.

6.2.4.2 Aromatic Structures

The predominant structures in coal are aromatic. There are a large number of substituted aromatic structures in the small-molecular phase of coal. The coal macromolecular structure is mainly composed of condensed aromatic nuclei. The aromaticity (f_a) directly reflects the distribution of carbon atoms in the aromatic nucleus structure and the lipid carbon structure; it is closely related to the chemical properties of hydrocarbons.

The C–H stretching vibration absorption peaks of aromatic rings, at 3000–3100 cm^{-1}, reflect the degree of depolymerization and sizes of the aromatic rings. In the range 1600–1450 cm^{-1}, the stretching vibration absorption peaks of aromatic skeletons appear at 1600, 1500, and 1450 cm^{-1}. The peak at 1600 cm^{-1} represents the stretching vibration of the benzene ring skeleton. Shenfu coal and its THF-soluble and –insoluble depolymerization products all have strong absorption peaks at 1600 cm^{-1}. Shenfu coal, which has an aromaticity of 0.77 and *R* bonding number of 3.75, contains a large number of aromatic ring structures, so these strong absorptions at 1600 cm^{-1} arise from aromatic C=C skeleton vibrations.

In summary, there are plenty of aromatic structure in Shenfu raw coal, the non-catalytic supercritical methanol extraction and BC II catalytic depolymerization have no evident influence on the C-H of aromatic skeleton. In the non-catalytic supercritical methanol extraction, the hydrogen bond are generated by the reaction of coal and methanol so the methanol is "adsorbed", which leads to the increase of oxygen in THF-insolubles. In the present of BC II catalyst, the absorption peak at 1437 cm^{-1} is enhanced, and the content of hydrogen and carbon are increased, suggesting alkylation reactions happen in the process.

6.2.4.3 Aliphatic C–H Structures

Aliphatic Structure of Shenfu Coal and Its Extracts

The absorption resonances in the 2800–3000 cm^{-1} region are attributable to C–H bonds of CH_3, CH_2, and CH groups. The IR spectra of the THF-soluble and -insoluble fractions of Shenfu coal have two strong absorption peaks, at 2920 and 2850 cm^{-1}, from the C–H symmetric and asymmetric stretching vibrations, respectively, of aliphatic hydrocarbons. A comparison of these peaks from the three samples shows that the absorption intensities of Shenfu coal and its THF-insoluble fraction are weak. There is a shoulder at 2950 cm^{-1} as well as the main peak at

2920 cm^{-1}, indicating the presence of methyl groups. The absorption intensities of these peaks for the THF-soluble fraction are significantly higher, and the shoulder at 2950 cm^{-1} is also larger. These observations suggest that Shenfu coal contains aliphatic structures, and the amount of aliphatic structures in the THF-soluble fraction increases during extraction.

Effects of Supercritical Pyrolysis and BC II Catalytic Reactions on C–H Aliphatic Structures

After coal depolymerization under supercritical methanol conditions and BC II catalysis, the 2920 and 2850 cm^{-1} peaks of both the THF-insoluble fractions are unchanged. However, the intensities of thee peaks in the THF-soluble fractions increase significantly, and the 2950 cm^{-1} shoulder peaks become strong. The similarities among the 2920 and 2850 cm^{-1} absorption peak of Shenfu coal, its THF-insoluble fraction, and the THF-insoluble fraction of its pyrolysis products show that extraction and the supercritical methanol process have little effect on depolymerization of the aliphatic structures of Shenfu coal. The reason for the increases in the intensities of the 2920 and 2850 cm^{-1} peaks of the THF-soluble depolymerization products is the same as that for the THF-soluble extracts, and both of them are caused by enrichment of the aliphatic compounds. The 2920 and 2850 cm^{-1} peaks of the THF-insoluble fraction from the BC II-catalyzed reaction arise from the C–H structures in the aliphatic structure of the raw coal, whereas these two peaks in the THF-soluble fraction are the result of enrichment of the aliphatic compounds or alkylation of aromatic structures. The peaks at 1375 cm^{-1} correspond to asymmetric vibrations of methyl groups connected to saturated carbon atoms. The 1375 cm^{-1} peaks of Shenfu coal, its THF-soluble and -insoluble fraction, and the THF-soluble and -insoluble fractions of the depolymerization products under supercritical methanol conditions are strong, but the effects of the extraction and pyrolysis on this peak are not obvious. The 1375 cm^{-1} peak of the THF-soluble and -insoluble fractions of the BC II catalytic reaction products are much stronger, confirming that alkylation occurs during the BC II catalytic reaction.

6.2.4.4 Carbonyl Structures

Carbonyl Peak Features

The C=O bond stretching vibrational frequency of aliphatic ketones appears at 1715 cm^{-1}. For ArCOR, the bond between the carbonyl and the aromatic ring can lead to p–π conjugation (conjugation between a π bond in benzene and the carbonyl C=O bond). The average cloud density in the conjugated system results in an average bond length, weakens the C=O double bond, decreases the bond force constant, and lowers the C=O frequency to 1695 cm^{-1}. If the R substituent has a high electronegativity, the positive charge on the carbon atom in the carbonyl increases as a result of electrostatic induction, i.e., the electrophilic reaction increases the positive charge, strengthens the C=O double bond, and increases the bond force constant. The inductive effect and conjugation lead to the C=O stretching vibration appearing at 1700 cm^{-1}.

Effects of Extraction, Pyrolysis, and BC II Catalytic Reaction on Carbonyl Group Peak

There are two shoulder peaks, at 1702 and 1682 cm^{-1}, in the FTIR spectrum of Shenfu coal. After extraction, there are no obvious changes in the 1700 cm^{-1} peaks of the THF-soluble and -insoluble fractions. The intensity of the peak at 1700 cm^{-1} for the THF-insoluble fraction obtained under supercritical methanol conditions increases significantly, especially compared with that of the THF-soluble fraction under depolymerization. However, the 1700 cm^{-1} peak of the THF-insoluble fraction of depolymerization products obtained in the BC II catalytic reaction is very small. The effects of different coal treatments on the 1700 cm^{-1} peak are as follows. Little effect is observed on the C=O structures of the compounds in the THF-soluble and -insoluble fractions on extraction with pentane, methanol, and THF. The intensities of the 1700 cm^{-1} peaks of the THF-soluble and -insoluble fractions increase on depolymerization under supercritical methanol conditions. The intensities of the 1700 cm^{-1} peaks of THF-insoluble fraction decrease or disappear for the BC II catalytic reaction, but the 1700 cm^{-1} peak of the THF-soluble fraction increases and becomes strong.

After treatment of the depolymerization products of coal with supercritical methanol, the 1700 cm^{-1} peak is strong, but there is no introduction of C=O structures during supercritical methanol treatment. The cause of the increase in the 1700 cm^{-1}peak is the C=O group in the reaction product, C=O···HO, between carbonyl and hydrogen bonds. Roy [39] reported that the absorption peak of this reaction product appears at 1600 cm^{-1}. There is a strong peak for the THF-soluble and -insoluble fractions, and this may be contributed by C=O···HO structures. After depolymerization, the products of the coal treated with supercritical methanol, i.e., $RC{=}O\cdots HOR^1$ structures, depolymerize. As a result of the differences between the structures of R and R^1, the larger-molecular-weight RC=O structure is present in the THF-insoluble fraction, but the smaller-molecular-weight $R^1C{=}O$ is extracted into the THF-soluble fraction. This is why the 1700 cm^{-1} peak is strong in both the THF-soluble and -insoluble fractions. We therefore concluded that the 1600 cm^{-1} peak is attributable to a C=O···HO structure, which is depolymerized under supercritical methanol conditions.

The 1700 cm^{-1} peak of the THF-insoluble fraction in the BC II catalytic reaction decreases or disappears as a result of C=O groups linking with ether bridges; the ether linkages break during depolymerization, and then the C=O compounds depolymerize to small compounds, which are extracted into the THF-soluble fraction. Direct depolymerization of the C=O structure may be another reason. GC-MS analysis of the water-soluble fraction in the BC II catalytic reaction demonstrates that acid and aliphatic compounds are present in the depolymerization products, proving that the above analysis is correct. The 1700 cm^{-1} peak of the THF-insoluble fraction disappears for the above reasons, and the intensity of the 1700 cm^{-1} peak of the THF-soluble fraction increases because small C=O groups are extracted into the THF-soluble fraction.

In summary, the effects of extraction on the C=O structures in the THF-soluble and -insoluble fractions is negligible. There are $RC{=}O\cdots OR^1$ structures in the coal, and these structures are easily depolymerized under supercritical methanol conditions. The 1700 cm^{-1} peak of the THF-insoluble fraction disappears as a result of depolymerization of the C=O structures to small-molecular species, which are extracted into the THF-soluble fraction in the BC II catalytic reaction.

6.2.4.5 Ether Structures

Features of Ether Peaks

The absorptions in the 1300–1100 cm^{-1} region are contributed by C–H (phenol), C_{ar}–O–C_{ar}, C–O (alcohols), C_{ar}–O–C_{al}, and C_{al}–O–C_{al}. The peak at 1250 cm^{-1} peak is attributed to the stretching vibrations of $-CH_2OCH_3$ and C=O structures. The 1060 and 1232 cm^{-1} peaks are the asymmetric stretching vibrations of $ROCH_3$. The 1120 cm^{-1} peak is attributed to the symmetric stretching vibration of $ROCH_3$.

Effects of Extraction, Pyrolysis, and BC II Catalytic Reaction on Ether Group Peaks

A peak at 1227 cm^{-1} appears in the 1300–1100 cm^{-1} range for Shenfu coal. The strong peak at 1227 cm^{-1} for Shenfu coal demonstrated the presence of –O– structures. After extraction with THF, the shape and intensity of the peak are unchanged, but the peak moves to 1266 and 1231 cm^{-1}, respectively. After supercritical methanol depolymerization, the shape and intensity of the peak are unchanged, but the intensity of the THF-soluble fraction peak increases significantly, and four strong peaks appear, at 1035, 1120, 1159, and 1232 cm^{-1}. The peaks of the distillate of the THF extract of the coal depolymerization product in the BC II catalytic reaction change significantly: for the THF-insoluble fraction, the peaks in the 1300–1100 cm^{-1} region disappear and only the 1081 cm^{-1} peak remains. Changes in the 1300–1100 cm^{-1} region for the THF-soluble fraction are also obvious: the peak intensity decreases, and a strong peak appears at 1219 cm^{-1}.

The similarity between the spectra of Shenfu coal and its THF extracts indicates that the extraction process has little effect on the –O– structures. The peak of the THF extract becomes a shoulder at 1232 cm^{-1}, indicating that the –O– groups in the THF extract are identical to the R_{al}–O–R_{al} structure. The effects of depolymerization and extraction on ether oxygen groups are almost same, but there are two types of ethers in the THF-soluble fractions: R_{al}–O–R_{al} (1232 cm^{-1}) and R_{al}–O–CH_3

(1060 and 1120 cm^{-1}). Cleavage of ether bonds in the BC II catalytic reaction leads to the peak in the 1300–1100 cm^{-1} region disappearing for the THF-insoluble fraction. The remaining peak, at 1081 cm^{-1}, indicates that some ethers are preserved in the BC II catalytic reaction.

6.2.5 *GC-MS Analysis of Extraction Fractions from Shenfu Coal Depolymerization*

Depolymerization and decomposition mainly occur at the bonds or side chains of bridges between the coal basic structural units. The formation of coal depolymerization liquefied products (such as coal tar and low-molecular-weight extracts) is mainly the result of chain cracking. In this section, the GC-MS characterization of small molecules in the extracted fractions of Shenfu coal depolymerization products under different reaction conditions is discussed.

6.2.5.1 Behavior of Model Compounds in Supercritical Methanol

The basic structural unit of coal consists mainly of aromatic nuclei with different degrees of condensation. The behavior of model compounds in supercritical methanol was investigated using toluene, ethylbenzene, phenol, ethyl benzoate, and methylnaphthalene as the model compound. Experiments were performed in a 50 mL reactor. A methanol solution (10 mL) containing 10 % of the model compound was depolymerized at 290 °C for 1 h and then cooled for GC analysis. Toluene, ethylbenzene, and ethyl benzoate were very stable at 290 °C and no pyrolysis products were detected. This indicates that the aromatic ring structure and some of the side-chain substituents (phenolic hydroxyl, methyl, ethyl, and ester) are stable in supercritical methanol at 290 °C. We assume that depolymerization does not crack the basic structural unit and side chains under these conditions.

6.2.5.2 Small Molecules Produced Under Different Depolymerization Conditions

Small Molecules Extracted Directly from Shenfu Coal

There are small molecules in coal, i.e., adsorbed CH_4, CO_2, and H_2O, and molecules that are part of high-molecular-weight compounds. Studies of pyridine extraction have shown that the average molecular weight is between 355 and 1025 [40–42]. The author used pentane, diethyl ether, methanol, and THF as solvents for coal extraction at low temperatures; no chemical reactions occurred. We propose that the extracts are small-molecular species present in coal. The extraction yields were 0, 1.70, 1.40, and 3.50 % for pentane, diethyl ether, methanol, and THF respectively.

Ketones are not stable, and their presence in coal is still under debate. The extraction results indicate the presence of 2- methylcyclohexanone and 1- amino-2-methyl-5,10-dianthrone; the ketone content is 6.65 %.

Table 6.4 shows the GC-MS analytical results for methanol extraction products. Hydrogenated aromatic structures are known to be hydrogen donor compounds in the initial stage of coal depolymerization and liquefaction; this is important for coal depolymerization and liquefaction.

Small Molecules in Supercritical Methanol Extraction Products from Shenfu Coal Depolymerization

Increasing the temperature in treatment with polar solvents can increase hydrogen-bond breaking. Fifty pentane-soluble components were detected by GC-MS after supercritical methanol depolymerization. These can be categorized into 12 groups, as shown in Table 6.5.

Small Molecules in Extracts of Alkali-Catalyzed Depolymerization Products from Shenfu Coal in Supercritical Methanol

The use of KOH in supercritical methanol coal pyrolysis increases depolymerization from 8.33 %, for the non-catalytic process, to 84.33 %, and the products are different. GC-MS analysis showed that the pentane extract of KOH catalytic

Table 6.4 GC-MS analytical data for methanol extracts of Shenfu coal

No.	Product	Formula
1.	2(3*H*)-dihydrogenated furanose	$C_4H_6O_2$
2.	1,1-Dimethylpropane	$C_5H_{12}O_2$
3.	2- Butyl tetrahydrogenated furan	$C_8H_{16}O$
4.	1,1-Terpineol	$C_{10}H_{18}O$
5.	3- Methyl-1- pentene-3- alcohol	$C_6H_{12}O$
6.	3- Halo-6-methoxypyridazine	$C_5H_5ClN_2O$
7.	3,3-Dimethoxy-1-pentene	$C_5H_{10}O_2$
8.	Hexadecane	$C_{16}H_{34}$
9.	*N*-(3-methoxyphenyl)-2,2-dimethylpropylamide	$C_{12}H_{17}NO_2$
10.	2,6,10,14-Tetramethyl 16-carbon alkanes	$C_{20}H_{42}$
11.	Dodecane	$C_{12}H_{26}$
12.	Nonadecane	$C_{19}H_{40}$
13.	16-Carbon alkyl halides	$C_{16}H_{33}Cl$
14.	Eicosane	$C_{20}H_{42}$
15.	3-Methyl 20-carbon alkanes	$C_{12}C_{21}H_{44}$
16.	Eighteen-carbon alkyl halides	$C_{18}H_{37}C_l$

Table 6.5 GC-MS analysis of pentane-soluble depolymerization products of supercritical methanol extraction

Product type	No. of compounds	Content (wt%)
Phenol	9	9.37
Methoxyphenol	3	2.09
Aromatic	15	39.31
Paraffin	1	4.75
Olefin	2	1.59
Heterocyclic	2	1.42
Hydrogenated heterocyclic	4	9.21
Alcohol	2	3.33
Ester	4	8.17
Ketone	4	4.43
Sugar	2	5.68
Quinone	2	5.17
Total	50	94.52

depolymerization contained cyclohexane, toluene, ethylbenzene, and ethyl benzoate, with contents of 51.36, 9.69, 18.63, and 8.20 % respectively.

The use of BC II instead of KOH as the catalyst increases the total conversion from 84.33 to 88.48 %. GC-MS analysis of the pentane extract showed that the main components were cyclohexane (32.59 %), toluene (6.63 %), ethylbenzene (9.43 %), and ethyl benzoate (1.62 %). Methanol can decompose under BC II catalytic conditions, and can also react with coal in this system. Alkylation and deoxygenation are the main reactions.

Statistical Average Structure of Small-Molecular Phase

The structural characteristics of coal extracts reflect the original structural units in the coal, so it is reasonable to deduce the coal structure from the structural characteristics of coal extracts. Many researchers have used the van Krevelen statistical method to investigate the statistical structures of coal, coal tar, and coal extracts. The average molecular structures of the pentane extracts of Shenfu coal depolymerization products in supercritical methanol, obtained using GC-MS analysis and the remove and replace method, are shown in Table 6.6 and 6.7

Table 6.6 Number of atoms in average molecule structure in pentane extracts from Shenfu coal depolymerization in supercritical methanol

Atom	C_{al}	C_{ar}	H_{al}	H_{ar}	H_{OH}	–O–
No. of atoms	3.731	9.801	9.553	5.796	0.364	0.060

Table 6.7 Average molecular structure of pentane extracts of Shenfu coal depolymerization in supercritical methanol

Parameter	Definition	Equation	Result
C_n	No. of carbon atoms	$C_{ar}+C_{al}$	13.53
H_n	No. of hydrogen atoms	$H_{ar}+H_{al}+H_{OH}$	15.71
S	Parameter of ring structure	$2C_n+2-H_n$	13.35
fa	Aromaticity	C_{ar}/C_n	0.72
x	Alkyl substitution H/C	H_{al}/C_{al}	2.55
$C_{p(un)}$	Unsubstituted peripheral aromatic carbon atoms	H_{ar}	5.80
$C_{p(s)}$	Substituted peripheral aromatic carbon atoms	$\sum C_{p(s),i*}$	2.39
C_{anb}	Aromatic non-bridging carbon atoms	$C_{p(un)}+C_{p(us)}$	8.19
C_{ab}	Aromatic bridging carbon atoms	$C_{ar}-C_{anb}$	1.16
R_a	Average number of aromatic ring structures	$(C_{ar}-C_{anb})/2+M_{an}$	1.91
M_{an}	No. of aromatic nuclei	$(C_{anb}/3)-(C_{ar}/6)$	1.10
R	Ring structures	$(S-C_{ar})/2$	1.78

*i denotes that the molecule was detected by GC-MS.

6.2.6 Reactivity of Shenfu Coal in Methanol–Alkali Systems

In this section, the effects of the reaction conditions on Shenfu coal depolymerization and liquefaction in methanol–alkali (BC II or KOH) systems are discussed.

The products of coal depolymerization and liquefaction in methanol–alkali systems can be classified as gaseous products, solid–liquid products, and water-soluble products. The yields of gaseous and water-soluble products are low (usually <6 %), and the main products are solid–liquid. The conversion of coal depolymerization and liquefaction can be expressed as

$$x = \frac{W_0 - W_{THF_i}}{W_0} = \frac{W_G - W_W + W_P + W_E + W_M + W_{THF_s}}{W_0}$$

where

W_0 is the weight of the coal sample (g, dry ash-free);
W_G is the weight of gaseous products (g);
W_W is the weight of water-soluble products (g);
W_P is the weight of pentane-soluble products (g);
W_E is the weight of ether-soluble products (g);
W_M is the weight of methanol-soluble products (g);
W_{THFi} is the weight of THF-insoluble products (g);
W_{THFs} is the weight of THF-soluble products (g).

Table 6.8 Composition of gaseous products of Shenfu coal depolymerization in methanol–alkali system (wt%)

Product	Reaction temperature (°C)	
	250	270
CH_4	0.02	0.08
C_2H_4	0.93	1.15
C_3H_8	0.01	0.03
C_3H_6	0.02	0.11
C_4H_8	0.05	0.15
C_4H_{10}	0.52	0.40
Total	1.55	1.92

6.2.6.1 Composition of Gaseous Products of Coal Depolymerization Under Supercritical Methanol and KOH Conditions

The gaseous products of depolymerization in supercritical methanol with KOH as the catalyst are CH_4, C_3H_8, C_4H_{10}, and other alkanes, olefins such as C_2H_4, C_3H_6, and C_4H_8, and dienes such as CH_2=CH-CH=CH_2. Table 6.8 shows the composition of the gaseous products of Shenfu coal depolymerization.

6.2.6.2 Composition of Water-Soluble Products of Coal Depolymerization and Liquefaction in Supercritical Methanol–Alkali System

After coal depolymerization in a methanol–alkali system at 290 °C, 2 N HCl was added, and the water-soluble products were extracted. GC-MS analysis showed that the water-soluble products of Shenfu coal depolymerization in supercritical methanol with KOH as the catalyst are cyclohexane (49.14 %), toluene (17.66 %), ethylbenzene (23.75 %), a methyl-substituted phenol (2.37 %), and methoxyphenol (1.71 %). Three esters (total 0.27 %), ketones (total 0.42 %), aldehydes, and naphthalene derivatives were also detected.

When the catalyst BC II was used, the product distribution varied significantly. Compared with the methanol–alkali system, the amounts of cyclohexane and ethylbenzene clearly decreased, but that of toluene increased; the amounts were 4.51, 6.60, and 45.66 %, for cyclohexane, ethylbenzene, and toluene, respectively. The product also contains phenolic, methoxyphenol, acid, ester, olefin, alcohol, poly-substituted naphthalene; their contents are listed in Table 6.9.

6.2.6.3 Effects of Reaction Time on Coal Depolymerization and Liquefaction

Shenfu coal depolymerization was performed with different reaction times. The experimental conditions were as follows: methanol 120 g, KOH catalyst 13.3 g,

Table 6.9 Water-soluble depolymerization products and contents (wt%) under different catalytic conditions

Product type	KOH		BC II	
	No. of Compounds	Content	No. of Compounds	Content
Cyclohexane	1	49.14	1	4.51
Toluene	1	17.66	1	45.66
Ethylbenzene	1	23.75	1	6.60
Phenolic	7	2.37	4	11.10
Methoxyphenol	4	1.71	1	0.48
Ester	3	0.27	2	7.19
Ketone	2	0.42	–	–
Aldehyde	1	4.95	–	–
Polysubstituted Naphthalene	1	0.15	1	0.23
Heteroatom compounds containing S and N	1	0.07	–	–
Acid	–	–	4	18.85
Olefin	–	–	1	0.04
Alcohol	–	–	1	0.25

Shenfu coal 15.0 g, reaction temperature 290 °C, and reaction pressure, i.e., self-generated pressure, 20.00–25.44 MPa. Figure 6.4 summarizes the effects of reaction time on Shenfu coal depolymerization and liquefaction in a methanol–alkali system over 2–60 min. The conversion increased with increasing reaction time. The conversions reached 83.20 and 84.33 % at 30 and 60 min respectively. The amounts of gaseous and water-soluble products (organics) also increased with increasing reaction time.

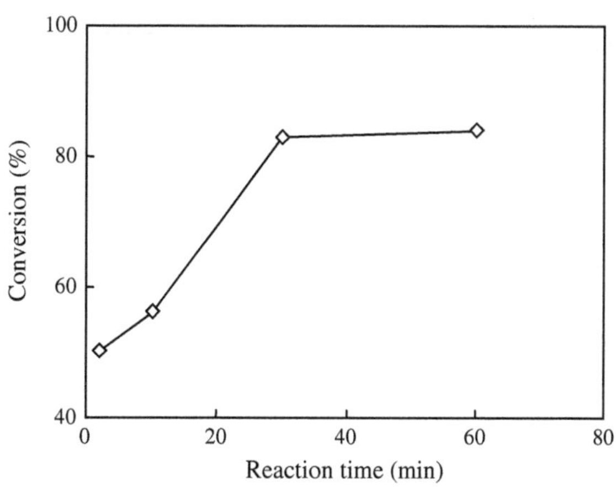

Fig. 6.4 Effects of reaction time on Shenfu coal depolymerization and liquefaction

Fig. 6.5 Effects of reaction time on extraction product distribution: *1*, pentane soluble; *2*, diethyl ether soluble; *3*, methanol soluble; *4*, THF soluble; and *5*, THF insoluble

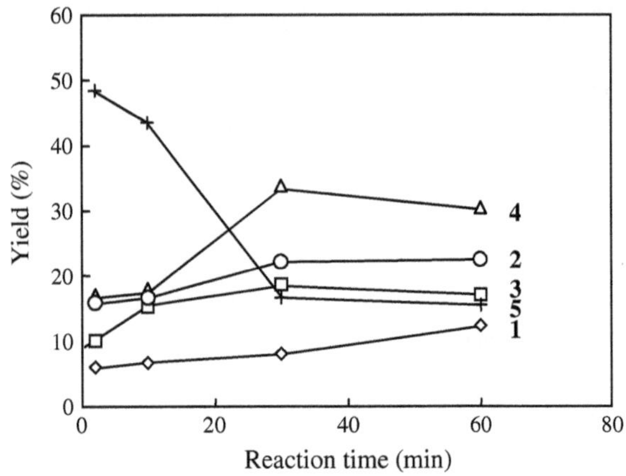

Figure 6.5 shows the distribution of soluble extracts of solid products at different reaction times. The amounts of pentane- and diethyl ether-soluble products increase with increasing reaction time. The amounts of methanol- and THF-soluble products are lowest, 18.70 and 33.77 %, respectively, at 30 min. This indicates that in the early stage of the depolymerization, the THF-insoluble material was converted to THF-, methanol-, and diethyl ether-soluble products.

6.2.6.4 Effects of Reaction Temperature on Coal Depolymerization

Shenfu coal depolymerization was performed at different reaction temperatures; the reaction time was fixed at 1 h. Figure 6.6 shows the conversion depending on reaction temperature. The extent of depolymerization in a methanol–BC II alkali system at 270 °C was 88.48 %. The conversion decreased with decreasing reaction temperature. At 250 °C, the conversion was only 36 %. At 305–310 °C, the

Fig. 6.6 Effects of reaction temperature on Shenfu coal depolymerization and liquefaction

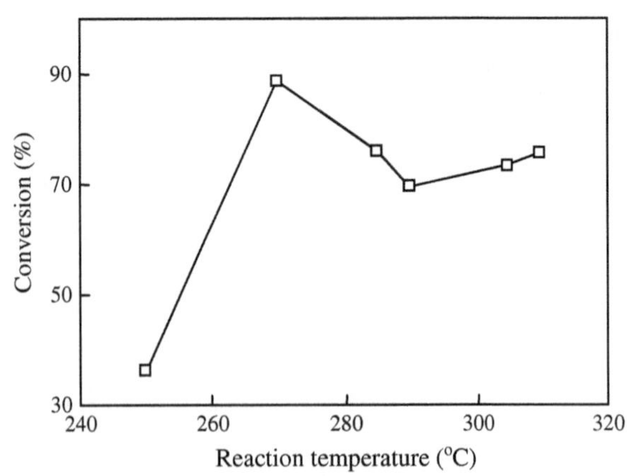

conversion rate was between 73.15 and 75.21 %.The maximum yield of solid–liquid products was obtained at 290 °C; the yield decreased to 17.28 % at 250–270 °C.

6.2.6.5 Effects of Coal Quantity on Depolymerization

The effects of coal quantity on depolymerization were studied under the following conditions: 290 °C, reaction time 1 h, methanol 38.5 g, and BC II as the catalyst. The results are shown in Table 6.10.

The distributions of solid products for depolymerization of different coal quantities are shown in Fig. 6.7 When the coal consumption is low, the conversions to pentane-, diethyl ether-, methanol- and THF-soluble products are high. Extraction decreases with increasing coal sample size. When the sample weighed more than 20 g, the product was tarry and insoluble in methanol, so it was not possible to use methanol to wash and collect the products; however, pentane was a good solvent for product collection. These results indicate that the chemical reactions of coal in methanol–alkali systems differ, depending on the sample size.

Table 6.10 Effects of coal quantity on depolymerization

	Coal quantity (g)		
	5	10	20
Pressure (MPa)	16.34	15.47	12.55
Conversion (%)	88.02	84.45	65.54
Solid, liquid weight (g)	4.74	9.31	18.33

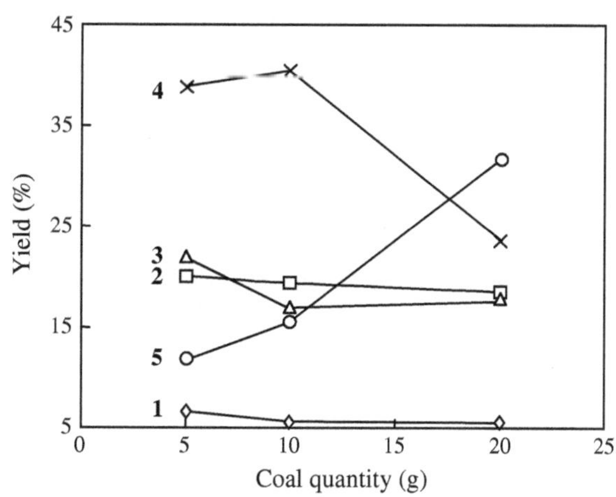

Fig. 6.7 Effects of coal quantity on extraction fraction distribution: *1*, pentane soluble; *2*, diethyl ether soluble; *3*, methanol soluble; *4*, THF soluble; and *5*, THF insoluble

References

1. Xie KC (1998) Fuel Chem 19(2):55
2. Heredy LA et al (1964) Fuel 43:414
3. Holy NL (1974) Chem Rev 74:243
4. Heredy LA (1979) Am Chem Soc Prepr Div Fuel Chem 24(1):142
5. Meriam JS et al (1981) Fuel 60:542
6. Ashida R et al (2008) Fuel 87:576
7. Ashida R et al (2009) Fuel 88:1485
8. Masaki K et al (2004) Energy Fuel 18:995
9. Morimoto M et al (2009) Energy Fuel 23:4533
10. Wang J et al (2005) Energy Fuel 19:2114
11. Sharma A et al (2008) Energy Fuel 22:3561
12. Sharma A et al (2008) Fuel 87:491
13. Sharma A et al (2008) Fuel 87:2686
14. Sharma A et al (2009) Energy Fuel 23:1888
15. Miura K (2000) Fuel Process Technol 62:119
16. Geymer DO (1974) US Patent 3,844,928, 5 Oct 1974
17. Bodily DM et al (1972) Am Chem Soc Prepr Div Fuel Chem 16(2):163
18. Bodily DM et al (1974) Am Chem Soc Prepr Div Fuel Chem 9(1):163
19. Tanner KI et al (1981) Fuel 60:52
20. Oblad HB (1982) Ph.D. Dissertation University of Utah
21. Mobley DP et al (1980) J Catal 64:494
22. Jensen RE (1981) MS Thesis University of Utah
23. Li CQ et al (2003) Energy Fuel 17:768
24. Takahashi K et al (2001) Energy Fuel 15:141
25. Opaprakasit P et al (2002) Energy Fuel 16:543
26. Makabe M et al (1981) Fuel Process Technol 5:129
27. Makabe M et al (1981) Fuel 60:327
28. Ouchi K et al (1981) Fuel 60:474
29. Lu HY et al (2011) Energy Fuel 25:2741
30. Anderson LL et al (1981) ACS Symp Ser 169:223
31. Lu HY et al (2010) J Wuhan Univ Sci Technol 33:83
32. Lei ZP et al (2011) Energy 36:3058
33. Sakbut PD et al (1988) Fuel Process Technol 18:287
34. Whitehurst DD et al (1976) Am Chem Soc Prepr Div Fuel Chem 21(5):127
35. Neavel RC (1976) Fuel 55:237
36. Wiser WH et al (1976) J Appl Chem Biotechnol 21:82
37. Wiser WH (1968) Fuel 47:475
38. Benjamin BM et al (1978) Fuel 57:267
39. Roy MM (1957) Fuel 35:926
40. Li XL et al (1999) 10th Int Conf Coal Sci, Taiyuan
41. Zhu SY et al (1994) Fuel Chem 22(3):427
42. Li F et al (1994) Fuel Sci Tech Int'l 12(1):151

Chapter 7
Coal Combustion

Abstract Combustion represents the earliest use of coal, and remains a fundamentally important application. However, burning coal directly generates pollution, and there is great interest in enhancing the efficiency and reducing the environmental impact of burning coal. The structure/property/reactivity relationships of coal are particularly important for understanding the complex multistage processes that govern coal combustion. In this chapter, we provide the relevant background for understanding the chemical and physical changes that occur during combustion, including pore models that have been developed to describe combustion. Finally, the results of investigations of the combustion parameters of four types of Chinese coal are presented to highlight the interplay between different coal structures and their combustion behaviors.

7.1 Introduction to Coal Combustion

As a result of technological limitations, about 80 % of coal in China is consumed by direct burning. This basic method of coal use and conversion has resulted in very serious pollution problems. China therefore urgently needs to improve coal combustion efficiency and to find cleaner methods for burning coal. Coal combustion is a very complex process. It is a multiphase chemical reaction, and is affected by various factors. Coal composition, combustion atmosphere, temperature, pressure, and even the combustion equipment all have important effects on the coal combustion process. This is a very attractive research area, and a range of in-depth studies have been performed to determine the key parameters that affect combustion. In this chapter, we focus on the changes in coal particle surfaces and the kinetics of coal combustion reactions.

K.-C. Xie, *Structure and Reactivity of Coal*, DOI 10.1007/978-3-662-47337-5_7

7.1.1 Combustion Reactions

Combustion is an oxidation reaction, in which combustible fuel and oxygen are chemically combined to release heat and light. The necessary conditions for combustion to occur and be sustained are the coexistence of combustible fuels and oxidizing agents such as air or oxygen, and attainment of the minimum ignition temperature. The minimum ignition temperatures for some solid fuels are shown in Table 7.1. There are many types of combustion reaction, depending on the completeness of oxidation, and combustion reactions can be classified as complete combustion and incomplete combustion. Based on the reactor type and control conditions, combustion can be classified as fixed-bed combustion, fluidized-bed combustion, and entrained-flow-bed combustion. Based on visible effects, combustion is classified as normal combustion, with a relatively stable combustion process and space, and abnormal combustion such as coal dust and flue gas explosions. Flamed and flameless combustions are also classification categories.

7.1.1.1 Coal Combustion Process

Coal combustion is a multistage process. If we focus on the changes in the coal, the process can be described as follows. First, the coal particles are heated to above 100 °C by an external source, and water is vaporized from the coal. Second, as the temperature increases further, volatiles are driven from the coal to form coal char. Third, when the ignition temperature is reached, the volatiles and coal char begin to burn. Finally, when all the organic matter has been burned out, the incombustible minerals form coal ash. These multiple stages may occur sequentially if the combustion space is sufficient. However, if the combustion space is limited or coal is continuously added, these combustion stages may occur simultaneously, e.g., volatiles may be released before water has been fully vaporized and coal char may start to burn before the volatiles have been fully released [1, 2]. The key combustible element in coal is fixed carbon, which is the major source of heat release, and takes the longest time to burn. The combustion of coal char is the most important stage, and determines the main characteristics of the coal combustion process. Generally, pyrolysis and volatile vaporization take up 10 % of the combustion time, and coal char combustion takes up to 90 % of the time.

If we focus on the movement of gases, the coal combustion process can be described as follows. First, oxygen passes through the boundary layer of the gas flow, and then diffuses into the micropores of the coal particles. Second, oxygen is chemically adsorbed and reacts on the coal surface. The reaction products enter the main gas flow by diffusion through multiple layers, starting from the micropores of

Table 7.1 Minimum ignition temperatures (°C) of some solid fuels

Fuel	Firewood	Lignite	Bituminous coal	Anthracite	Coke
Ignition temperature	280–300	250–410	400–500	550–600	>700

Fig. 7.1 Schematic diagram of coal combustion kinetics

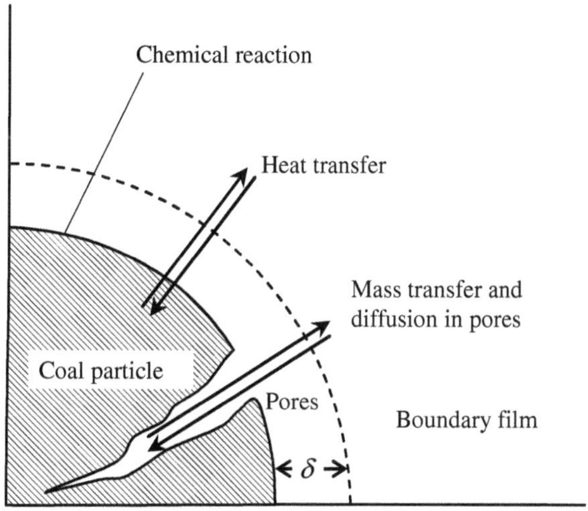

coal particles, to the coal particle surface, and then from the ash layer to the gas flow boundary layer. The total coal combustion time consists of the chemical reaction time plus the gas diffusion time. The chemical reaction time is determined by the inherent properties of the coal such as composition and degree of metamorphism, and the external temperature. The gas diffusion time is determined by the gas flow speed and the thickness of the ash layer, as shown in Fig. 7.1.

There are two typical reaction scenarios in coal combustion. One is that the combustion status is dominated by the chemical reaction rate; in this case, the rate of combustion is controlled by the rate of oxidation on the coal surface. The combustion time depends on the chemical reaction time. This scenario often occurs when the coal surface temperature is low. The most effective method for increasing the combustion reaction rate is to increase the external temperature. In the other scenario, the combustion status is mainly controlled by the rate of gas diffusion; in this case, the combustion rate is determined by how fast oxygen is diffused into the coal reaction surface, and is not dependent on the external temperature. The most effective ways of increasing the combustion reaction rate are to decrease the coal particle size to reduce the coal ash thickness, to increase the gas flow rate to reduce the gas layer thickness, and to increase the concentration of reactants on the coal particle surface. It is worth mentioning that the actual combustion status may fall between these two scenarios, in which case increasing the external temperature and the gas flow rate both increase the coal combustion rate.

7.1.1.2 Chemical Changes During Coal Combustion

Carbon and hydrogen are two of the main elements involved in coal combustion. The following are some of the major reactions and their reaction heats under standard conditions:

Complete combustion of carbon:

$$C + O_2 \rightarrow CO_2 \quad \Delta_r H_m^0 = -393.51\,\text{kJ mol}^{-1}$$

Incomplete combustion of carbon:

$$C + \frac{1}{2}O_2 \rightarrow CO \quad \Delta_r H_m^0 = -110.52\,\text{kJ mol}^{-1}$$

Combustion of CO:

$$CO + \frac{1}{2}O_2 \rightarrow CO_2 \quad \Delta_r H_m^0 = -282.99\,\text{kJ mol}^{-1}$$

Combustion of hydrogen:

$$H_2 + \frac{1}{2}O_2 \rightarrow H_2O(g) \quad \Delta_r H_m^0 = -241.83\,\text{kJ mol}^{-1}$$

These basic reaction equations can be used to predict the theoretical values for the required oxygen, and the composition and amount of flue gas produced in a combustion reaction. Different carbon allotropes have different chemical reaction heats. The reaction heat values in the above equations are based on the graphite form of carbon. The reaction heat of amorphous carbon is higher than that of graphite.

Not all the chemical reactions in coal combustion are combustion reactions. Other reactions such as carbon gasification and the water gas shift reaction often occur simultaneously. The following are some major reactions and their reaction heats under standard conditions:

Gasification of CO_2:

$$C + CO_2 \rightarrow 2CO \quad \Delta_r H_m^0 = 172.47\,\text{kJ mol}^{-1}$$

Gasification of steam:

$$C + 2H_2O(g) \rightarrow CO_2 + 2H_2 \quad \Delta_r H_m^0 = 90.15\,\text{kJ mol}^{-1}$$

Water gas shift reaction:

$$CO + H_2O(g) \rightarrow CO_2 + H_2 \quad \Delta_r H_m^0 = -41.16\,\text{kJ mol}^{-1}$$

Methanation:

$$CO + 3H_2 \rightarrow CH_4 + H_2O(g) \quad \Delta_r H_m^0 = -206.16\,\text{kJ mol}^{-1}$$

Coal combustion can be described as a two-stage process. The first stage is the chemical adsorption of oxygen on the coal surface to produce intermediate substances, C_xO_y. The second stage is decomposition of the intermediate substances to two combustion products, i.e., CO_2 and CO.

$$xC + \frac{1}{2}yO_2 \rightarrow [C_xO_y] \rightarrow mCO_2 + CO$$

At different combustion temperatures, the chemical reaction modes and concentration of gases on the coal particle surfaces are different [3]. The CO/CO_2 ratio near the coal particle surface changes as a function of temperature. Below 1200 °C, this ratio is less than 1. In this case, the concentration of CO_2 is greater than that of CO, and the process tends to be an oxidation reaction. Above 1200 °C, this ratio is greater than 1, and the process tends to be a reduction reaction. At around 1200 °C, CO/CO_2 is close to 1.

If steam is present in the combustion gas flow, hydrogen will be produced in the combustion process. The reaction rate of the hydrogen molecule is higher than that of CO, and the steam molecule has a higher reaction rate than CO_2. As a result, while CO_2 gasifies one carbon atom, hydrogen gasifies multiple carbon atoms. The presence of steam therefore significantly accelerates coal combustion. This is why coal gasification cannot be neglected in analysis of the coal combustion process. The products of combustion and gasification are different. The process is controlled by the amount of oxygen available.

7.1.1.3 Brief Review of Changes in Surface Structure of Coal During Combustion

Changes in surface structure during pyrolysis. Coal pyrolysis occurs before combustion. The changes in the surface structure during pyrolysis have a direct impact on the coal combustion state [4, 5]. Maria et al. [6] studied the surface area changes during coal pyrolysis at low temperatures. The results are shown in Table 7.2. The data in the table show that the surface areas covered by N_2 and CO_2 both increase with increasing temperature and degree of pyrolysis. However, the rate at which the surface area covered by CO_2 changes is much greater than that for N_2. Before the

Table 7.2 Surface area changes during pyrolysis. Reprinted from Ref. [6], Copyright 1983, with permission from Elsevier

Temperature (°C)	Weight loss ratio (%)	Surface area (m^2 g^{-1})		
		N_2 (77 K)	CO_2 (195 K)	CO_2 (298 K)
350	4.6	<1	21	118
375	8.2	<1	22	127
400	13.0	<1	27	147
500	22.4	2.4	233	304
600	24.9	10.0	206	383
Untreated	–	1.8	12.5	90

temperature reaches 500 °C, the surface area covered by CO_2 increases significantly, whereas that covered by N_2 changes little.

Note that N_2 can only detect medium to large pores, and CO_2 can detect micropores. These observations indicate that most micropores are formed at the beginning of the temperature increase as volatiles are vaporized. The amount of medium to large pores does not change much at this stage. As the temperature is further increased, volatiles are strongly vaporized. At this stage, more micropores are created, and some micropores develop into medium to large pores.

Maria et al. also studied the pyrolysis of coal samples with high volatile concentrations. The results suggest that at the beginning of coal pyrolysis, the surface area covered by CO_2 is increasing, whereas that covered by N_2 is decreasing. This may contribute to a high concentration of volatiles, causing vigorous pyrolysis in the initial stage. In this case, the formation of small pores and the collapse of large pores occur simultaneously. This is why the surface areas in these two cases are different.

Studies were performed on four types of Chinese coal. Up to 500 °C, the pore volumes and pore surface areas do not change much, which is consistent with the observations of Maria et al. When the pyrolysis temperature reaches 500–700 °C, significant amounts of volatiles, e.g., C_nH_{2n}, CO, and CO_2, are vaporized. These volatiles originate from inside the coal particles, so their vaporization leads to increases in the pore volume and pore surface area. Above 700 °C, the coal mobility and tar precipitation clog some pores, causing decreases in the pore surface area and pore volume. Above 800 °C, lighter substances such as H_2 escape, leaving behind many micropores and causing an increase in the pore surface area [7]. Maria et al. [6] also studied changes in the fractal dimensions of Victoria coal. The results show that the changes in fractional dimensions differ slightly depending on the gaseous atmosphere and treatment method. Generally, the fractional dimensions do not change much at low temperatures, but increase significantly at high temperatures. In a N_2 atmosphere, the fractional dimension increase is not very obvious. For acid-washed coal samples, the fractional dimension is stable, at around 2, indicating that acid washing makes the coal surface smoother, leading to the formation of Eulerian planes [8].

Generally, as more volatiles are released during pyrolysis, the pore structure becomes better developed, leading in particular to the formation of a large number of micropores. The formation of micropores increases the inner surface area and surface roughness, and the fractional dimensions increase accordingly. This creates favorable conditions for coal combustion.

Early studies by the author [9] also indicated that the inherent minerals and major mineral compositions all have favorable influences on pore diffusion and new surface creation during coal coking.

Changes in surface structure during combustion. The coal surface structure changes significantly during combustion [10, 11]. Davini et al. [12] studied the surface area changes during coal and coke combustion. The main results are shown in Figs. 7.2 and 7.3. The general trends in the surface area changes are similar. The surface area first increases rapidly, followed by a gradual decrease. The time at which the surface area is maximum corresponds well to the time at which the combustion rate is maximum. The surface change curves also show that the higher the coking temperature is, the shorter the time taken to reach the peak surface area, and the faster the maximum combustion rate is achieved. The inner surfaces of coal particles provide the required contact areas for two-phase gas–solid reactions. An increase or decrease in the inner surface area is therefore directly related to an increase or decrease in the combustion reaction rate. An increase in the coking temperature causes an increase in the extent and speed of volatile vaporization. This increases the inner surface area, which provides favorable conditions for the combustion rate to reach its maximum value. The results of research by Ghetti et al. [13] also support the observations made by Davini et al. Ghetti et al. noted that after making necessary corrections to the coal surface area, the coal combustion rate increases monotonically with increasing coal surface area, and these two parameters reach their maximum values at the same time.

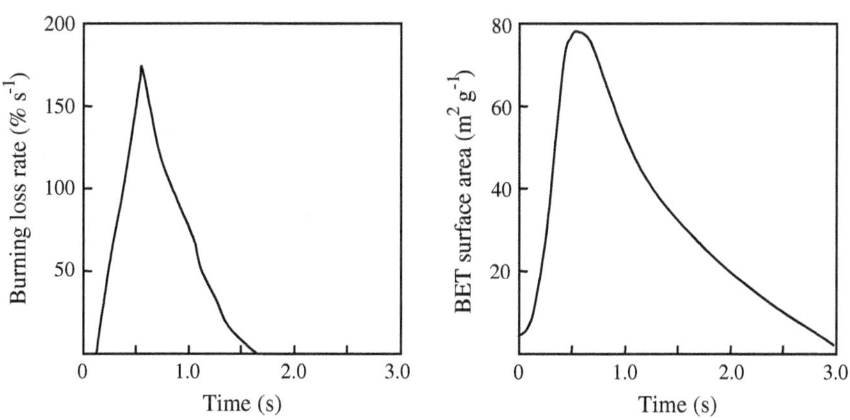

Fig. 7.2 Burning loss and Brunauer–Emmett–Teller (BET) surface area change curves for coal combustion. Reprinted from Ref. [12], copyright 1996, with permission from Elsevier

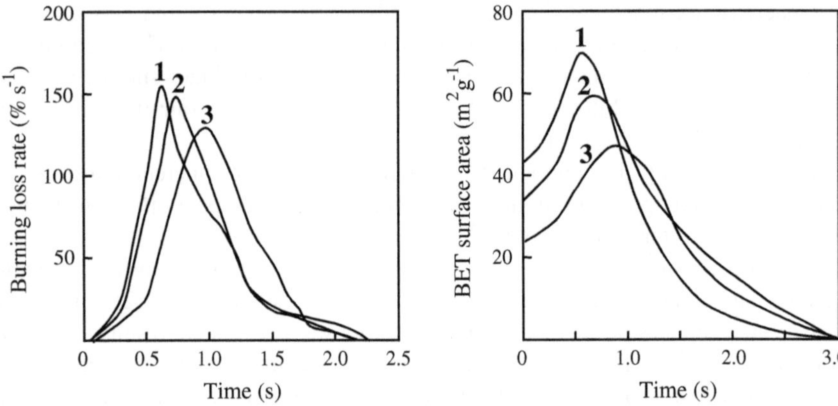

Fig. 7.3 Burning loss and BET surface area change curves for coke combustion at different coking temperatures: *1* 1000 °C; *2* 800 °C; and *3* 500 °C. Reprinted from Ref. [12], copyright 1996, with permission from Elsevier

Adams et al. [14] investigated three types of coal at three different combustion temperatures, and identified the changes in the coal surface areas. The results showed that the specific surface area was increasing during most of the combustion process. After the coal conversion rate had reached above 80 %, the specific surface area started to decrease. In contrast, the total coal surface area reached its maximum value when the conversion rate was 50 %. Adams et al. suggested that the 50 % conversion rate was the key turning point. This is the point at which the reactants have the maximum total surface area, which provides the maximum contact surface for the two-phase gas–solid reaction.

Gu et al. [15] studied the coal pore changes of three coals during coal combustion, and obtained data on the pore size distributions at different combustion stages at the same combustion temperature. Their results show that various coals, from lignite to anthracite, have similar trends in pore distribution changes. Before combustion, the pore size distribution has a single peak value, centered at around 30 nm. As combustion progresses, bituminous coal shows the most significant changes in pore size. The peak sizes become larger, and double peaks are formed. After combustion for 25 s, all three coals have double peaks or even multiple peaks in their pore distribution curves. Consequently, the pore network becomes well developed, which also shows that the pores are constantly changing during the combustion process.

Salatino et al. [16] measured the fractional dimensions of coal surfaces during combustion. Their results show that the fractional dimension is in the range 2.7–2.8, and this value does not change significantly at different combustion stages. It is believed that once the coal coke has started burning, its surface roughness has reached some degree of saturation. Although pores are opened up and collapse during combustion, the overall surface roughness is not further developed. This is why the fractional dimension does not show much variation.

7.1.2 Pore Models of Coal Combustion

A well-connected pore system and large specific surface area are favorable conditions for coal combustion. Different pore networks and specific areas lead to different combustion characteristics. The coal surface characteristics change as combustion progresses.

Comparing to homogeneous reactions, the multiphase reactions involved in combustion have two important characteristics features, i.e., the diffusion of reactant molecules and reaction boundary conditions. Coal particles have well-connected pore structures. This has two implications. The first is that the pore structure provides diffusion paths for the gaseous oxygen molecules. The second is provision of a large specific surface area, which is needed for gas–solid reactions to occur on the contact surface.

There have been some studies aimed at understanding the pore structure changes in coal combustion, and some theoretical models have been proposed. Thiele and Wheeler [17] modeled the pore structures in gasification reactions as disconnected uniform hollow structures. This model describes certain experimental observations, but clearly the pore distribution in coal cannot be uniform. Based on this work, Hashimoto et al. [18] extended this model to a random pore model, but this extended model did not become popular because it included too many parameters that cannot be easily determined experimentally. Recent models are the tree-shaped model developed by Simons et al. [19] and the random pore model developed by Bhatia et al. [20].

7.1.2.1 Single-Pore Model and Tree-Shaped Model

The single-pore model was proposed by Peterson et al. [21]. In this model, it is assumed that the pore structure in the porous material consists of hollow cylinders of constant diameter but variable height. When the gas–solid reaction takes place, the surface area occupied by the solid-phase reactant is assumed to be the same as the inner surface area of the hollow cylinder, and the reaction space is taken to be the volume enclosed by the hollow cylinder. The advantage of this model is that only two independent parameters are involved, i.e., the pore diameter and the pore height. However, this model has obvious problems. First, it does not take the pore distribution into consideration: clearly pores are not uniform hollow cylinders. Second, during the gas–solid reaction, the pore structure is constantly changing, but this model assumes that the pore structure is the same after the reaction. In addition, this model only considers the kinetic boundary conditions; it only applies to a special case of uniform reactant concentrations. These limitations led to the development of a better pore structure model, i.e., the tree-shaped pore model.

The tree-shaped pore model is a semi-empirical equation. In this model, a pore is defined as a hollow cylinder of radius r_p and height l_p; l_p is proportional to r_p, with a proportionality constant of about 10. The pore structure in the coal resembles an inverted tree structure. Large pores have openings on the coal surface, and small pores have opening into large pores, similar to tree branching. The pore radius distribution is described by a semi-empirical equation. For radii between r_p and $r_p + dr_p$, the pore radius distribution is

$$g(r) \propto \frac{1}{r_p^3}.$$

This model can be used to compute the total surface area of a pore structure with a root branch radius of R_T. It is assumed that a kinetic-controlled reaction and a diffusion-controlled reaction are in equilibrium at a pore of radius r_c. For pores of radius $r_t > r_c$, the reactions inside pores are diffusion controlled, whereas for pores of radius $r_t < r_c$, the reactions inside the pores are kinetically controlled. The pore size distribution function has to be considered in determining the rate of the gas–solid reaction for the entire pore structure. The tree-shaped model was the first to introduce the concept of pore size distribution. In addition, it distinguishes between kinetic-controlled and diffusion-controlled reactions. This model is one step closer to reality. However, the results predicted by this model for gasification reactions still do not agree well with the experimental data.

7.1.2.2 Random Pore Model

On the basis of previous pore theories, Bhatia et al. [20] proposed a new pore model, i.e., the random pore model. Bhatia assumed four pore structure models. In the first model, all pores (large and small pores) are randomly distributed throughout the relevant space. Pores of any size can overlap in an arbitrary manner. In the second model, large pores are distributed randomly in the solid, whereas small pores are randomly distributed in the space outside large pores. In the third model, sets of small pores are randomly distributed in the solid, but different sets do not overlap. In the fourth model, sets of small pores are randomly distributed in the solid, and different sets can overlap. Bhatia compared the adsorption values predicted by the models and the experimental data, and concluded that the third model was the most accurate. Based on the third model, Bhatia derived the conversion rate in a gasification reaction:

$$x = 1 - \left(1 - \frac{\tau}{\sigma}\right)^3 \exp\left[-\tau\left(1 + \frac{\varphi\tau}{4}\right)\right],$$

where

φ is the pore structure parameter;
σ is the pore size parameter;
τ is time.

The random pore model includes various parameters such as the pore volume, pore surface area, and pore structure. It also enables derivation of an explicit expression for the conversion rate. All these parameters can be easily obtained, so this model is more accurate and of practical use. However, this model still has some minor shortcomings, e.g., the effect of minerals on reactions is not addressed.

7.1.2.3 Shapeless Pore Model

Chen et al. [22] proposed a shapeless pore model. This model avoids the problem of pore shapes and sizes. It does not deal with the actual pore structure, does not treat each cell as a computational unit cell, and does not perform an integration along the perimeter of each cell. Instead, the integration is performed along the radius, based on the overall characteristics of all the pores. The sum of areas of the openings of large pores on the coal particle surface is defined as the diffusion flux area, S_d; S_d, the initial specific surface area S_r, and the actual combustion specific area are expressed as follows:

$$S_d = 4\pi r^2\theta,$$

$$\theta = \theta_0 + (1 - \theta_0)(1 - A)B,$$

$$S_r = 4\pi r^2 A_r \rho(1 - \theta_0),$$

$$S_{re} = S_r(1 - A)(1 - B),$$

where

θ is the porosity;
A is the ash content;
B is the burning degree;
R is the coal particle radius (m);
A_r is the specific area of the raw coal (m^2);
ρ is the density of the raw coal (kg cm^{-3}).

The overall reaction rate of coal combustion can be calculated using this model combined with mass transfer, heat transfer, and kinetic equations.

A more thorough study of pore structure modeling is still needed, to gain a better understanding of the coal combustion process and to obtain better agreement between model predictions and experimental results.

7.1.3 Studies of Changes in Surface State During Combustion of Four Coal Samples

This section presents results from the study of four different Chinese coal types at different combustion temperatures. The aim is to give readers a clear picture of how coal surface structures change during combustion. The four coal samples are Fuxin long flame coal, Xinzhi fat coal, Fengfeng meager coal, and Jincheng anthracite; their proximate and ultimate analytical data are shown in Table 2.9 in Chap. 2.

7.1.3.1 Determination of Coal Surface Structural Parameters During Combustion

In our experiments with 100 μm coal samples, isothermal combustion conditions were maintained at relatively low temperatures, i.e., 550, 600, and 650 °C, and the results were compared. According to published reports, for coal particles of such size, inner–outside diffusion plays a negligible role in combustion reactions.

The surface structure data for the four types of coals at three different temperatures are given in Tables 7.3, 7.4, 7.5, 7.6. The most probable pore size and average size are indicators of the pore size distribution in the pore structure. These two parameters do not change significantly as combustion progresses, i.e., the overall pore size distribution does not change greatly, indicating that the pore volume and specific surface area are changing simultaneously in the same direction.

Table 7.3 Surface structure data for combustion of Jincheng anthracite

Temperature (°C)	Conversion rate (%)	Specific surface area (m^2 g^{-1})	Most probable pore size (Å)	Average pore size (Å)	Pore volume (mL g^{-1})	Fractional dimension
550	11.48	128.0	8.71	18.86	0.09	2.87
	19.29	228.2	8.89	16.57	0.17	2.87
	28.92	269.0	8.54	16.52	0.20	2.87
	49.76	309.4	9.00	17.82	0.26	2.77
	69.39	306.0	8.89	18.63	0.26	2.74
	81.28	138.2	8.19	18.08	0.17	2.66
600	10.06	99.5	15.62	23.11	0.09	2.79
	18.68	114.1	8.31	13.24	0.07	2.93
	31.48	204.6	8.29	15.07	0.13	2.96
	59.86	250.7	8.36	15.95	0.20	2.81
	73.39	164.0	9.42	17.53	0.17	2.78
	83.90	156.0	8.49	18.49	0.16	2.78
	94.50	140.6	–	–	0.16	2.86
650	11.00	93.0	9.25	16.29	0.08	2.73
	22.41	155.9	8.21	16.12	0.10	2.93
	30.20	185.8	8.27	17.50	0.14	2.82
	44.35	188.8	–	–	0.18	2.82
	67.47	179.2	8.75	15.51	0.11	2.85
	91.70	106.6	8.32	15.99	0.10	2.77

Table 7.4 Surface structure data for combustion of Fengfeng meager coal

Temperature (°C)	Conversion rate (%)	Specific surface area (m^2 g^{-1})	Most probable pore size (Å)	Average pore size (Å)	Pore volume (mL g^{-1})	Fractal dimension
550	10.00	101.3	8.87	18.27	0.11	2.58
	22.24	184.6	9.68	17.70	0.13	2.93
	37.88	329.3	8.36	14.94	0.23	2.93
	66.90	379.3	8.67	17.26	0.54	2.85
	82.58	491.6	9.05	18.57	0.30	2.68
	94.37	304.6	9.43	17.74	0.21	–
600	10.42	134.4	8.30	14.95	0.09	2.85
	19.14	177.6	8.37	16.20	0.13	2.85
	38.44	250.2	–	–	0.18	2.91
	57.95	346.6	8.47	16.25	0.27	2.78
	75.65	431.8	9.33	20.87	0.41	2.76
	96.41	263.6	–	–	0.38	2.70
650	17.86	116.9	8.71	19.11	0.11	2.74
	27.88	194.4	8.13	18.33	0.18	2.84
	52.82	222.7	8.83	17.40	0.16	2.87
	66.11	357.9	–	–	0.30	2.81
	78.90	400.0	8.29	18.69	0.44	2.69
	91.73	272.8	8.20	19.95	0.26	2.67

Table 7.5 Surface structure data for combustion of Xinzhi fat coal

Temperature (°C)	Conversion rate (%)	Specific surface area (m^2 g^{-1})	Most probable pore size (Å)	Average pore size (Å)	Pore volume (mL g^{-1})	Fractal dimension
550	10.46	174.9	–	–	0.20	2.87
	18.70	206.5	8.80	19.74	0.25	–
	29.05	312.9	8.19	17.84	0.31	2.79
	44.05	420.5	8.68	17.75	0.39	2.65
	70.13	437.8	8.62	17.18	0.43	2.60
	83.07	315.5	11.49	24.61	0.29	–
600	17.06	113.3	8.76	22.08	0.20	2.54
	36.60	175.0	10.81	18.93	0.21	2.82
	52.97	321.0	–	–	0.32	2.72
	75.04	359.7	8.17	16.77	0.34	2.76
	91.35	228.7	–	–	–	2.60
650	22.43	139.3	8.36	15.16	0.11	2.96
	33.81	275.3	8.93	21.45	0.30	2.79
	47.96	278.8	8.93	19.13	0.23	2.78
	63.19	289.0	8.29	18.37	0.31	2.73
	77.48	292.3	–	–	0.54	2.45
	91.46	130.5	–	–	–	–

Table 7.6 Surface structure data for combustion of Fuxin long flame coal

Temperature (°C)	Conversion rate (%)	Specific surface area ($m^2\ g^{-1}$)	Most probable pore size (Å)	Average pore size (Å)	Pore volume ($mL\ g^{-1}$)	Fractal dimension
550	28.58	178.3	9.35	17.47	0.15	2.86
	41.87	247.7	–	–	0.18	–
	56.53	303.7	8.29	16.84	0.23	2.83
	73.55	322.2	8.54	15.86	0.23	2.88
	88.28	446.5	9.88	18.25	0.74	2.58
600	15.40	88.6	8.43	16.77	0.13	2.60
	27.90	162.6	8.62	18.41	0.13	2.83
	45.01	237.5	8.84	15.32	0.16	2.92
	60.51	278.8	–	–	0.27	2.89
	74.52	351.3	8.40	16.36	0.28	2.74
	86.12	491.8	10.35	18.42	0.63	2.77
650	29.52	100.9	–	–	0.19	2.68
	46.51	225.3	8.31	18.71	0.19	2.80
	57.18	259.1	8.27	14.92	0.19	2.80
	73.85	260.9	8.35	17.25	0.21	2.82
	81.60	271.7	8.80	15.86	0.24	2.82

The specific surface area and fractional dimension are the two most important parameters reflecting changes in the surface structure.

7.1.3.2 Changes in Specific Surface Area During Coal Combustion

Figure 7.4 shows the changes in the specific surface area during combustion of these four coal samples. The curves show common trends, but each coal has its own unique features. At the beginning of combustion, the specific surface areas of all four coals start to increase rapidly. At around 10 % conversion, the specific surface area has increased from 10 $m^2\ g^{-1}$ for the raw coal to 500 $m^2\ g^{-1}$. The surface area then continues to increase.

Jincheng coal reached its maximum specific surface area at 50 % conversion. For the three tested temperatures, the maximum specific area, 300 $m^2\ g^{-1}$, was achieved at 550 °C. The specific surface area then started to decrease. Fengfeng meager coal and Xinzhi fat coal reached their maximum specific surface areas, close to 500 $m^2\ g^{-1}$, at 80 % conversion at 550 °C.

A comparison of the results for these three coals shows that the general trends in the surface area changes are similar, i.e., the specific surface area initially increases rapidly, followed by a decrease as combustion proceeds. The differences are that different coals reached their maximum surface areas at different conversion degrees.

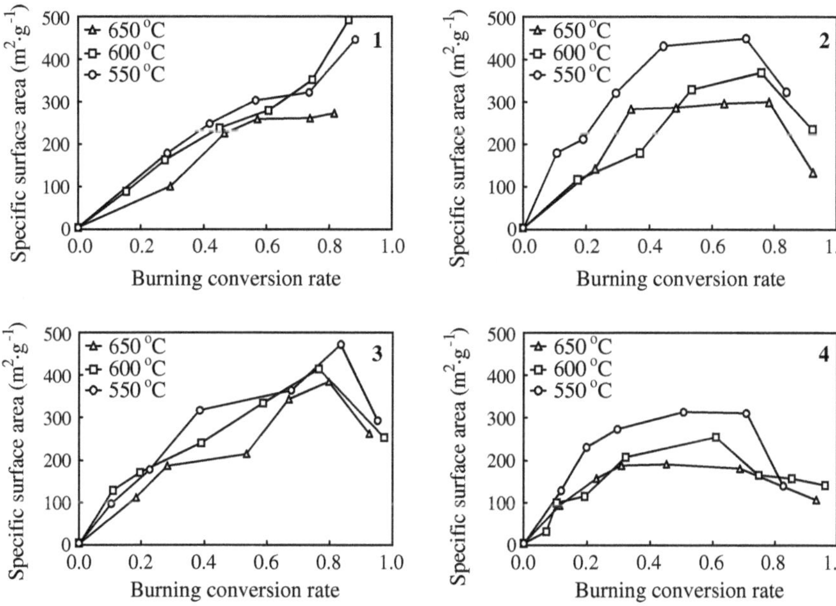

Fig. 7.4 Changes in specific surface area in combustion of coal samples: *1* Fuxin long flame coal; *2* Xinzhi fat coal; *3* Fengfeng meager coal; and *4* Jincheng anthracite

The behavior of Fuxin long flame coal is different from those of the others. Its specific surface area continues to increase up to a 90 % conversion. This might be because of its low degree of metamorphism. The uniform distribution of ashes may also play an important role in supporting growth of the pore network.

The surface area changes can be explained as follows. At the beginning of combustion, volatiles start to escape, leaving many new micropores on the coal surface. The consumption of organic carbon causes growth of the existing micropores. As a result of new pore creation and old pore enlargement, the specific surface area of coke starts to increase rapidly. When the combustion reaction reaches a certain degree, all the volatiles have escaped, and the process becomes dominated by a two-phase gas–solid reaction between carbon and gaseous oxygen. The consumption of organic carbon continues, and pores start to merge, weaken, and finally collapse. This process causes the specific surface area to decrease until it reaches a minimum value.

The porosities of the four coals were also measured using the N_2 adsorption method. The results show that at the tested combustion temperatures, neither the average radius nor the most probable radius changed significantly; their values are around 1.8 and 0.9 mm, respectively. The test equipment can measure pore sizes in the range 0.4–30 nm, so we deduced that combustion does not significantly change pore sizes in this range. This confirms that the rapid growth in the surface area at the beginning of combustion mainly contributes to new pore creation. This is further confirmed by the pore volume changes observed in the experiments. As the combustion reaction proceeded, the mass specific pore volume changed simultaneously with the coke specific surface area.

The structural characteristics of coal have an important effect on the surface structural changes of coke during combustion. Our results indicate that at three different combustion temperatures, the degrees of conversion corresponding to the maximum specific surface areas are the same, i.e., for different combustion temperatures, the boundary conditions for these combustion reactions are similar, and these boundary conditions are directly related to the coal rank.

For a coal, a higher metamorphic degree corresponds to a lower conversion degree at the point of maximum specific surface area. A lower metamorphic degree corresponds to a higher conversion degree. This may be because the degree of coalification deepens and the number of aromatic units on the coal surface increases, strengthening the interconnections between different units.

As a result, it becomes easier for pores to collapse as combustion progresses. The specific surface areas of coals with lower metamorphic degrees tend to decrease more during coal combustion. Comparisons of the specific surface areas at different temperatures also indicate that for coke of the same coal, the specific surface area at 550 °C is greater than that at 600 °C. The specific surface area reaches its minimum at 650 °C. In other words, a lower combustion temperature corresponds to a higher specific surface area. This is because at lower temperatures, the reaction conditions become milder, facilitating the creation of new micropores. The vigorous reactions at high temperatures can easily cause the pore structures to rupture, which does not facilitate the formation of pore structures with large surface areas.

7.1.3.3 Fractal Features During Combustion

In adsorption experiments, when $p/p_0 < 0.37$, we consider that a monolayer of gas molecules is adsorbed on the micropore surfaces. The adsorption features reflect the characteristics of a solid surface. Based on this assumption, Avnir et al. [23] derived a new adsorption equation that is related to the fractal structure of a solid surface:

$$\theta = k\left[\ln\left(\frac{p_0}{p}\right)\right]^{-(3-D)},$$

where

- θ is the relative adsorption;
- k is the adsorption constant;
- p, p_0 are the equilibrium and saturation pressure (Pa);
- D is the surface fractal dimension of burned coke.

Jincheng anthracite was burned for different times at 650 °C. Figure 7.5 shows a logarithmic plot of the measured adsorption data. It can be seen from this plot that the experimental data agree well with the modeled predictions. The correlation

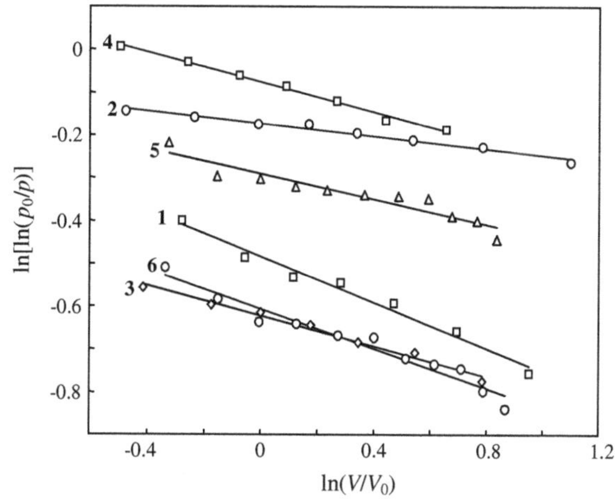

Fig. 7.5 Jincheng anthracite coke adsorption at 650 °C for different combustion times: *1* 2 min; *2* 4 min; *3* 5 min; *4* 6 min; *5* 7 min; and *6* 10 min

coefficients between these two sets of data are higher than 99 %. By fitting the data to a straight line, the fractal dimensions at different degrees of coal conversion can be obtained from the line slope. This method was also used to obtain adsorption data for the three other coals at different combustion temperatures. These data can be used to determine the dependence of the specific surface area and fractal dimension on the carbon conversion rate.

Figure 7.6 is a plot of the fractal dimension data for four coals at three different temperatures. Because of the differences in coal type, composition, and chemical structure, these samples underwent different surface structural changes during combustion. However, they also showed some common features. For instance, the specific surface areas of all these coal samples changed from 10 m^2 g^{-1} before burning to a maximum value of 500 m^2 g^{-1} after burning. Before burning, the fractal dimensions of all these coals were smaller than 2.5. After burning, the fractal dimensions increased to almost 3. This indicates that during combustion the surface structures of the coal changed significantly. It can also be seen from the plot that the fractal dimensions of these coals first increased and then decreased. However, the conversion rate at which the fractal dimension was maximum was much smaller than the conversion rate at which the specific surface area reached its maximum value.

Jincheng anthracite had the highest degree of degradation, and showed the most regular trend in surface structural changes. At all three temperatures, the specific surface area first increased rapidly, reached its maximum at 45–50 % conversion, and then decreased slowly. This is because at the beginning of combustion, escape of volatiles and the consumption of organic solids create new pores and enlarges the existing pores, leading to a rapid increase in the coke specific surface area. When the conversion degree is above 50 %, most of the organic solids have been consumed. As a result, the surface area of the micropores starts to decrease and large pores start to collapse, causing a decrease in the specific surface area of the coke. Raw Jincheng coal has a surface fractal dimension of about 2.48. When combustion

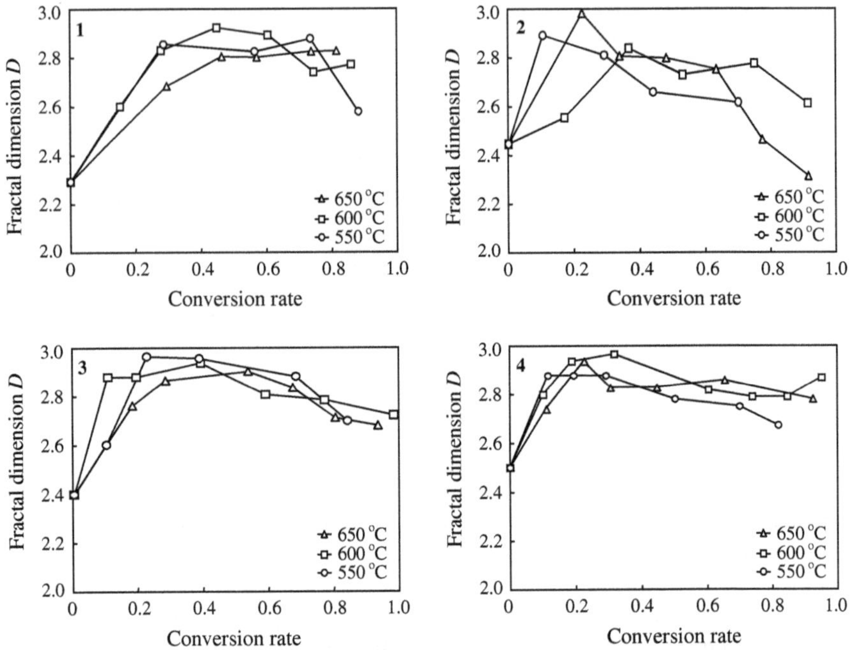

Fig. 7.6 Changes in surface fractal dimensions during coal combustion: *1* Fuxin long flame coal; *2* Xinzhi fat coal; *3* Fengfeng meager coal; and *4* Jincheng anthracite

starts, the fractal dimension increases quickly. When 30–40 % of the organic matter has been consumed, the coke surface has a three-dimensional network of pores at all three temperatures. At this point, the two-dimensional pore theory is no longer applicable. The fractal dimension then starts to decrease until it reaches about 2.75. The reason for the decrease is that the flowability of the coke decreases the pore roughness.

The fractal dimension changes for Fengfeng meager coal are similar. The fractal dimension increases quickly at the beginning of combustion. When the combustion rate reaches 25–40 %, the fractal dimension is close to 3. It then starts to decrease. The situation for Xinzhi fat coal is slightly different: its rate of increase is not so consistent, but its maximum value is also 2.8. The fractal dimension then starts to decrease at a faster rate. The fractal dimension increase for Fuxin long flame coal is slow, and the maximum value is not reached until the combustion degree is more than 50 %. These results show that the fractal dimension of Jincheng anthracite increases most rapidly, followed by Fengfeng meager coal, Xinzhi fat coal, and Fuxin long flame coal. This order is the same as that of the coal ranks, indicating that the higher the coal rank is, the faster the fractal dimension increases. This can be explained by the way in which coal burns. Coal combustion consists of three stages: escape of volatiles, combustion of volatiles, and combustion of organic solids. These three stages overlap. The volatile contents of low-degradation coals are much higher than those of high-degradation coals. The initial combustion stage is dominated by the vaporization and combustion of volatiles. The main combustion reaction is a homogeneous gas-phase reaction. High-degradation coals have much lower volatile contents. The initial combustion reaction is dominated by a two-phase gas–solid reaction between organic matter and oxygen. As a result, the surface structures of high-degradation coals undergo more vigorous changes than those of low-degradation coals, causing the fractal dimensions to increase more rapidly. When the fractal dimension passes its maximum value, the subsequent decrease is mild.

The final values of the fractal dimensions for these four coals are all around 2.7–2.8; these are much higher than the values for the corresponding raw coals. The entire combustion process can be regarded as a process that increases the fractal dimensions of the coal surface, causing the coal surface morphology to become more complex and irregular. It is difficult to analyze such complex reaction boundary conditions using conventional two-phase gas–solid reaction kinetics, and the results are inaccurate. Fractional theory and the introduction of fractional kinetics enables a better understanding and a more realistic picture of this complex coal combustion process to be achieved.

The current author studied coke surfaces using scanning electron microscopy (SEM) to clarify the correlation between surface morphology and fractional features. The coke sample was made using Jincheng anthracite at 650 °C. The coke pore structure is clearly revealed using SEM. The surface morphology of the pore structure continuously changes as combustion progresses.

When the coal conversion is about 10–20 %, the coke surface has very sharp wavy structures; this is clear evidence of the presence of pore structures. In comparison with previous analysis, the fractal dimension at this stage increases quickly until it reaches a maximum value. The fractional features on the coke surface are clearly seen in SEM images. When the coal conversion is between 30 and 45 %, the wavy structures become much smoother, but the density of the surface pores increases significantly. In comparison with previous data, the fractal dimension at this stage decreases slightly, but the specific surface area continues to increase until it reaches a maximum value. This trend is consistent with the information provided by the SEM images.

When the coal conversion is between 65 and 90 %, the SEM images show that most organic carbonates have been burned out, leaving inorganic salt ashes. At this stage, the pore structures starts to collapse and the surface morphology becomes much simpler. The specific surface area and fractal dimensions start to decrease. Finally, the coal reactivity approaches zero.

7.1.3.4 Change in Total Reaction Area During Combustion

The coke specific surface area at different burning percentages is an important parameter in measuring the coke surface morphology. However, this parameter does not indicate the contact surface area used in the two-phase gas–solid reaction.

This is because as combustion proceeds, the amount of reactants decreases. Converting the specific surface area to total reaction area makes more practical sense in studies of the coal combustion rate [24].

Figure 7.7 shows the total reaction areas for the four Chinese coals. These curves are very similar to those for the coke reaction rates; these will be discussed in later sections. Both parameters first increase and then decrease, and the curve shapes are almost identical, with the peak values occurring at conversion rates of about 50 %. This is clear evidence that the combustion reaction rate is directly related to the total reaction area. This point will be further discussed in the section on the kinetics of coke combustion.

The above discussion shows that the coal surface structure undergoes vigorous changes during combustion. The specific surface area initially increases and then decreases. The conversion rates corresponding to the peak surface areas differ for different coal ranks. The higher the coal degradation is, the lower the conversion rate is, and vice versa. The specific surface area also depends on the combustion temperature and coal rank. Coals with higher ranks have smaller specific surface areas, and lower temperatures facilitate the formation of larger specific surface areas. Coal combustion involves fractal growth of surface structures, and the fractal growth rate is faster than the surface area growth rate. Combustion reactions take place in a well-connected three-dimensional pore network, which is different from

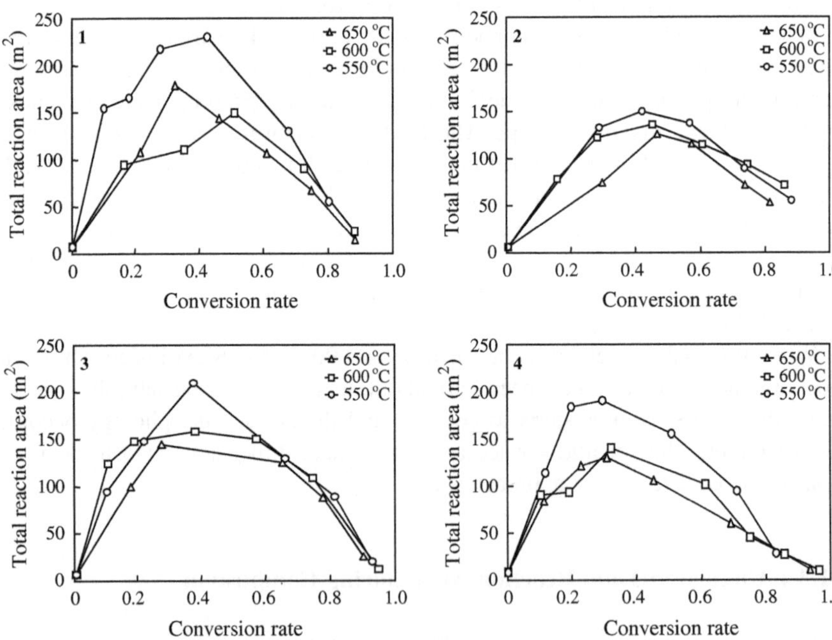

Fig. 7.7 Changes in total reaction areas during combustion of coal samples: *1* Fuxin long flame coal; *2* Xinzhi fat coal; *3* Fengfeng meager coal; and *4* Jincheng anthracite

the reaction surface for common two-phase gas–solid reactions. The rate of the fractal dimension increase is also related to the coal class. The higher the degradation degree is, the faster the fractal dimension growth rate is.

7.2 Kinetics of Coal Combustion

Coal combustion kinetics is closely related to the combustion conditions. The differences in the reaction kinetics under different conditions are quite complex [25, 26]. Sometimes the differences are caused by reaction conditions such as the combustion method, and the heating and temperature control methods. Other factors may also play an important role, such as systematic errors in kinetic experiments, numerical-processing method, and model selection. In this section, studies of the combustion kinetics of four types of coal, i.e., Fuxin long flame coal, Xinzhi fat coal, Fengfeng meager coal, and Jincheng anthracite, are presented.

7.2.1 Studies of Kinetics of Coal Combustion Using Temperature-Programmed Thermogravimetric Analysis (TGA)

The differences among coal classes can result in fundamental differences among combustion reactions. In this section, experimental studies of coal combustion using temperature-programmed thermogravimetry are presented. The coal samples were milled and screened. The TG curves obtained can be used to determine various characteristic temperatures such as the ignition and after-burn temperatures. The experimental data can be further used in the analysis of combustion kinetics.

7.2.1.1 Characteristic Combustion Temperatures

The ignition temperature is the temperature at which combustion starts. On a TG curve, it is the point at which the sample starts to lose mass, and its value reflects the ease with which coal combustion starts. The temperature at which the reaction rate is maximum is the temperature at which the combustion reaction rate is maximum. On a TG curve, this is the temperature at which the differential thermal gravimetry curve reaches its peak value. It is a good indicator of the combustion characteristics of a coal. The 50 % conversion temperature is the temperature at which the mass loss is half of the total organic mass loss. This parameter is similar to the half-life of a radioactive material. It is an important indicator of the combustion characteristics of a coal. The after-burn temperature corresponds to the temperature at which the combustion reaction completely stops. On a TG curve, this is the temperature at which the coal sample mass is constant. The ignition temperature and after-burn temperature are the two end points of the reaction.

These characteristic temperatures can be obtained from TG curves. Table 7.7 lists the characteristic temperatures for four coals. Fengfeng meager coal has the highest ignition temperature, i.e., 437 °C, followed by Jincheng anthracite, with an ignition temperature of 430 °C. Fuxin long flame coal has a much lower ignition temperature, 271 °C. The maximum difference between the ignition temperatures of these four coals is about 160 °C, indicating that the different coal classes have very different combustion behaviors. The large differences among the ignition temperatures can be explained by the coal combustion process.

Initially, the coal mainly undergoes pyrolysis, during which volatiles escape. This stage can be considered as the start of combustion. On the TG curve, this is indicated by a continuous and clear loss of mass. Different coal classes have very different coal compositions and structures. In general, a higher degree of degradation corresponds to a denser structure, which means a lower volatile content. When fewer volatiles are available to escape at the beginning of combustion, it is more difficult to start the process.

The ignition temperatures of these four coals confirm this. As the degree of degradation increases, the ignition temperature also increases. The only exception here is Jincheng anthracite, which has an ignition temperature slightly lower than that of Fengfeng coal. This is understandable because coals have complex compositions and structures, and the ignition temperature is affected by many factors.

We can also see from Table 7.7 that other characteristic temperatures such as the temperature of the maximum reaction rate and the temperature at which 50 % of organic matter has been burnt fall within a range of 100 °C, which is smaller than the differences among the ignition temperatures. With the exception of Fuxin long flame coal, the temperatures of the maximum reaction rate and 50 % organic burning are almost the same, with differences smaller than 5 °C. This indicates that the time at which 50 % of the organic matter has been burned corresponds to the time at which the reaction rate reaches its maximum. The temperature of the maximum reaction rate is slightly lower than the temperature for 50 % organic

Table 7.7 Characteristic coal combustion temperatures (°C)

Coal sample	Ignition temperature	Temperature of maximum reaction rate	50 % organic burning temperature	Burn-Out temperature
Fuxin long flame coal	271	514	497	636
Xinzhi fat coal	362	566	571	693
Fengfeng meager coal	437	566	571	664
Jincheng anthracite	430	589	593	710

burn-out, which means that the time taken to reach the maximum reaction rate is smaller than the time taken to burn out 50 % of the organic matter.

The burn-out temperatures for these four coals are relatively close to each other. Jincheng anthracite has the highest burn-out temperature, 710 °C; the burn-out temperature of Fengfeng meager coal is 693 °C. This shows that the burn-out temperature is not related to the ignition temperature. Coals with higher ignition temperature do not necessarily have higher burn-out temperatures.

7.2.1.2 Calculation of Reaction Activation Energy

Once the reaction model has been selected, the function $f(x)$ in the reaction kinetic equation can be determined. Based on the Arrhenius equation, $f(x)$ is integrated to give $F(x)$, and then the logarithm of $F(x)$ is plotted against the inverse of the temperature, $1/T$. The slope of the line corresponds to the reaction activation energy.

The reaction model was selected based on the methods described in Chap. 5, Sect. 5.2.1. Nine models of two-phase gas–solid reactions were considered: four diffusion models, three nucleation models, and two interphase reaction models. After a careful comparison, we determined that the shrink core model is the most appropriate for describing the combustion process to be discussed in this section.

The numerical results are listed in Table 7.8. Jincheng anthracite has the highest activation energy. The activation energies for the other three coals are nearly the same.

A comparison of the reaction activation energies and characteristic temperatures shows that Jincheng flameless coal has the highest ignition, maximum reaction rate, 50 % burn-out, and full burn-out temperatures. It is therefore more difficult to start the combustion reaction; therefore it has the highest reaction activation energy. It is worth noting that although the reaction activation energy has a clear physical meaning in reaction kinetics, and can be used to measure the level of difficulty of certain reactions, it has limitations in dealing with coal combustion reactions. This is because coal combustion is complex and affected by many factors, so the superficial reaction activation energy values obtained here can only be used for qualitative analysis.

Table 7.8 Activation energies obtained using temperature-programmed thermogravimetry ($kJ\ mol^{-1}$)

Coal type	Fuxin long flame coal	Xinzhi fat coal	Fengfeng meager coal	Jincheng anthracite
Activation energy	74.09	74.49	73.19	86.70

7.2.2 Coal Combustion Kinetics Under Fixed-Bed Conditions

Fixed-bed reactors are widely used in coal combustion and gasification, so the study of coal combustion characteristics under fixed-bed conditions is of practical significance. Fixed-bed experiments are relatively simple to conduct, and have fewer influencing factors to consider, so they are ideal for investigating reaction mechanisms. Studies of fixed-bed combustion reactions for various coals at different temperatures are presented in this section.

Low-temperature isothermal combustion conditions were used in the experiments. The conversions as a function of time are shown in Fig. 7.8.

7.2.2.1 Kinetic Calculations

Calculation of Reaction Rate

If the initial mass of the coal sample is m, and the mass after burning is m', the conversion rate is

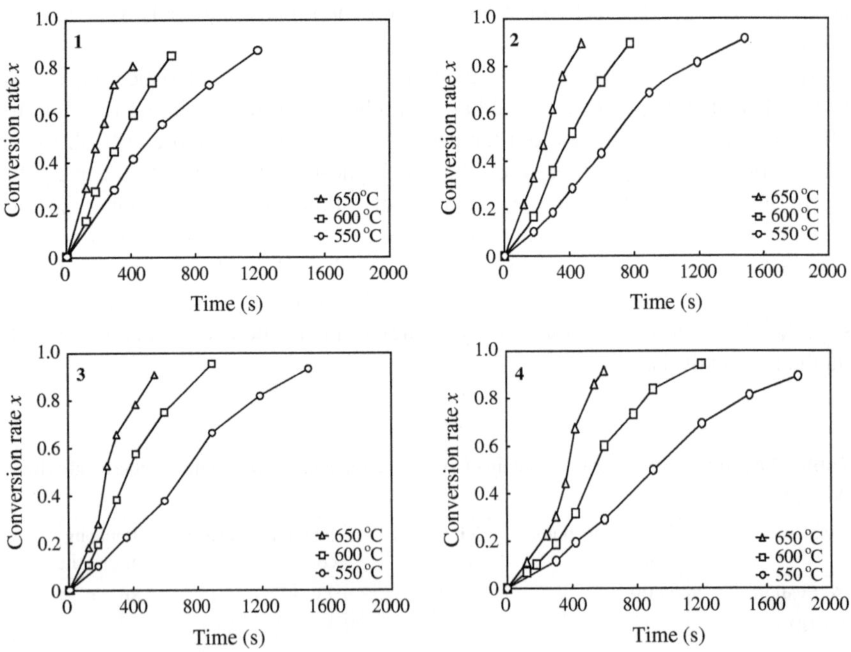

Fig. 7.8 Combustion conversion curves of coal samples: *1* Fuxin bituminous coal; *2* Xinzhi fat coal; *3* Fengfeng meager coal; and *4* Jincheng anthracite

$$x = \frac{m - m'}{m}.$$

The $x - t$ plot can be fitted to a polynomial:

$$x = c_0 + c_1 t + c_2 t^2 + \cdots + c_n t^n,$$

where n is the polynomial degree; it is often taken to be 3–6 in experiments. Differentiation gives the reaction rate:

$$\frac{\mathrm{d}x}{\mathrm{d}t} = c_1 + 2c_2 t + \cdots + nc_n t^{n-1}.$$

If the mass loss curve cannot be well fitted well to a polynomial [27], the reaction rate can be obtained by the central differential derivation equation, as follows:

$$\frac{\mathrm{d}x_i}{\mathrm{d}t} = \frac{(x_{i+1} - x_i)}{2(t_{i+1} - t_i)} + \frac{(x_i - x_{i-1})}{2(t_i - t_{i-1})}.$$

This equation can be used to obtain the reaction rate at the ith point; combustion rate curves can also be obtained, as shown in Fig. 7.9.

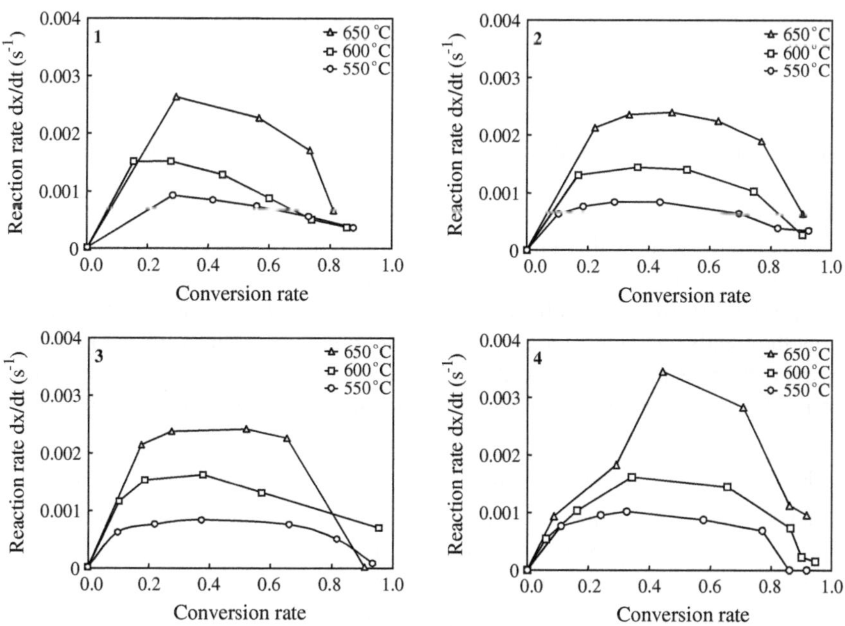

Fig. 7.9 Combustion rate curves of coal samples: *1* Fuxin bituminous coal; *2* Xinzhi fat coal; *3* Fengfeng meager coal; and *4* Jincheng anthracite

Selection of Reaction Model and Calculation of Activation Energy

The reaction model was selected using the method described in Table 5.2 in Chap. 5 to identify the reaction mechanism. In a comparative study, the author found that the random nucleation model, with $n = 2$, is suitable for coal combustion conditions, as discussed above. Because the fixed-bed combustion conditions used were isothermal, the procedure for finding the reaction activation energy is slightly different from the TG method.

Once the reaction model has been selected, the $F(x) - t$ plot can be obtained from the $x - t$ combustion curve at a given temperature. The reaction rate of the fixed-bed combustion at this temperature can be obtained from the slope of the linear line plot of $F(x) - t$. The chosen kinetic model gives linearity of $F(x) - t$, with a linearity correlation coefficient greater than 99 %. According to the Arrhenius equation, when the Arrhenius plot of 1 nk versus $1/T$ is linear, the slope of the line gives the superficial reaction activation energy. The calculated results for the reaction activation energies are listed in Table 7.9. The data in the table show that the activation energy order is anthracite > fat coal > long flame coal > lean coal. This order is consistent with the results in the previous section, shown in Table 7.8. The reaction activation energies obtained from fixed-bed combustion are slightly smaller than those obtained under TG conditions, which indicates that fixed-bed conditions facilitate combustion.

7.2.2.2 Features of Combustion Reactions of Four Types of Coal

Figure 7.8 shows the plots of conversion rate versus time for four types of coals at three different combustion temperatures. It can be seen from the plots that the burn-out time at 650 °C is about one-third of that at 550 °C. This indicates that the combustion reaction is fairly sensitive to temperature, and that at a relatively low temperature, the combustion reaction is controlled by chemical kinetics. The combustion plots show that Fuxin long flame coal has the shortest combustion time and Jincheng anthracite has the longest combustion time.

Figure 7.9 shows plots of the reaction rates for these four coals. The plots clearly show the trends and characteristics of the reaction rates. In the combustion process, the reaction rate first increases, then decreases, and reaches a maximum value at around 50 % conversion. This trend indicates that once the ignition temperature is

Table 7.9 Calculation results for kinetics of coal combustion under fixed-bed conditions

Coal type	Reaction rate (s^{-1})			Activation energy (kJ mol^{-1})
	550 °C	600 °C	650 °C	
Fuxin long flame coal	0.0011	0.0020	0.0031	65.58
Xinzhi fat coal	0.0010	0.0019	0.0032	73.55
Fengfeng meager coal	0.0011	0.0019	0.0029	61.32
Jincheng anthracite	0.0008	0.0015	0.0026	74.48

reached, the organic matter in the coal begins to react vigorously with oxygen. As the reaction progresses, the reaction rate increases until it reaches the maximum value. When the conversion rate is above 50 %, organic carbonates are exhausted and the amount of combustible matter decreases. This causes the reaction to slow down until it reaches zero. For the same class of coal at three different temperatures, the peak values appear at roughly the same conversions on the reaction rate curves. For flameless coal, the maximum reaction rate occurs at 50 % conversion, for Fengfeng meager coal and Xinzhi fat coal, the maximum reaction rates occur at 40 % conversion, and for Fuxin long flame coal it occurs at 30 % conversion.

These results indicate that the combustion characteristics have a close correlation with the coal class structure. Differences among coal class structures lead to each coal type having unique combustion behavior. The maximum reaction rate is also related to the coal degradation. The lower the degree of degradation is, the faster the reaction rate reaches its maximum. Coals with higher degrees of degradation reach their maximum reaction rates more slowly. This is because coals with higher degradation degrees have denser and closer structures; this does not facilitate oxygen diffusion, and fast propagation of the reaction is deterred. The plots also show that at lower temperatures, the maximum reaction rate decreases. This indicates that at lower temperatures, the combustion reaction is milder, and the coal class structure has a more profound effect on the combustion rate. However, at higher temperatures, the combustion reaction is fast and vigorous. The reaction rate is mainly controlled by the temperature, and the coal class structure plays a minor role.

Figure 7.7 shows plots of the total reaction area versus conversion rate. The figure shows that for Fuxin bituminous coal, Xinzhi fat coal, and Fengfeng meager coal, the reaction rate reaches its maximum at the same time as the total reaction area reaches a maximum. This timing corresponds to a conversion rate of about 40 %. Jincheng anthracite behaves differently. Its reaction rate reaches a maximum at 50 % conversion, whereas the total reaction area reaches a maximum at 40 % conversion. This is because when the conversion reaches 40 %, Jincheng anthracite has already achieved the maximum reaction interface for a two-phase gas–solid reaction. However, the reaction activation of Jincheng anthracite is much higher than those of the other three coals, so more energy needs to be accumulated to reach the maximum reaction rate.

This is why the reaction rate cannot keep pace with the total reaction area. The conversion rate at which the reaction rate reaches the maximum lags behind the conversion rate at which the total reaction area reaches its peak value. The reaction activation energies for the other three coals are smaller, therefore once the surface area conditions are satisfied, the two-phase gas–solid reaction can easily progress. In this case, the reaction rate and total reaction area are changing at the same pace.

Based on the above analysis, it can be concluded that the total reaction area provides the basic conditions for the two-phase gas–solid reaction in the combustion process. In general, if the reaction activation energy is small, then the reaction rate increases along with the total reaction area. When the reaction reaches a critical point, the available amount of combustibles and the total reaction area start to decrease. This causes the reaction rate to decrease, until all the reactants have been burned out. There are exceptions for some coals that have high reaction activation energies. For these coals, the reaction is not only determined by the interface, but is also related to energy transfer during combustion. In this case, the increase in the reaction rate lags behind the increase in the total reaction area: the increases in these two parameters are out of phase.

7.2.3 Isothermal TG Studies of Combustion Kinetics

To better understand the combustion process under different conditions, isothermal TG experiments were performed to simulate fixed-bed conditions on the TG scale. Five temperatures were studied: 480, 510, 540, 570, and 600 °C. When the required temperature was reached, N_2 was replaced by air to start combustion until the sample reached a stable mass.

The previously discussed model selection method showed that the random nucleation model, with $n = 2$, is the most suitable for the current combustion reaction. This model is the same as the fixed-bed combustion model, as discussed in Sect. 7.2.2 of this chapter, but is different from the program-controlled combustion model discussed in Sect. 7.2.1 of this chapter. A comparison of the conditions used in these three combustion reactions shows that the isothermal TG and program-controlled TG reactions belong to the same reaction system. However, as one is isothermal and the other is temperature increasing, these two combustions have different reaction models.

Fixed-bed and isothermal TG combustions belong to different reaction systems. However, because these two reactions have the same temperature conditions during combustion, they have the same reaction model. This shows that the temperature plays a critical role in coal combustion. Once the model has been selected, the reaction activation energies can be calculated for these four coals under the test conditions. The calculation results are listed in Table 7.10. A comparison of the reaction activation energies obtained using program-controlled TG experiments and fixed-bed combustion experiments shows that the results obtained using isothermal TG are more credible, and are consistent with the current test conditions.

Fixed-bed combustion experiments have relatively high errors because there are more parameters that are affected by the operator's experience. For program-controlled TG experiments, it is difficult to compare the reaction activation energies because different coals have different ignition and burn-out temperatures, so combustion occurs at different temperature ranges. The reaction activation energies

Table 7.10 Activation energies obtained using isothermal TG (kJ mol^{-1})

Coal type	Fuxin long flame coal	Xinzhi fat coal	Fengfeng meager coal	Jincheng anthracite
Activation energy	68.83	74.64	72.42	78.93

obtained from the isothermal TG experiments show that Jincheng anthracite has the highest reaction activation energy, and Fuxin long flame coal has the lowest. The order is almost the same as that for the coal class.

7.2.4 Fractal Analysis of Combustion Kinetics

7.2.4.1 Application of Fractal Theory to Coal Combustion

The introduction of fractal geometry has provided a useful tool for understanding the complexity of coal particle shapes and structures, enabling quantitative studies to be performed [28]. Gu et al. [29] studied coal particle shapes, and defined a shaping parameter for determining coal particle shapes:

$$\chi = \frac{4\pi A}{L^2},$$

where

χ is the coal particle shaping parameter;
L is the projected perimeter of the particle (m);
A is the projected area of the particle (m^2).

It is found that the coal particle shape does not change significantly during combustion. This is contrary to the common belief that coal particles become spherical at high temperature, as a result of melting. Based on this finding, the following expression can be obtained:

$$\log A(r) = \frac{2}{D}\log L(r) - \log C^2,$$

where

r is the fractal measurement unit;
D is the fractal dimension;
C is a constant.

This expression defines the relationship between the projected perimeter $L(r)$ and the projected area $A(r)$ as a function of the fractal measurement unit r. Studies have shown that for a given coal type, the fractal dimensions do not change at any combustion stage, regardless of which combustion method is used. However, different types of coal have different coal particle fractal dimensions. The fractal shaping parameter is therefore inherent to a coal type, providing a unifying standard for analyzing combustion processes for different coal types.

Many different models for obtaining quantitative descriptions of coal pore structures have been proposed, such as the tree-type and cylinder-type models.

Although these models have interesting features, they are not adequate or satisfactory. Fractal theory provides a direct method for determining pore structures, while avoiding the issue of pore shape. Earlier studies [30] compared the pore structures in SEM images at magnifications of ×2000 and ×5000. We found that the pore structures at different scales are self-similar. Histogram analysis of SEM intensity images has also shown that coal pore structures are self-similar.

This observation has opened the door for the application of fractal theory to the experimental study of coal combustion. Based on the classical theory of multiphase combustion together with some reasonable assumptions, an empirical model has been established to describe the combustion rate, q_c. This model takes many factors into consideration, including the effect of fractional dimension growth on combustion, the different cases of diffusion-controlled and kinetic-controlled reactions, the kinetic-driven fractal growth coefficient α_1 and fractal dimension D_S, the diffusion-driven fractal growth coefficient α_2 and fractal dimensions D_d, and the area reduction factor β, resulting from pore collapse. The combustion rate q_c from this model is expressed as

$$q_c = \left(\alpha_1 V^{\frac{D_S}{3}} + \alpha_2 V^{\frac{D_d}{3}} + 1\right)(1 - V)^{\beta} \times K P_{O_2}.$$

Figure. 7.10 shows the calculated results obtained using this model. The model predictions agree well with the experimental data, indicating that this model can be used to predict changes in the combustion rate.

Coal surface structures and the coal combustion process have been studied for many years. However, there is still no theory that combines and unifies the research results from these two areas, enabling breakthroughs. The introduction of fractal kinetic theory may bridge the gap between these two areas. It may be possible to make breakthroughs in this field by choosing the right test method and identifying

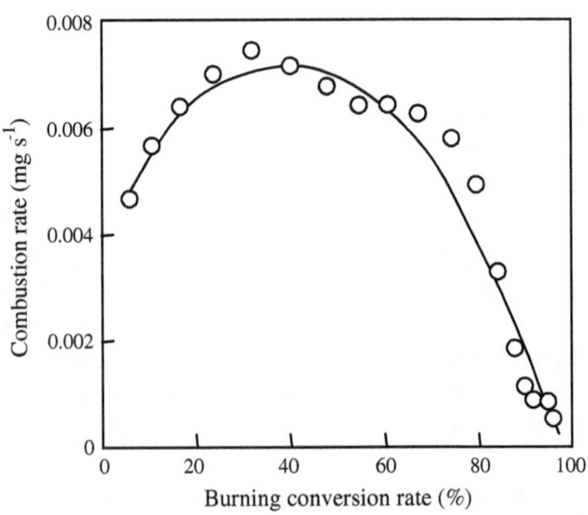

Fig. 7.10 Model predictions and experimental values for coal combustion rates; splashes are experimental values and curve is model predictions

the common critical factors. It is worth pointing out that coal combustion is a very complex process. Fractal kinetic theory has only recently been introduced in this field, so much more research work is required to make the theory better fit practical applications.

7.2.4.2 Fractal Kinetic Model of Coal Combustion

Chemical kinetics as an independent scientific field has a long history of over 100 years. However, until now, chemical kinetic theory has generally only been applicable to homogeneous chemical reactions [31]. Current chemical kinetic theory is unsatisfactory for inhomogeneous reactions. This is mainly because inhomogeneous reactions have different spatial dimensions. Unlike homogeneous reactions, inhomogeneous reactions depend closely on the diffusion behaviors of the reactant molecules. When the spatial dimension of the reaction medium becomes fractional, or when the inhomogeneous reaction medium begins to show fractional characteristics, the diffusion of reactant molecules does not follow the behavior described by classical theory. In contrast to homogeneous reactions, inhomogeneous reactions exhibit non-conventional chemical kinetic behavior [32].

The media structures in fractional reactions are complex, so reactions at the contact surface boundary are believed to be random and orderless. Farin and Avnir [33] listed 31 inhomogeneous-phase media with fractal characteristics. Coal is one of them. For coal in particular, the combustion reaction is highly non-symmetric and non-linear on different layers. Therefore, to better understand the kinetics of the multiphase combustion of coal, the fractional characteristics need to be considered, and fractal kinetic theory should be applied. Only in this way can we obtain predicted results that are in agreement with experimental data.

Coal combustion is affected or controlled by many internal and external factors. When all these factors are taken into consideration simultaneously, obtaining a clear understanding of the actual process is tedious and complex. In establishing the fractal kinetic model, the current author has focused on the most important factors, i.e., time effects in the reaction process, the reaction fractional-order effect, and restriction effects of the surface structure and ash layer.

Effect of Time on Reaction

The effect of time on the reaction refers to the changes in the reaction rate coefficient as a function of time. In classical reaction kinetics, the rate coefficient is a constant in three-dimensional space. It does not change as a function of time. This is not the case in fractal kinetics. The reason is that the reactant molecules have special diffusion features in the fractal medium. In ordinary Eulerian space, when a particle is performing a random walk in the medium for a period of time t, the total number of points the particle has passed through can be expressed as

$$S(t) \propto t.$$

The above equation does not hold for a fractal medium. Instead, the expression is

$$S(t) \propto t^{\gamma}.$$

where γ is a quantity that depends on the fractional dimension. The non-classical diffusion behavior of reactant molecules in a fractal medium results in peculiar chemical reaction kinetics.

Klymko et al. [34] found that the reaction rate coefficient, k, is closely related to the random walk behavior of particles in the reaction medium. This is expressed as

$$k \propto \frac{dS(t)}{dt}.$$

Therefore, in a fractal reaction medium, the rate coefficient should satisfy the following expression:

$$k \propto t^{\gamma-1}.$$

This indicates that in a fractal reaction medium, the rate coefficient is no longer a constant. Instead, it is a function of time and can be expressed as

$$k = k_f t^{-\tau}.$$

Many studies have shown that, in general, τ can be taken to be 1/3. This is the well-known "AO guess" [35]. This guess has been verified by many computer simulations. Based on this assumption, we derive the following expression for the chemical reaction kinetics:

$$\frac{dx}{dt} \propto k_f t^{-\frac{1}{3}},$$

where k_f is the kinetic coefficient for the fractal reaction.

Reaction Fractional-Order Effect

According to chemical reaction kinetics, the kinetic equation for coal combustion can be expressed as the carbon combustion rate:

$$\frac{dx}{dt} = kf_1(x)f_2(P_{O_2}).$$

Here f_2 is a function of oxygen concentration.

During the reaction, f_2 is fixed as a result of the constant influx of air. The carbon combustion rate can therefore be expressed as

$$\frac{\mathrm{d}x}{\mathrm{d}t} = k' f_1(x).$$

In addition, we assume that f_1 has the form

$$f_1(x) = x^{\alpha}.$$

We can then expect that the reaction order α may not be a whole integer. Instead, it might be a fraction for a fractal medium reaction. This is because coal combustion is so complex that it cannot be described as a basic reaction or a mixture of several basic reactions. Coal combustion is therefore influenced by many factors.

Restriction Effects of Surface Structure and Ash Layer

The effects of reaction time and reaction order suggest that the reaction rate increases as the conversion rate increases during combustion. However, this is not the case in reality. As the reaction proceeds, the amount of organic matter in the coal continues to decrease, and the ash content keeps increasing. At the same time, the reaction area is decreasing, which causes the reaction to slow down. Simons et al. [36] proposed that the efficient reaction area is proportional to $(1-\theta)$, where θ is the porosity, i.e., the reaction rate changes in the same direction as $(1-\theta)$. Based on this concept, we introduce a component in the reaction rate equation that suppresses combustion. The organic matter is consumed during combustion; therefore the coal porosity increases as a result of thinning of the pore walls and an increase in the pore volume, resulting in the following expression:

$$\frac{\mathrm{d}x}{\mathrm{d}t} \propto (1-x)^{\beta},$$

where β is an exponent to be determined. A larger value of β indicates a stronger hindrance of the reaction rate. A lower value of β indicates a weaker effect.

By combining the above three factors, the following equation for modeling the fractal kinetics of coal combustion is obtained:

$$\frac{\mathrm{d}x}{\mathrm{d}t} = k_{\mathrm{f}} x^{\alpha} (1-x)^{\beta} t^{-\frac{1}{3}}.$$

In the above equation, α and β can be obtained by fitting the equation with the experimental data.

7.2.4.3 Primary Treatment of Combustion Reaction Using Fractal Kinetic Model

We used the above modeling equation to recalculate the combustion rates obtained in the isothermal TG experiments discussed in Sect. 7.2.3 of this chapter. By comparing to the experimental data and making adjustments, we obtained the best fitted values for the parameters in the modeling equation. We found that $\alpha = 3/5$ for all four types of coal under all combustion conditions; this indicates that the coal reaction order is between 0 and 1. As a result of mixed control by many factors, the superficial reaction order is 3/5. The β value depends on the coal type and combustion conditions. Table 7.11 lists the β values for each type of coal at five combustion temperatures. The data in the table show that β is a fraction between 1/2 and 1. The lower the temperature is, the higher the β value is, suggesting that the ash layer and surface structure do constrict the reaction, and the lower the temperature is, the higher the restriction effect is.

A comparison of the modeled predictions with the experimental data for all the coal types shows that the modeling results are in good agreement with the experimental data.

Figure 7.11 shows the theoretical and experimental values for Fengfeng meager coal at a combustion temperature of 600 °C. Comparisons for the other types of coal show similar results. The combustion rate initially increases rapidly, followed by a gradual decrease after the peak value. The current theory performs well in modeling the experimental changes in the combustion reaction rate.

We used a fractal kinetics model to calculate the combustion rate coefficient and reaction activation energy for four types of coals at various combustion temperatures; the results are listed in Table 7.12. The data in the table show that the combustion rate coefficients obtained using fractal kinetics are higher than those obtained using conventional kinetic theory. The reaction activation energy obtained using fractal kinetic theory is lower than that obtained using conventional kinetic theory. These differences are the result of using different computational methods. The rate coefficient, k, in conventional kinetics is a function of both the fractional rate coefficient k_f and of time t; k decreases as the reaction time t increases. The longer the reaction time is, the more the rate coefficient k varies. In the computation using conventional kinetic theory, because the reaction time is longer at lower temperature, and the time has a large effect on k value variations, higher predictions

Table 7.11 β values for combustion of four types of coal

Temperature (°C)	Fuxin long flame coal	Xinzhi Fat coal	Fengfeng Meager coal	Jincheng anthracite
480	3/4	19/20	2/3	19/20
510	10/9	8/9	7/8	7/8
540	19/20	8/9	7/8	3/4
570	7/8	8/9	7/8	3/4
600	3/4	3/4	2/3	4/7

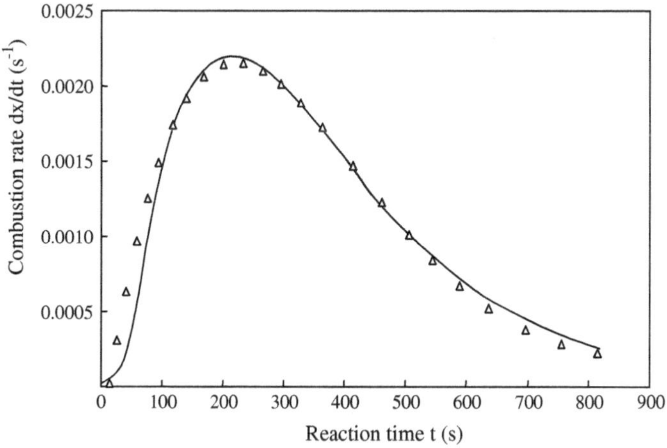

Fig. 7.11 Model predictions and experimental values for combustion rate of Fengfeng meager coal; splashes are experimental values, and curve is model prediction

Table 7.12 Fractal kinetic model results for combustion of four types of coal

coal type	Combustion rate coefficient (s^{-1})					Activation energy (kJ mol^{-1})
	480	510	540	570	600	
Fuxin long flame coal	0.011	0.023	0.026	0.031	0.035	58.01
Xinzhi fat coal	0.015	0.022	0.028	0.037	0.035	52.80
Fengfeng meager coal	0.014	0.021	0.036	0.038	0.040	62.29
Jincheng anthracite	0.026	0.035	0.040	0.054	0.047	40.42

are obtained for the reaction activation energy. It is also worth pointing out that the fractal kinetic theory also has shortcomings. The fractal kinetic model has only one indeterminate parameter β, so it is more simplified than other combustion models. However, because β depends on the coal type and combustion conditions, it is difficult to apply this model to practical problems. More in-depth studies are needed to achieve a better understanding of β, to make this model more meaningful and applicable. These are the current tasks and challenges in using fractal kinetic theory to solve the complex problem of coal combustion.

References

1. Gurgel Veras CA et al (1999) Combust Flame 116(4):567
2. Senneca O et al (2006) Combust Flame 144(3):578
3. Chen P (2001) Nature, classification and utilization of Chinese coal. Chemical Industry Press, Beijing, p 57
4. Indrek K et al (2007) Proc Combust Inst 31(2):1897

5. Wang B et al (2012) Energy Fuels 26(3):1565
6. Maria M et al (1983) Fuel 62:1393
7. Qiu JH (1994) J Fuel Chem Technol 22:316
8. Johnston PR et al (1993) Interface Sci 155:146
9. Xie KC et al (1990) Gasification kinetics and minerals role. Shanxi Science & Education Press, Taiyuan p303
10. Alvarez D et al (2007) Energy Fuels 21(2):1085
11. Lee DW et al (2012) Ind Eng Chem Res 51(42):13580
12. Davini P et al (1996) Fuel 75:1083
13. Ghetti P et al (1985) Fuel 64:950
14. Adams KE et al (1989) Carbon 27(1):95
15. Gu F et al (1993) J Fuel Chem Techn 21:425
16. Salatino P et al (1993) Carbon 31:501
17. Wheeler A (1951) Advances in catalysis 3:249 42 Chapter 7
18. Hashimoto K, Silveston PL (1973) AIChE J 19:259
19. Simons GA et al (1979) Combust Sci Technol 19:217
20. Bhatia SK (1987) AIChE J 33:1707
21. Peterson EE (1973) AIChE J 19:268
22. Chen H et al (1994) CIESC J 45(3):327
23. Avnir D et al (1989) Langmuir 5(6):1431
24. Senneca O et al (2011) Proc Combust Inst 33(2):1763
25. Gil MV et al (2012) Appl Energy 91(1):67
26. Senneca O et al (2010) Fuel 102:751
27. Wang FM et al (1992) Applied numerical calculation methods. Science Press, Beijing
28. He W et al (2013) Combust Sci Technol 185(11):1624
29. Gu F et al (1994) Sci China (A) 24(9):1001
30. Ren YZ (1997) J Eng Thermophys 18(1):103
31. Wan R et al (1996) Chemistry 7:1
32. Avnir D (1989) The fractal approach to heterogeneous chemistry. Wiley, Chichester, p 295
33. Farin D et al (1987) J Phys Chem 91:5517
34. Klymko PW et al (1983) J Phys Chem 87:4565
35. Alexander S et al (1982) J Phys Chem 43:1625
36. Simons GA (1983) Prog Energy Combust Sci 9:269

Chapter 8
Coal Swelling

Abstract Coal swelling is a phenomenon associated with physical and morphological changes that occur when coal is steeped in a particular solvent. Coal may be considered as a cross-linked polymeric macromolecule that may allow particular solvent molecules to penetrate the flexible framework. Analysis of the changes to the coal and solvent during swelling can provide valuable structural information on the coal, and facilitate more efficient use of this coal. This chapter describes coal swelling from a phenomenological viewpoint. We discuss prior studies in this area and the models that have been used to describe the swelling behavior. We compare the behavior of raw coal against that which has been swelled in pyrolysis and liquefaction processes. We also study the changes in properties (thermal, pore, and surface effects), char properties that occur and demonstrate the utility of swelling as a technique to probe the coal structure. We discuss the relationship between swelling and structural parameters, and present an investigation of the swelling behaviors of eight Chinese coals.

8.1 Introduction to Coal Swelling

The study of coal swelling began in 1961 after van Krevelen [1] proposed that coal has a polymeric structure. The simple techniques can provide a large amount of structural information and has a wide range of applications in coal science. The three major advances in swelling studies were as follows. Sanada et al. [2] used the Flory–Huggin lattice model to calculate the average molecular weight of coal, and determined its extent of crosslinking; Solomon et al. [3] determined the swelling ratio during coal tar pyrolysis, which reflects the beginning of the crosslinking reaction in coal pyrolysis, predicted the generation of gases such as CO_2 and CH_4, and described the coal network structure qualitatively; and the use of swelling as a pretreatment in coal liquefaction. In polymer science, the swelling of crosslinked polymers results from the presence of polymer chains because chemical bonds in the polymer form crosslinks to give a three-dimensional polymeric network

K.-C. Xie, *Structure and Reactivity of Coal*, DOI 10.1007/978-3-662-47337-5_8

structure. Crosslinks have some flexibility, so they can be bent and stretched, but low-molecular-weight solvents can enter the macromolecular mesh to cause swelling of the polymer. In the swelling process, the solvent molecules spread throughout the polymer, making the network swell. Penetration by low-molecular-weight solvents increases the distance between crosslinking points in the cross-linked polymer, so the elastic restoring force increases. When the elastic restoring force of the polymeric network and the diffusion of the solvent reach equilibrium, the polymer volume no longer increases. Different experimental conditions such as the swelling temperature or the nature of solvent lead to different equilibrium states. If the experimental conditions are fixed, different swelling equilibrium states indicate different extent of crosslinking, i.e., the densities of the crosslink bonds are different. When the conditions are fixed, the swelling ratio of a crosslinked polymer is closely related to its extent of crosslinking.

Polymer swelling thermodynamics is based on the Flory–Huggin lattice model of the structural features of macromolecular chains. This theory is not appropriate for determining coal structure because coal swelling is accompanied by rupture of weak bonds and dissolution of some small molecules. The overall process is therefore irreversible, and it is impossible for the coal to return to the original state.

8.1.1 Measurement and Calculation of Swelling Ratio

There are two types of swelling: swelling in liquid solvents and swelling in solvent vapors. When coal swelling occurs in a solvent vapor, the solvent is adsorbed on the coal surface and a phase change occurs; thus this is not suitable for thermodynamic studies. The swelling extent of the void and coal stacking must also be considered. However, this method has the advantage of avoiding solvent extraction, and post-treatment of the sample is easy. Coal swelling in a liquid solvent is a relatively simple experiment that is performed in a tube, so that the swelling volume can be measured by reading the scale before and after swelling [4–7].

The ratio of swelling is given by

$$Q_d = \frac{V_0 + V_s}{V_0} = \frac{h_2}{h_1},$$

where

Q_d is the swelling ratio of the dry coal sample;

V_0 is the coal sample volume before swelling (mm^{-3});

V_S is the increase in the coal sample volume when swelling reaches equilibrium (mm^{-3}); and

h_1, h_2 are the coal sample stack heights before and after swelling reaches equilibrium in the tube (mm).

If we assume that minerals occupy certain volumes before and after swelling, but the minerals do not swell, the ratio of swelling taking the mineral effect into consideration is:

$$Q_{\mathrm{dmmf}} = \frac{(Q_{\mathrm{d}} - 1) + \left(1 - \chi \cdot \frac{\rho_{\mathrm{d}}}{\rho_{\mathrm{mm}}}\right)}{\left(1 - \chi \cdot \frac{\rho_{\mathrm{d}}}{\rho_{\mathrm{mm}}}\right)},$$

where

χ is the mineral content (wt%);
ρ_{d} is the dry coal density (g cm^{-3}); and
ρ_{mm} is the mineral density (g cm^{-3}).

If minerals are removed using HCl/HF, the equation can be simplified to

$$Q_{\mathrm{dmmf}} = Q_{\mathrm{d}} = \frac{h_2}{h_1}.$$

8.1.1.1 Solvent Diffusion in Coal

An understanding of the solvent diffusion mechanism in coal is essential for coal use; for example, in the pretreatment process in direct coal liquefaction, the swelling process, in which the solvent loosens the coal structure, determines whether the hydrogenating agents can quickly enter the coal structure.

In polymer chemistry, there are two solvent diffusion mechanisms for glassy polymers: Fickian diffusion, which is caused by the concentration gradient inside and outside the particles [8, 9], and Case II diffusion, which is controlled by the looseness of the structure, i.e., there is a transition from the glassy to a rubbery state [10]. An understanding of the diffusion mechanism during swelling therefore contributes to understanding of the coal structure.

In general, the diffusion rate is not closely related to the coal rank. The larger the molecular volume of the solvent, the lower the diffusion rate. The diffusion mechanism changes from Fickian diffusion to Case II diffusion with increasing solvent alkalinity. The solvent diffuses through the highly irregular coal structure, the aperture changes during swelling, and interactions between polar solvents and coal can change the adhesive properties and elasticity of the coal; therefore the traditional glassy polymer diffusion mechanism cannot be used to explain diffusion in coal. The coal-swelling dynamic equation is usually given as

$$\frac{M(t)}{M_{\mathrm{e}}} = \frac{Q(t) - 1}{Q_{\mathrm{e}} - 1} = kt^n,$$

where

$M(t)$ is the amount of solvent absorbed in a given time;
M_e is the amount of solvent absorbed at swelling equilibrium;
$Q(t)$ is the swelling ratio at a given time;
Q_e is the swelling ratio at swelling equilibrium; and
k, n are diffusion parameters related to the coal properties.

8.1.1.2 Swelling Process and Coal Structure

Coal structure and reactivity are very important properties in coal science and technology. Study of the relationship between coal structure and reactivity is complicated by the non-uniformity of coal and its components, and many aspects remain unclear. van Krevelen [1, 11] investigated coal structure from the perspective of a colloid. Solomon et al. [12] and Wolf [13] used model compounds to simulate coal pyrolysis to clarify the pyrolysis mechanism. Coal structure is thought to consist of a covalently bonded three-dimensional macromolecular network. Non-covalent hydrogen bonds and van der Waals forces interactions, which are categorized as secondary forces, also play important roles. The secondary forces determine many aspect of the physical structure of the coal, in terms of the small-molecule phases that are embedded in the macromolecular network through secondary forces. During extraction processes, the secondary forces in the macromolecular network are disrupted.

Nearly all coal ranks, from lignite to anthracite, partially dissolve in suitable solvents accompanied by swelling occurs, making this a versitile technique in the study of coal. Dryden [14] systematically studied the swelling properties of coal macromolecules in various solvents. This method is used for quantitative studies of coal macromolecular networks to estimate the extent of crosslinking in the raw coal. Polymers that do not pyrolyze easily but swell readily, are known to be crosslinked substances. Therefore, swelling gives an indication of crosslinking in coal. A higher degree of swelling leads to a lower extent of crosslinking. Different coal particles swell at different rates, so that swelling processes are kinetically controlled. The macerals and minerals in coal affect swelling properties, and their hardness and shape affect the swelling process. For these reasons, 20–30-μm-thick coal wafers are used to study coal–solvent systems, to overcome the problems of irreversible swelling and coal heterogeneity [15]. This method can determine geometric and size changes in the sample surface during swelling. The results show that propylamine and pyridine have different impacts on coal. For example, if propylamine is used in the swelling process, the coal size changes through swelling, but the shape is unchanged. After removal of the solvent, the original shape is restored. Pyridine causes coal volume to double during swelling, but the volume then contracts after removal of the solvent. These differences are attributed to dissolution of some materials from the coal, and structural rearrangement of macromolecular chains.

Cody et al. [16] investigated the thin-layer swelling of coal of different ranks and found anisotropic swelling behavior in all solvent types. The swelling ratio is larger in the vertical direction than in the parallel direction, and the times needed to reach swelling equilibrium are different, depending on the coal rank. When the solvent is removed, the coal cannot be completely restored to the original state. In general, the thin-layer volume increases, and the coal is transformed from the glassy to the rubbery state.

Swelling is also commonly used to determine the extent of crosslinking during coal pyrolysis, together with functional group changes and the escape of gases; it can accurately reflect the crosslinking reactions during coal pyrolysis and the controlling factors. Solomon et al. [17] measured the swelling ratio at different heating rates, and the results indicate that crosslinking is closely related to coal rank. For lignite, the crosslinking starts before tar precipitation; for high-volatile bituminous coal, the crosslinking temperature is higher than that for tar precipitation.

There are diverse views on the swelling process because of differences among the test methods and approaches used. Typically, studies of the relationship between the swelling process and coal structure examine the solvent diffusion mechanism, the solvent electron-donor number (EDN), the solvent basicity (pK_b), the size of the molecules, coal rank, coal oxidation process, and the influence of the coal:solvent ratio on the swelling ratio.

8.1.2 Model of Coal Physical Structure and Swelling Process

8.1.2.1 Two-Phase Model of Swelling Process

According to the two-phase model, the main body of the coal structure is a three-dimensional crosslinked macromolecular network containing small molecules; the network and small molecules are connected through electron donators and acceptors. The model is based on the viscosity and elasticity exhibited by the coal structure when the swelling reaches equilibrium in a particular solvent. Coal has swelling characteristics, indicating the presence of crosslinks; the crosslinks lead to formation of a macromolecular network. In swelling tests, the main structure of the coal is not destroyed; the small portion that is dissolved is the small-molecule phase.

^{1}H nuclear magnetic resonance (NMR) spectroscopy was used to examine coal before and after swelling with fully deuterated pyridine. The results show that the amount of hydrogen increases by 40–50 % in the fluidized phase [18, 19]. This is because the added pyridine destroys the non-covalent bonds in the coal structure, the small-molecule phase decreases, and the amount of fluidized hydrogen increases. Brenner studied the optical properties of swollen coal and raw coal, and concluded that when raw coal is immersed in a polar solvent such as pyridine, the coal rapidly swells and is transformed from a glassy state to materials similar to highly crosslinked plastics. The coal structure becomes loose, proving that the coal consists of highly flexible macromolecules [20].

The two-phase model of coal suggests the presence of small molecules, and that the small molecules and macromolecular networks are connected by electron donors and acceptors, which are present in both the macromolecular phase and as small molecules. There are many advantages to the two-phase model, but it also raises many questions, such as the following. What are the relative contents of small molecules and macromolecules in the coal? How do these contents change during coalification? Are there clear boundaries between the two phases? The main body of the coal is a macromolecular network structure, similar to a polymer, suggesting that the swelling process should be reversible, when actually, it is not.

8.1.2.2 Association Model of Swelling Process

Nishioka [21] proposed the association model, based on swelling and extraction experiments. Brenner [22] observed that the coal structure is unchanged after nitrogen purging at a particular temperature and removal of the solvent, suggesting swelling is a reversible process. Cody used Brenner's method and repeated many experiments; the results show that the volume changes from −27 to 6 % after removal of the solvent. Moreover, the solvent is not completely removed, i.e., 5–15 % remains, suggesting the swelling process is irreversible [16]. Sanada et al. [2] found that the swelling ratio of the same coal at 298 and 323 K differ significantly, which is clearly incompatible with elastic swelling.

Nishioka studied the effect of the coal to solvent ratio on swelling. It was found that a lower ratio leads to a higher swelling ratio. It is suggested that at low concentrations, more coal associates, so the swelling ratio is greater, i.e., the coal associative structure determines the swelling ratio. Nishioka proposed a coal association model (Fig. 8.1), based on coal extraction experiments. The structural model stresses the importance of non-covalent bonds in the coal, and these determine the main physical and chemical properties of the coal.

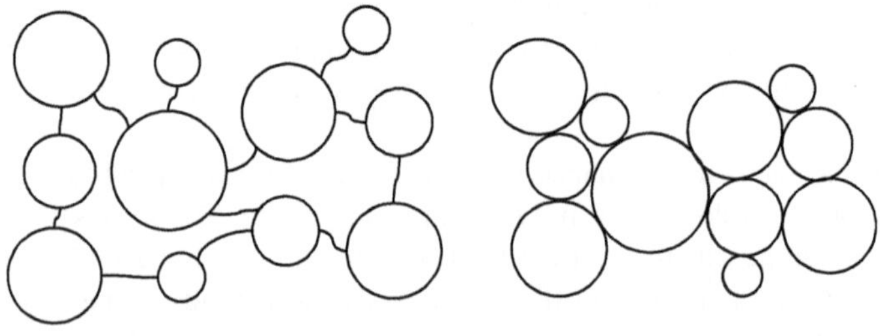

(1) (2)

Fig. 8.1 Two-Phase Model (*1*) and Association Model (*2*). Reprinted from ref. [21]. Copyright 1993, with permission from American Chemical Society

Table 8.1 Descriptions of swelling process

Experiment	Two-phase model	Association model
Swelling process	Reversible	Irreversible
Coal:solvent ratio	Unrelated	Important
Swelling ratio (Q) of phases	$Q_{Solubles} > Q_{Insolubles}$	$Q_{Solubles} < Q_{Solubles}$

It should be noted that the two-phase model and the association model give different descriptions of the swelling process because they emphasize different contents from different perspectives (Table 8.1).

8.1.3 Factors Affecting Swelling Ratio

8.1.3.1 Effect of Solvent on Swelling Ratio

Szeliga studied the swelling of three types of coal using 20 different solvents with different EDNs [23]. The results show that the swelling ratio is clearly dependent on the solvent. The coal swells slightly when the EDN is 0–16, possibly because of diffusion of the molecules. The coal swelling ratio increases sharply when the EDN is 16–30, mainly as a result of interactions between the solvent and electron-donating and electron-accepting functional groups.

8.1.3.2 Effect of Solvent Alkalinity on Swelling Ratio

The solvent alkalinity (pK_b) essentially reflects the hydrogen-donating ability of the solvent. Hall et al. [24] concluded that the solvent pK_b has a strong correlation the with the coal swelling ratio. There is a critical pK_b, above which the swelling ratio starts to decrease, above this critical value, the alkalinity has no influence on the swelling ratio. The solvents used are listed in Table 8.2. If the steric effect of the solvent is not considered, the interactions of hydrogen bonds in the coal with the solvent increase with increasing solvent alkalinity, the hydrogen bonds that connect the coal macromolecular structures are broken, and the solvent forms a stable

Table 8.2 Solvent pK_b values and molecular volumes. Reprinted from ref. [24], Copyright 1988, with permission from Elsevier

Solvent	pK_b	Molecular volume ($cm^3\ mol^{-1}$)
Tetramethyleneimine	2.7	84.1
2-Methylpyridine	8.0	98.8
Pyridine	8.6	80.9
3-Chloropyridine	11.5	95.1
2-Chloropyridine	13.5	94.1
3-Fluoropyridine	14.4	86.1

system with the coal body. Swelling kinetic studies show that the solvent basicity has important effects on the coal swelling ratio and solvent diffusion; the swelling rate changes with changes in solvent alkalinity [25].

8.1.3.3 Solvent Steric Effect on Swelling Ratio

Aida et al. concluded that the swelling ratio increases with increasing solvent molecular volume, but stops at a certain critical value. When the coal and solvent interact, hydrogen bonds break, the extent of crosslinking decreases, and solvent molecules diffuse into the coal pores, so the network structure expands [26, 27]. Brenner [28] confirmed that after diffusion of amine molecules into the coal macromolecular structure, the structure changed from a glassy state to a rubbery state, mainly as a result of fracture of hydrogen bonds by the amine. When hydrogen bonds are broken, the coal molecular chains are reoriented; the extent of the reorientation depends on the number of hydrogen bonds fractured and the extent of crosslinking of the covalent bonds. The size of the solvent molecules has an important effect on the swelling ratio and the swelling kinetics, but it does not change the solvent diffusion mechanism. The solvent pK_b and the solvent steric volume therefore determine the swelling ratio and are important in the study of coal structure. The solvent EDN plays an auxiliary role in coal swelling.

8.1.3.4 Effect of Coal Rank on Swelling Ratio

The swelling ratio is significantly affected by the crosslinking degree of covalent bonds and the functional group distribution in the coal; these factors are closely related to the coal rank. When the carbon content is larger than 85 %, the hydrogen bonds in coal almost disappear. Many functional groups such as hydroxyl and carbonyl groups readily interact with polar solvents to form hydrogen bonds and destroy the original structure of the coal. The numbers of such functional groups decreases with increasing coalification. Sanada et al. [2] studied various coals with different carbon contents, and found that coal swelling decreases with increasing carbon content. The swelling ratio decreases significantly when the carbon content is higher than 85–87 %, The coal type therefore affects solvent diffusion, and is important in studies of the swelling mechanism.

8.1.3.5 Effect of Oxidation on Swelling Ratio

Ndaji et al. [29] heated coal at 200 °C for 24 h and studied the effect of coal oxidation on the swelling ratio; the experimental data are listed in Table 8.3. The results show that the swelling ratio increases after oxidation of the coal. Fourier-transform infrared (FTIR) spectroscopy shows that oxidation destroys some of the covalent bonds, decreases the ratio of aliphatic chains to aromatic rings, increases

Table 8.3 Effect of oxidation on swelling ratio. Reprinted from ref. [29], Copyright 1993, with permission from Elsevier

Coal	Species 1		Species 2		Species 3		Species 4	
	Raw	After	Raw	After	Raw	After	Raw	After
Oxygen content (%)	13.9	21.3	8.2	16.7	3.7	9.2	4.2	5.6
Swelling ratio	2.17	2.37	2.1	2.33	1.09	1.87	1.01	1.81

the ratio of hydroxyl groups to aromatic rings, and increases the number of carbonyl groups. These factors lower the extent of crosslinking, forming a more open coal structure, resulting in an increase in the swelling ratio.

8.1.3.6 Effect of Coal to Solvent Ratio on Swelling Ratio

The effect of the coal:solvent ratio is often overlooked in swelling experiments. Turpin et al. [30] attributed changes in the swelling ratio caused by the coal:solvent ratio to experimental error. This ratio has an important effect on the swelling ratio. Larsen also observed this effect in his experiments [31]. He proposed that extraction occurs when a solvent is used for coal treatment, causing the solvent activity to decrease, and reducing the coal swelling ratio to some extent. This effect depends on the coal:solvent ratio, but the relationship between the swelling ratio and this ratio has not been clarified. Nishioka [32] noted that the swelling ratio should be correlated with the coal:solvent ratio in studies of coal structure. Coal-swelling experiments should be performed at low coal:solvent ratios, otherwise it is difficult for the coal and the solvent to achieve association equilibrium, so the results do not clearly reflect the swelling mechanism.

8.1.3.7 Effect of Moisture on Swelling Ratio

Coal is a porous solid and contains moisture. The moisture is present in either the free or bound state, but these states cannot be completely distinguished. The coal structure has similar characteristics to colloids: when moisture is lost, the coal structure contracts; when moisture is gained, the coal structure expands. Suuberg et al. [33] studied different coal ranks, ranging from lignite to bituminous coals, and found that there is a linear relationship between volume shrinkage and coal water loss:

$$\text{Volume shrinkage rate } (\%) = a + b \times \text{dehydration rate } (\%),$$

where a and b are constants related to the coal at a certain temperature.

Coal moisture affects the adsorption and diffusion rate of the solvent in the coal, thus affecting the swelling ratio. The solvent diffusion rate of lignite is higher before drying, and the solvent diffusion rate of bituminous coal is almost the same before

Table 8.4 Swelling ratios for different drying conditions. Reprinted from ref. [33], Copyright 1993, with permission from American Chemical Society

Drying condition	Sample 1	Sample 2	Sample 3	Sample 4
No drying	1.48	1.62	1.33	1.60
Drying in concentrated H_2SO_4 (1 h, 300 K)	2.20	2.01	1.90	2.10
Drying in concentrated H_2SO_4 (30 d, 300 K)	2.22	2.06	2.10	2.14
Drying in concentrated H_2SO_4 (1 h, 373 K)	2.34	2.50	2.05	2.47

and after drying. Swelling ratio studies show that for pretreatment above 373–423 K, the swelling ratio significantly increases. However, drying is not effective for all solvent types. The swelling ratio does not increase when the solvent is tetrahydrofuran (THF). This is because when the coal loses moisture, new non-covalent bonds are formed in the coal. THF is weakly polar, so it cannot destroy these bonds as pyridine does, leading to small changes in the swelling ratio. When the solvent is pyridine, the swelling ratio significantly increases after removal of moisture (Table 8.4). To achieve comparable swelling ratios for studying diffusion mechanisms, the coal must be pretreated to prevent interference by water in the coal.

8.1.3.8 Effect of Minerals on Swelling Ratio

A solution of HCl/HF is often used to remove minerals before coal use. As the data in Table 8.5 show, the coal swelling ratio generally increases after removal of minerals [34]. FTIR spectroscopy shows that the absorption of the C–H bond vibrations of aromatic rings does not change significantly, but the intensity of the carbonyl group absorption increases, and that of the hydroxyl group decreases. These changes might be caused by carbonyl group conversion to carboxyl groups, which are strong hydrogen-bond acceptors. The swelling effect is therefore strengthened in solvents such as pyridine and *N*-methyl-2-pyrrolidone (NMP). However, the changes may simply be caused by removal of materials (such as clay) that are strongly bound in the solvent, leading to an increase in the swelling ratio.

Table 8.5 Swelling ratios before and after removal of minerals. Reprinted from ref. [34], Copyright 1989, with permission from American Chemical Society

Solvent	Species 1		Species 2		Species 3	
	Raw	After	Raw	After	Raw	After
Pyridine	2.4	2.5	2.0	2.9	2.2	3.1
NMP	2.3	3.1	1.8	2.9	2.1	3.0
THF	1.7	2.3	1.7	2.3	1.6	2.2
Nitrobenzene	1.2	1.7	1.2	1.6	1.2	1.5
Ethanol	1.6	1.7	1.6	1.7	1.5	1.7

8.1.4 Swelling During Pyrolysis

8.1.4.1 Effect of Swelling on Tar Yield During Pyrolysis

Solomon [35] studied the crosslinking reactions during coal pyrolysis and proved that –C=O, –COOH, and –OH functional groups participated in crosslinking reactions at 573–673 K, forming CO_2 and H_2O, and decreasing the tar yield. If the amounts of volatile and liquid products are to be increased, it is necessary to suppress crosslinking reactions. As a result of mass transfer limitations in the coal pyrolysis, coal tar precursors remain in the coal for some time. The precursors are broken into smaller molecules, and require more hydrogen to achieve stability. If the coal tar consists of a large number of small molecules, hydrogen will not be used effectively, so the tar yield decreases. These are the main reasons for low tar yields.

If a solvent is used in the pretreatment of coal, the rate and extent of crosslinking reactions decrease. ^1H NMR spectroscopy was used to study a mixture of coal and pyridine; the addition of pyridine improved the hydrogen fluidity. Barton et al. [36] used ^1H NMR spectroscopy to prove that only 10 % of the hydrogen in untreated coal was fluid, and deuterated pyridine increased the hydrogen fluidity by 40–50 %; the hydrogen transfer efficiency in coal swelling increased and the tar yield was improved. Additionally, solvent molecules diffuse into the microporous structure of the coal, expanding the pore volume, which reduces the residence time of tar in the coal. The possibility of molecules repolymerizing to form semi-coke is reduced, improving the tar yield. The solvent also produces hydrogen radicals during pyrolysis and the radicals are transferred to the coal tar, leading to high yields.

8.1.4.2 Effect of Swelling on Pyrolysis Mechanism

Kazuhiro [37] used pyridine vapor to study coal swelling during pyrolysis, and proved that the tar yield increased during low-temperature pyrolysis, and formation of CO, CO_2 and H_2O was suppressed.

Miura [38] proposed that coal swelling leads to fast pyrolysis. It is believed that at 760 °C, coal phenolic hydroxyl groups undergo condensation reactions to form H_2O:

$$\mathrm{Ar{-}OH + Ar'{-}OH \rightarrow Ar{-}O{-}Ar' + H_2O}$$

$$\mathrm{Ar{-}OH + Ar'{-}OH \rightarrow \underset{\diagdown O \diagup}{Ar{-}Ar'} + H_2O}$$

Ar–OH pyrolysis may occur as follows:

$$\mathrm{Ar - OH \rightarrow Ar = O + H\cdot},$$

where Ar and Ar′ are aromatic groups connected to coal macromolecules

There are hydrogen bonds between Ar–OH and Ar′–OH in coal:

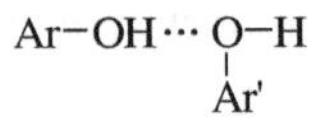

Swelling breaks the hydrogen bonds, regenerating Ar–OH and Ar′–OH; it also suppress the condensation reaction of phenolic hydroxyl groups in the coal, promoting the pyrolysis reaction, and generating a large amount of hydrogen radicals; these stabilize the coal tar precursors, and increase the tar yield.

8.1.4.3 Effect of Solvent Type on Coal Pyrolysis Products

Miura et al. used various solvents for pretreatments of coal before pyrolysis. It was found that the swelling ratio and tar yield are highest for coal samples pretreated with phenol [38]. The swelling treatment does not change the CO and CO_2 yields or the composition and yield of hydrocarbons. The tar yield increases, and the rate of increase is related to the formation of H_2O, H_2, and semi-coke. Swollen coal pyrolysis tar contains more oxygen than raw coal tar.

After pyridine vapor treatment, the CO, CO_2, and H_2O yields decrease significantly and the tar yield increases during pyrolysis. The amount of char decreases, but the numbers of hydrogen bonds and oxygen functional groups increase significantly. Pyridine treatment of coal samples increases the yields of volatile components and tar at 386 °C. Treatment with tetralin increases the volatile component yield above 590 °C. At 764 °C, tetralin treatment results in a higher tar yield than pyridine treatment. These results indicate that pyridine is more effective at promoting pyrolysis at low temperatures, whereas tetralin is more efficient at high temperatures.

The changes in the swelling ratio during coal pyrolysis can be used to study differences between coal macromolecular structures. In low-temperature pyrolysis, as a result of CO_2 formation, the swelling ratio of low-rank coal decreases [39, 40]. Suuberg et al. studied the swelling ratios of four lignites at 373, 473, and 573 K, and found that the swelling ratio decreased with increasing temperature [41]. At 373 K, the amount of moisture removed decreased with increasing swelling ratio; –COOH groups in the coal reduced the swelling ratio by 5–15 % above 473 K. These results indicate that, even under mild conditions, swelling treatment is disadvantageous when lignite is used for the production of low-molecular-weight products.

Yun used differential scanning calorimetry (DSC) combined with a swelling method to study pyrolysis of a bituminous coal, namely Pittsburgh 8 coal, and found two distinct transition temperature regions: 250–300 °C and 350–400 °C [42]. The low-temperature transition region is caused by low-temperature tar formation and the loose coal structure, and the high-temperature transition region is related to covalent bond crosslinking reactions. These transition points do not exist for low-rank coals and anthracite.

8.1.4.4 Swelling During Coal Liquefaction

Currently, the main problems in coal liquefaction are that large amounts of light hydrocarbons are generated, hydrogen consumption is very high, and side reactions reduce the yield of heavy oil. These problems can be overcome by methods such as lowering the reaction intensity, changing the reaction reagent, or appropriate pretreatment of the coal. For any pretreatment, the pore structure of the coal should first be considered because the reaction of the reagent with the coal particles is controlled by diffusion in the initial stage of liquefaction. Swelling is a good method for maximizing contact between the solvent and coal surface. Swelling of the coal macromolecular network might result in rearrangement of the pore structure, which would increase contact between hydrogenating agents and the active sites. These effects are beneficial for the transfer of free radicals, reduce side reactions, and increase coal liquefaction yields and product quality. The data in Table 8.6 show the effect of swelling on coal liquefaction [43].

The swelling ratio is determined by the solvent properties and the nature of the coal. Swelling has two drawbacks. First, a small amount of solvent remains in the coal structure, affecting the coal liquefaction product. Second, the solvent used is generally very expensive, and therefore not suitable for commercial purposes.

Table 8.6 Effect of swelling on coal liquefaction. Reprinted from ref. [43], Copyright 1991, with permission from Elsevier

Sample	Conversion (%)			
	Heavy oil	Asphalt	Pre-asphalt	Amount
Coal 1, raw	22	31	16	69
Coal 1, swollen	37	33	13	83
Coal 2, raw	32	22	13	67
Coal 2, swollen	50	22	6	78
Coal 3, raw	33	15	8	56
Coal 3, swollen	51	12	5	68

8.2 Study of Swelling Behaviors of Coals

Coal macromolecular networks, i.e., the aromatic ring sections, play crucial roles in coal mechanical properties, mass transfer during pyrolysis, and the swelling characteristics and other physical properties; they therefore determine coal chemical reactivities. The network structure is formed by intertwined crosslink bonds and intermolecular forces, to form a three-dimensional structure with specific arrangements. The most important interactions are the crosslinks between aromatic clusters. The crosslinks contain covalent bonds such as aromatic carbon–carbon bonds, and there are many non-covalent interactions such as hydrogen bonds, van der Waals forces, and electron-donator–electron-acceptor interactions. Different crosslink distributions lead to reactivity differences with changes in the coalification degree. However, research on the role of non-covalent bonds in the coal structure and their impacts on coal reactivity is rare. We investigated the swelling ratios and properties of eight different ranks of coal to study the distribution of weak bonds in the coal and analyzed the roles of these weak bonds in the coal structure, to establish a swelling model and gain further understanding of the coal structure.

To investigate the swelling mechanism and establish the model, it is necessary to choose suitable experimental methods and solvents.

The analytical data for Fushun bituminous coal are listed in Table 8.7. For other samples, readers are referred to Table 2.4 in Chap. 2.

8.2.1 Degree of Swelling

8.2.1.1 Measurement of Swelling Ratio

Five solvents were used in the experiments, i.e., pyridine, THF, NMP, tetralin, and CS_2; the properties of the solvents are shown in Table 8.8. The ratios (v/v) of pyridine to CS_2 and NMP to CS_2 were both 1:1. Minerals were removed by pretreating the coal with a mixture of 40 % distilled water, 45 % HCl (36 %), and 15 % HF, to reduce experimental errors caused by minerals and to reduce their impact on solvent diffusion. The coal sample was pulverized to less than 75 μm to ensure a narrow uniform particle size distribution, otherwise small particles embedded in the spaces between large particles will change the bulk density and affect the accuracy of the swelling ratio measurements. The coal sample was dried in a vacuum oven at

Table 8.7 Proximate and ultimate analyses of coal sample

Coal sample	Proximate analysis			Ultimate analysis				
	M_{ad}	A_d	V_d	C_{daf}	H_{daf}	O_{daf}	N_{daf}	S_{daf}
Fushun bituminous coal	3.41	10.65	39.39	73.35	5.26	18.69	1.95	0.87

Table 8.8 Physical properties of solvents

Solvent	Molecular weight	Density (g cm^{-3})	Properties	Solubility
Pyridine	79.10	0.938	Colorless or slightly yellow	water, ethanol, ether, and petroleum ether
NMP	99.15	1.028	Colorless	water, ethanol, and ether, in any ratio
CS_2	76.13	1.263	Colorless, combustible, toxic	ethanol and ether, in any ratio
Tetralin	132.20	0.973	Colorless	ethanol and ether
THF	72.10	0.889	Colorless	ethanol

373 K for 3–4 h to avoid experimental errors caused by moisture. The coal sample mass was 600 mg, and the solvent volume was 6 mL. Typically, the coal sample was placed in a test tube of diameter 6 mm and height 5 cm. The test tube was centrifuged at 4500 r min^{-1} for 15 min, and the height, h_1, was recorded. The solvent was slowly added to the test tube to prevent rapid swelling of the coal. The sample was continuously stirred to guarantee contact between the coal surface and the solvent. The sample was left at room temperature for 2 days, until swelling equilibrium was reached. The tube was again centrifuged at 4500 r min^{-1} for 15 min, and the height, h_2 was recorded. The swelling ratio Q_d was calculated using the equation in Sect. 8.1.1 of this chapter.

The swelling measurement results are shown in Table 8.9. As can be seen from the table, the order of the swelling ratios of the eight coal samples is Pingzhuang lignite > Fushun bituminous coal > Fuxin flame coal > Pingshuo gas coal > Zaozhuang coking coal > Xinzhi fat coal > Fengfeng meager coal ≈ Jincheng anthracite. The swelling ratio substantially decreases with increasing coal rank.

8.2.1.2 Effect of Temperature on Swelling Ratio

The temperature determines the swelling ratio at equilibrium and the swelling rate. We investigated the effect of temperature on the swelling ratio by measuring the

Table 8.9 Coal swelling ratios in various solvents

Coal	Pyridine	NMP	Tetralin	THF	Pyridine + CS_2	NMP + CS_2
Pingzhuang lignite	2.66	2.80	1.75	1.85	2.25	2.40
Fushun bituminite	2.50	2.72	1.68	1.82	2.10	2.00
Pingshuo gas coal	2.07	2.38	1.48	1.52	2.00	2.40
Fuxin long flame coal	2.44	2.61	1.64	1.79	2.10	2.20
Xinzhi fat coal	1.40	1.50	1.10	1.28	2.00	2.22
Zaozhuang coking coal	1.94	2.20	1.51	1.45	1.64	1.89
Fengfeng meager coal	1.00	1.00	1.00	1.00	1.18	1.04
Jincheng anthracite	1.00	1.00	1.00	1.00	1.11	1.27

Table 8.10 Swelling ratios at different temperatures

Coal	Solvent	25 °C	40 °C	60 °C
Pingzhuang lignite	Pyridine	2.66	2.60	2.62
	NMP	2.80	2.76	2.75
Pingshuo gas coal	Pyridine	2.07	2.02	2.05
	NMP	2.38	2.20	2.24
Zaozhuang coking coal	Pyridine	1.94	1.98	1.92.
	NMP	2.20	2.22	2.26
Fuxin long flame coal	Pyridine	2.44	2.45	2.40
	NMP	2.61	2.65	2.66

swelling ratio at 25, 40, and 60 °C. The experimental results are listed in Table 8.10. The swelling ratio does not change significantly at the different temperatures studied, but the swelling rate accelerates. The solvent diffuses into the coal structure, occupies weak bonding sites, and causes elastic deformation of the coal framework, thereby increasing the coal volume. As the temperature increases, the number of weak bonds in the coal does not change, so the swelling ratio does not change significantly at the swelling equilibrium. Suuberg et al. [44] suggested that the enthalpy change is zero when the coal–solvent system is close to equilibrium, so temperature changes do not affect the swelling ratio in this temperature range.

8.2.1.3 Swelling Ratios of Coal Macerals

The non-uniformity of coal must be taken into consideration in studies of coal structure and reactivity. Coal is a complex mixture of different macerals at different contents, and the overall structure and properties are the result of interactions among various macerals. The swelling properties of three coal macerals were studied to further our understanding of the swelling process. We chose three macerals of Pingshuo gas coal, and vitrinite and resin macerals of Fushun bituminous coal for the swelling experiments. The mass of each sample was 600 mg, and the solvent volume was 6 mL. The ultimate analytical data for Pingshuo gas coal macerals are listed in Chap. 3 Table 3.11. The ultimate analysis data for vitrinite and resin of Fushun bituminous coal are shown in Table 8.11. The swelling ratios are listed in Table 8.12.

Table 8.11 Fushun bituminous coal maceral ultimate analysis (dry ash-free basis, %)

Maceral	C	H	N	$S+O$
Vitrinite	72.08	5.66	1.68	20.58
Resin	73.12	5.28	1.72	19.88

Table 8.12 Swelling ratios of different macerals

Coal	Maceral	Pyridine	NMP	Pyridine + CS_2	NMP + CS_2
Pingshuo gas coal	Vitrinite	2.11	2.61	1.80	2.15
	Liptinite	1.86	1.70	1.75	1.86
	Inertinite	1.11	1.15	1.18	1.15
Fushun bituminite	Vitrinite	2.27	2.71	1.94	2.27
	Resin	2.04	2.43	1.97	2.13

As the data in Table 8.12 show, the order of the swelling ratios of the three coal macerals is vitrinite > liptinite > inertinite. It is known that the inertinite group has the highest aromaticity, crosslinking degree, and degree of graphitization, and is very similar to the structure of anthracite. Inertinite therefore has the lowest swelling ratio.

8.2.2 Swelling Mechanism

8.2.2.1 Weak Bonding Interactions Between Solvent and Coal

To determine the weak bonding interactions between the coal and the solvent, it is assumed that the number of weak bonds that interact with the solvent is η_s (mol). When swelling equilibrium is achieved, the adsorbed solvent volume of η_s, $V_{m,S}$, and the swelling ratio Q have the following relationship:

$$Q = \frac{n_S V_{m,S} + V_0}{V_0},$$

$$\frac{(Q-1)}{V_{m,S}} = \frac{n_S}{V_0},$$

where

$V_{m,S}$ is the solvent molar volume (L mol^{-1}); and

V_0 is the initial volume of the coal sample (L).

The unit weak interaction number (η_s/V_0) reflects the interaction degree between the coal and the solvent, which is an indication of the strength of the non-covalent-bonding forces in the coal structure. Table 8.13 lists the calculated weak bonding interaction numbers.

Figure 8.2 shows the weak bond number trend when pyridine is used as the solvent. As can be seen from the figure, the weak bond number gradually decreases with increasing coalification. For high metamorphic grade lean coal and anthracite, some non-chemical interactions such as hydrogen bonds and van der Waals forces almost disappear because of the high degree of aromatic structure condensation, and they are replaced by delocalized π bonds, so swelling does not occur. The FTIR

Table 8.13 Weak bonding interaction numbers in unit volume (mol L^{-1})

Coal	Pyridine	NMP	Tetralin	THF
Pingzhuang lignite	19.7	18.6	5.5	10.5
Fushun bituminite	17.8	17.8	5.0	10.1
Pingshuo gas coal	12.7	14.3	3.8	6.4
Fuxin long flame coal	17.1	16.7	5.8	9.7
Xinzhi fat coal	4.7	5.2	4.7	3.5
Zaozhuang coking coal	11.1	12.4	3.8	5.6
Fengfeng meager coal	0.0	0.0	0.0	0.0
Jincheng anthracite	0.0	0.0	0.0	0.0

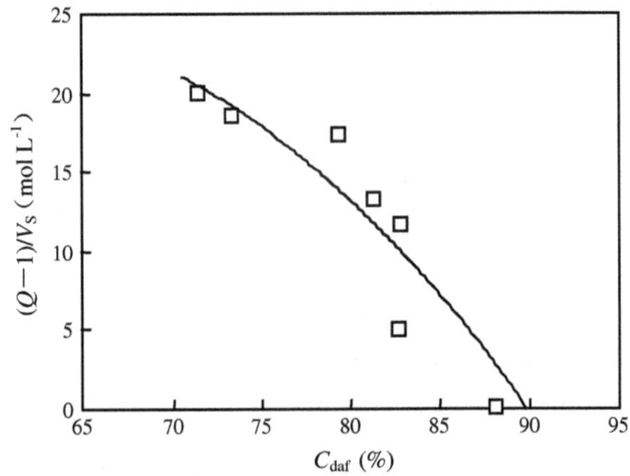

Fig. 8.2 Changes in number of weak bonds in pyridine solvent with coal rank

spectra in Chap. 2 Table 2.5 show that from lignite to anthracite, the numbers of polar hydroxyl functional groups in the coal significantly decrease, so hydrogen-bond formation by association of hydroxyl groups also decreases. The C–H vibration absorption peak gradually weakens with increasing coalification degree, indicating that the aliphatic hydrogen content and branching decrease, the cross-linking degree and number of aromatic C–C bonds increase, and the structure becomes graphitic. The differences among the swelling ratios of the eight coal samples are strong evidence that the swelling ratio is inversely proportional to the crosslinking degree of the macromolecular network. Xinzhi fat coal has unusual properties compared with the other coal samples studied. Its carbon content is about 83 %, and it is a volatile bituminous coking coal, but its swelling ratio is unusual; this requires further study.

8.2.2.2 Residual Solvent in Coal

Residual solvent affects the results of swelling experiments. Attempts should be made to remove it completely. A commonly used method is to repeatedly wash the coal with distilled water and then dry the sample in a vacuum oven at 100 °C. The ultimate analytical data for the coal samples after washing are shown in Table 8.14. The data show that the nitrogen content of the samples washed with only distilled water was higher than that of raw coal samples, indicating the presence of residual solvent. The solvents used in the experiments are soluble in ethanol, and ethanol has a lower boiling point than these solvents, so the swollen coal was repeatedly washed with ethanol and vacuum dried at 373 K for 24 h. The data in Table 8.14 show that this is a more effective method for rinsing coal samples, but that some solvent cannot be removed completely.

8.2.2.3 Relationship Between Solvent Properties and Swelling Ratio

The swelling ratio is determined not only by the nature of coal, but also by the physical properties of the solvent used. The strength of the solvent–coal interaction is reflected by factors such as the EDN, the solubility parameter δ, and the solvent molar volume $V_{m,S}$. The important physical parameters of some common solvents are listed in Table 8.15.

The EDN and the electron-acceptor number (EAN) are indicators of the electron-donating and electron-accepting abilities of a solvent. Coal can donate and accept electrons simultaneously, so the EDN and EAN are related to the strength of the coal–solvent interaction. During swelling, the solvent destroys hydrogen bonds, van der Waals forces, and other non-covalent-bonding forces in the coal, occupies interaction sites, replaces the weak bonds in the coal structure, and expands

Table 8.14 Ultimate analysis (%) of swollen pingshuo gas coal

Coal sample	Rinsed with distilled water			Rinsed with ethanol		
	C_{daf}	H_{daf}	N_{daf}	C_{daf}	H_{daf}	N_{daf}
Pyridine-swollen	79.89	5.38	2.72	79.82	5.29	1.93
NMP-swollen	79.93	5.29	2.32	79.76	5.19	2.06
Pyridine + CS_2 swollen	79.74	5.43	2.75	79.64	5.24	2.04
NMP + CS_2 swollen	79.87	5.33	2.34	79.92	5.22	1.97
Raw coal	79.84	5.27	1.79	79.84	5.27	1.78

Table 8.15 Solvent physical properties

Solvent	EDN	EAN	Molar volume, $V_{m,S}$ (mL mol^{-1})	Solubility Parameter, δ	Dipole moment, μ (D)	Viscosity, η (Pa s)
Pyridine	33.1	14.2	84.32	10.7	2.37	0.974
NMP	27.3	13.3	96.45	11.0	4.09	1.69
CS_2	0	0	60.28	10.0	0	0.362
Tetralin	–	–	135.87	9.50	–	2.26
THF	20.0	8.0	135.87	9.10	1.75	–

agglomerated macromolecules, leading to a volume increase. The data in Table 8.9 show that the coal swelling ratios are highest in pyridine and NMP because these solvents have high EDN and EAN values. Although mixed solvents are good extraction agents, they are usually inferior to pyridine and NMP in swelling reactions. Addition of CS_2 reduces the viscosities of pyridine and NMP (e.g., the viscosity of NMP + CS_2 decreases to 0.615 Pa s), but it also reduces the swelling function of the solvent. The addition of CS_2 can increase solvent diffusion, improve the mass transfer efficiency in the coal structure, and accelerate the swelling rate, but the swelling ratio after swelling equilibrium is lower. In this respect, Xinzhi fat coal is again unusual because its swelling ratio in a mixed solvent system is significantly higher than that in a single solvent.

8.2.2.4 Effect of Coal to Solvent Ratio on Swelling

For most coal samples, under the conditions used in the experiments described here, a portion of the small molecules dissolve during swelling (Table 8.16). The coal–solvent system can therefore be regarded as a suspension. In this system, there is a balance between the insoluble and soluble components. The experimental results for swelling of various coal samples in pyridine and NMP at different coal:solvent ratios are shown in Table 8.17. As the data in the table show, the swelling ratio Q and coal:solvent ratio are closely related; when the ratio is low, non-covalent-bonding forces are disrupted and weak bonding interactions are replaced, so the swelling ratio increases. The number of weak bonding interactions is fixed, so there is a limit to the value of $(Q\text{-}1)/V_S$.

When the weak bonding interaction sites in the coal are completely occupied, the swelling ratio reaches a constant value. Dissolution of a small amount of material is inevitable during swelling, but the coal weight loss has a limit, and does not increase with increasing solvent amount. At low coal:solvent ratios, this small amount of dissolution does not affect the solvent activity during swelling. If we assume that each weak bonding interaction site is an adsorption site, the swelling and adsorption processes are similar at low coal to solvent ratios, so the Freundlich equation [45] can be used to express the relationship between the coal:solvent (C/S) ratio and weak bond adsorption sites:

Table 8.16 Coal sample weight losses (%) during swelling

Coal	Pyridine-swollen	NMP-swollen
Pingzhuang lignite	11.83	9.82
Fushun bituminous coal	9.10	8.16
Fuxin long flame coal	6.30	5.80
Pingshuo gas coal	8.20	6.92
Zaozhuang coking coal	4.97	4.73

Table 8.17 Swelling ratios and weak bonding interaction at different coal:solvent (*C*/*S*) ratios

Coal Sample	Pyridine			NMP		
	C/*S* (wt%)	*Q*	$(Q\text{-}1)/V_{m,S}$ (mol L^{-1})	*C*/*S* (wt%)	*Q*	$(Q\text{-}1)/V_{m,S}$ (mol L^{-1})
Pingzhuang lignite	0.180	2.11	13.16	0.165	2.28	13.27
	0.151	2.35	16.01	0.144	2.32	13.69
	0.127	2.45	17.20	0.119	2.39	14.41
	0.114	2.54	18.26	0.104	2.54	15.97
	0.106	2.67	19.81	0.093	2.56	16.17
	0.082	2.80	21.35	0.075	2.71	17.73
Fuxin long flame coal	0.211	1.60	7.12	0.185	1.72	7.47
	0.143	1.82.	9.72	0.130	1.88	9.12
	0.106	2.00	11.86	0.097	2.18	12.23
	0.082	2.28	15.18	0.075	2.32	13.69
Pingshuo gas coal	0.249	1.61	7.23	0.221	1.89	9.23
	0.151	1.92	10.91	0.193	2.00	10.37
	0.111	2.20	14.23	0.146	2.22	12.65
	0.106	2.21	14.35	0.097	2.54	15.97
	–	–	–	0.090	2.75	18.14
Zaozhuang coking coal	0.211	1.60	7.12	0.185	1.72	7.47
	0.143	1.82	9.72	0.130	1.88	9.12
	0.106	2.00	11.86	0.097	2.18	12.23
	0.082	2.28	15.18	0.075	2.32	13.69

$$q = kc^{1/n},$$

$$q = \frac{Q-1}{V_{m,S}} = \frac{n_S}{V_0},$$

$$c = C/S,$$

where k and n are parameters related to the solvent and the coal sample.

If we take logarithms of both sides,

$$\ln q = \ln k + \frac{1}{n}\ln c.$$

If the coal:solvent ratio and weak bonding interaction sites match the above assumptions, the number of sites can be calculated using the Freundlich equation.

However, the swelling ratio of Xinzhi fat coal does not change with the coal: solvent ratio (Table 8.18), and the swelling ratios of various coal samples do not increase for tetralin and THF as the solvent volume increases, indicating that the effect of the coal:solvent ratio on weak bonds requires further study.

Table 8.18 Swelling ratios and weak bonding interaction of Xinzhi fat coal at different coal: solvent (*C*/*S*) ratios

Pyridine			NMP		
C/*S* (wt%)	*Q*	$(Q\text{-}1)/V_{m,S}$ (mol L^{-1})	*C*/*S* (wt%)	*Q*	$(Q\text{-}1)/V_{m,S}$ (mol L^{-1})
0.154	1.40	4.74	0.165	1.52	5.39
0.106	1.40	4.74	0.145	1.53	5.50
0.077	1.42	4.98	0.097	1.50	5.18
0.192	1.45	5.34	0.070	1.56	5.81

8.2.2.5 Calculation of Average Molecular Weight of Coal

Swelling is the primary method used to determine the macromolecular structural features of coal, and enables estimation of the molecular weight of coal connected by the crosslink bonds. Solvent absorption causes the macromolecular structure of a material to expand, so the swelling ratio is determined by the crosslinking degree. The crosslinking degree can be determined by measuring the swelling ratio (Q) and the interaction coefficient (β). Although coal swelling is not identical to polymer swelling, the information obtained using this method is still valuable. The three most commonly used equations are the Flory–Rehner [46], Kovac [47], and Barr–Howell–Peppas equations [48]. All these average-molecular-weight calculations are based on the assumption that the free energy of the coal–solvent mixture is equal to the free energy of the elastic deformation of the macromolecular network when thermodynamic equilibrium is reached. The specific equations are as follows.

Flory–Rehner equation:

$$M_C = \frac{\rho_C M_S}{\rho_S} \frac{\frac{1}{2}Q - \left(\frac{1}{Q}\right)^{1/3}}{\ln\left(1 - \frac{1}{Q}\right) + \beta \cdot \frac{1}{Q} + \frac{1}{Q}}.$$

Kovac equation:

$$M_C = \frac{\rho_C M_S}{\rho_S} \frac{\left(\frac{1}{Q}\right)^{1/3} + \frac{1}{N} \cdot Q^{1/3}}{-\left[\ln\left(1 - \frac{1}{Q}\right) + Q + \beta \cdot \left(\frac{1}{Q}\right)^2\right]}.$$

Barr–Howell–Peppas equation:

$$M_C = \frac{\rho_C M_S}{\rho_S} \frac{\left[\frac{1}{2} \cdot \frac{1}{Q} - \left(\frac{1}{Q}\right)^{1/3}\right]\left[1 + \frac{1}{N} \cdot \left(\frac{1}{Q}\right)^{1/3}\right]}{\ln\left(1 - \frac{1}{Q}\right) + Q + \beta \cdot \left(\frac{1}{Q}\right)^2}.$$

In these equations,

M_C is the average molecular weight between coal macromolecular crosslinking points;

M_S is the solvent molecular weight;

ρ_C is the coal density, usually 1.3 g cm^{-3};

ρ_S is the solvent density (g cm^{-3});

β is the interaction coefficient between coal and the solvent (0.3 is used for most polymers) and

N is the number of aromatic clusters at coal macromolecular crosslinking points, and can be calculated as $N = M_C/M_r$, where M_r is the average molecular weight of aromatic clusters; their numerical values are listed in Table 8.19.

The molecular weights obtained using the above equations are shown in Table 8.20. The data show that the molecular weights obtained using different methods are very different, but they show the same trends. The obtained absolute values of the molecular weights, M_C, which are determined by the swelling ratio of the coal macromolecules, are not necessarily reliable, but their relative values are reasonable. Information on the coal macromolecular network configuration can be obtained from the trend in the average molecular weight. As can be seen from Table 8.20, regardless of the solvent and calculation method, M_c and the aromatic

Table 8.19 Average molecular weights of aromatic clusters in coal samples

Coal	Pingzhuang lignite	Fushun Bituminite	Fuxin long flame coal	Pingshuo gas coal	Xinzhi fat coal	Zaozhuang coking coal
M_r	370.5	360.1	373.1	384.8	389.6	329.2

Table 8.20 Average molecular weights (M_c) of coal samples

Solvent	Coal	Flory–Rehner		Kovac		Barr–Howell–Peppas	
		N	M_c	N	M_c	N	M_c
Pyridine	Pingzhuang lignite	2.97	1100	4.83	1791	5.34	1980
	Fushun bituminous	2.60	936	4.38	1577	4.92	1772
	Fuxin flame	2.35	878	4.05	1511	4.69	1749
	Pingshuo gas coal	1.45	557	2.82	1085	3.72	1432
	Xinzhi fat coal	0.39	152	1.20	467	2.44	950
	Zaozhuang coking coal	1.40	461	2.80	921	3.72	1224
NMP	Pingzhuang lignite	3.87	1433	6.29	2331	6.30	2335
	Fushun bituminite	3.70	1332	5.83	2099	6.08	2190
	Fuxin long flame coal	3.21	1198	5.20	1940	5.61	2093
	Pingshuo gas coal	2.44	939	4.20	1617	4.81	1851
	Xinzhi fat coal	0.58	227	1.55	604	2.73	1063
	Zaozhuang coking coal	2.30	757	4.07	1339	4.72	1553

cluster number, N, are lowest for the fat coal, indicating that it has fewer weak bonds and the material is highly crosslinked. Low metamorphic grade lignite and Fushun coal have higher M_c and N values, indicating that their macromolecular structures are relatively loose and the extent of crosslinking are low. Rough estimations of the structure parameters of the coals used in our experiments (except fat coal) suggest that the average number of aromatic clusters between crosslinks is four to six, with an average molecular weight of 1500–2500 mass units.

8.2.2.6 Nature of the Swelling Process

Coal has hydrogen-donating and hydrogen-accepting functional groups. Swelling breaks the weak bonds among small molecules and the coal structural units, which expands the coal volume, resulting in structural changes and rearrangements by reaction with electrophilic and nucleophilic reagents. The swelling process is the opposite of elastic contraction of the network structure, and the swelling equilibrium is decided by the number of crosslinks, which consist of covalent bonds, i.e., coal–solvent interactions cause bond redistribution to give a new equilibrium state. The volume expansion observed in swelling experiments is the result of coal macromolecular structural rearrangements; the driving force is the change in the free energy of coal–solvent mixing. Larsen, Liotta, and Hombach [49–51] all reported that hydrogen bonds are the most important non-covalent bonds, and are the main factor controlling the swelling process. Marzec [52] suggested that electron acceptors determine the swelling ratio, and hydrogen bonds are simply electron acceptors. During swelling, the coal structural units are connected to each other by crosslinking bonds, and this prevents dissolution of the main coal structure. The coal swelling ratio is generally greater than 1, indicating a flexible, elastic structure among the crosslinks. Low-rank coal has a low extent of crosslinking and long molecular chains, whereas high-rank coal has a highly crosslinked structure and short molecular chains.

The swelling process in a coal–solvent system is shown in Fig. 8.3. In state A, the raw coal and the solvent are not mixed. The solvent gradually penetrates into the coal structure, the macromolecular structure begins to loosen, and coal–solvent interactions gradually replace coal intramolecular weak bonds, to give the metastable state, B. This state has the largest swelling ratio, but swelling equilibrium is not yet achieved. This is because the rate at which the solvent replaces weak bonds and the coal structural rearrangement rate are not equal. As a result of formation of new associations in the coal structure and molecular structural rearrangements in the new system, the structure may contract, and deformation occurs, generating resistance to solvent diffusion into the coal; equilibrium state C is then reached. After the coal has been rinsed with ethanol and dried in a vacuum, it forms a new state, D, which has a lower energy than the raw coal, but still maintains some of the changes caused by the swelling process. There are constraint relationships for the swelling energy of the process shown in Fig. 8.3: $E_{\text{raw coal}} + E_{\text{solvent}} = E_A$ and $E_{\text{swollen coal}} + E_{\text{solvent}} = E_D$, and it is assumed that the E_{solvent} does not change after

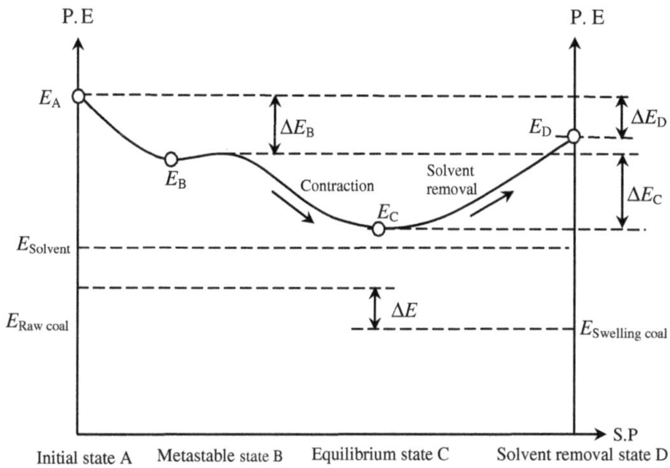

Fig. 8.3 Scheme of swelling mechanism (*P.E* Potential Energy; *S.P* Swollen Process)

swelling. The primary swelling effect is expansion of the coal macromolecular network to a state resembling that before coal formation, in which non-covalent bond distribution plays an important role.

8.2.3 Changes in Coal Caused by Swelling

It is interesting to determine whether the coal surface structure changes after swelling. The thermal properties of coal are important indicators in coal processing and use, and directly reflect structural differences after swelling. We used small-angle X-ray scattering (SAXS) and DSC to investigate the changes in the pore structure and thermal properties caused by swelling, and used the results to describe the main physical properties of the swollen coal.

8.2.3.1 Thermal Properties of Coal

DSC is a very sensitive technique for investigating thermal effects by studying the relationship between sample heat flow input and sample temperature. The basic principle of the technique is described by

$$\frac{dH}{dt} = mC_p \frac{dT}{dt},$$

where

dH/dt is the heat power (mW);

C_p is the heat capacity of the sample (J g^{-1} K^{-1}); and

M is the instrument constant, related to the measuring conditions and unit.

The measurements were performed using a Seteram DSC III calorimeter. The masses of the raw and swollen coal samples were both 20 mg. The nitrogen flow rate was 45 mL min^{-1}. The ramp rate was 20 °C min^{-1}, in the temperature range 25–600 °C. The DSC results show that the swellings of various coals followed the same trend as that of the raw coal. The DSC heat flow and heat capacity curves of raw Pingshuo gas coal and coal swollen in NMP + CS_2 are shown in Figs. 8.4 and 8.5, respectively.

There is an endothermic peak at about 100 °C for the raw and swollen coals, corresponding to dehydration. Weight loss caused by pyrolysis occurs between 200 and 400 °C. The raw coal shows significant heat flow and heat capacity changes, mainly because of phase transfer and removal of small molecules by the breaking of hydrogen and other non-covalent bonds, low-temperature crosslinking reactions, and gas evolution, which cause significant heat fluctuations. For the swollen coal, because the heated solvent destroys the majority of the weak bonds

Fig. 8.4 DSC heat flow curves of Pingshuo gas coal: *1* Raw and *2* Swollen. Reprinted from ref. [53]. Copyright 2000, with permission from Elsevier

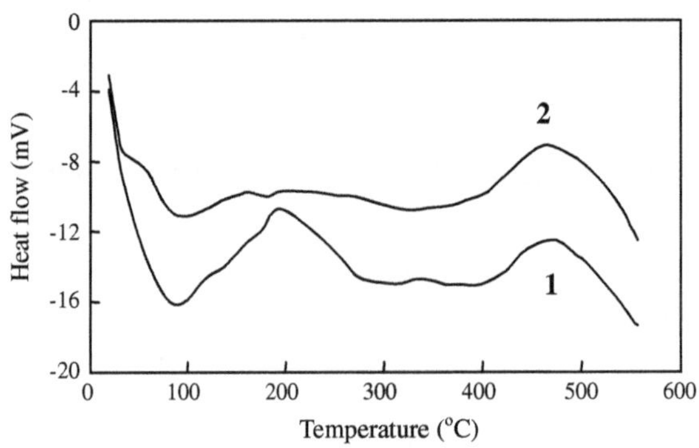

Fig. 8.5 DSC heat capacity curves of Pingshuo gas coal: *1* Raw and *2* Swollen. Reprinted from ref. [53]. Copyright 2000, with permission from Elsevier

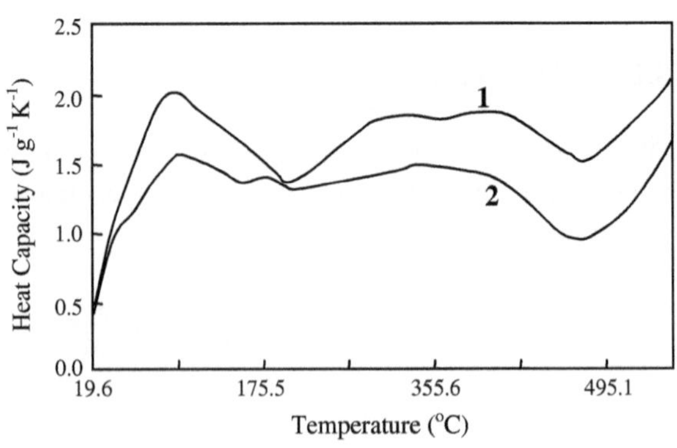

during swelling, and increases the fluidity of the small-molecule phase, the heat flow change is very slight in this temperature range.

8.2.3.2 Changes in Coal Surface Properties by Swelling

We used SAXS to examine the fractal properties of the surfaces of Fuxin flame, Pingshuo gas, and Zaozhuang coking coals before and after swelling. We know that the scattering intensity $I(\theta)$ of the X-ray scattering at an angle θ follows the relationship

$$\ln I(\theta) = A + (D - 6)\ln\theta,$$

where

D is the fractal dimension on the sample surface;
A is a constant related to the intensity of the incident X-ray; and
θ is the scattering angle (rad).

The measured intensities of the X-rays at different scattering angles can be used to obtain the relationship between $\ln I(\theta)$ and $\ln\theta$. A linear relationship confirms that the fractal characteristics and the slope can be used to estimate the fractal dimension D. Our experimental results show that good linear relationships between $\ln I(\theta)$ and $\ln\theta$ are observed for raw and swollen coals. The experimental results for Zaozhuang coking coal are shown in Fig. 8.6. The calculated fractal dimensions are shown in Table 8.21.

As can be seen from Table 8.21, the fractal dimension increases with increasing coal rank. The fractal dimensions of swollen and raw coals increase significantly, indicating that there are major differences between their surfaces. During swelling, the solvent expands the pores in the coal structure. Gether [54] showed that a large number of oxygen-containing functional groups are attached to the coal surface

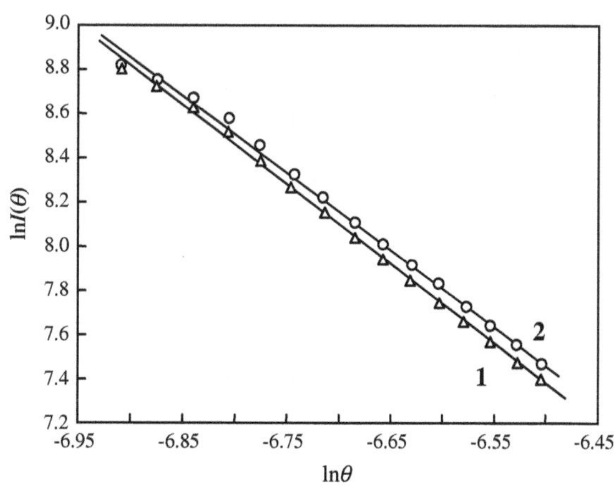

Fig. 8.6 Relationship between scattering intensity and scattering angle for Zaozhuang coking coal: *1* Raw and *2* Swollen

Table 8.21 Raw and swollen coal surface fractal dimensions

Coal	Raw	Swollen
Fuxin long flame coal	2.33	2.60
Pingshuo gas coal	2.36	2.49
Zaozhuang coking coal	2.41	2.55

through hydrogen bonds. The addition of pyridine and NMP solvents destroys the weak bonds at the coal surface, and new associations are formed. When the solvent is removed, part of the pore structure is damaged, increasing the coal surface roughness.

8.2.3.3 Effect of Swelling on Coal Char Structure

Char Surface Area and Pore Volume

Raw and swollen coals were each placed in a crucible and heated at 900 °C for 7 min in the absence of air. The crucible was rapidly removed from the furnace, cooled in air for 5 min, and weighed at room temperature. This procedure corresponds to the proximate analysis of coal volatiles. The surface area was obtained using the Brunauer–Emmett–Teller equation and the pore volume was calculated using the Dubinin-Radushkevich equation; the results are given in Table 8.22.

As the data in Table 8.22 show, the char surface area and pore volume decrease with increasing coal rank. In the pyrolysis process, during which volatiles are removed, the structure of low-rank coal becomes more open because it has more

Table 8.22 Coal char surface areas and porosities

Coal	Surface area (m^2 g^{-1})	Pore volume (mL g^{-1})
Pingzhuang lignite	484.5	0.31
Pingzhuang lignite, pyridine swollen	150.4	0.11
Pingzhuang lignite, NMP swollen	236.4	0.17
Fushun bituminous	228.7	0.40
Fushun bituminous, pyridine swollen	173.7	0.21
Fushun bituminous, NMP swollen	177.6	0.20
Fuxin flame	236.3	0.18
Fuxin flame, pyridine swollen	159.3	0.16
Pingshuo gas coal	245.6	0.27
Pingshuo gas coal, pyridine swollen	192.7	0.26
Pingshuo gas coal, NMP swollen	197.4	0.20
Zhozhuang coking coal	45.37	0.10
Zaozhuang coking coal, pyridine swollen	33.39	0.05

volatiles. In contrast, high-rank coals, which contain less volatiles, show crosslink destruction and crystal ordering, resulting in loss of part of the surface area, which lowers the surface area and pore volume. After swelling pretreatment, the surface area and pore volume of the swollen coal are significantly lower than those of raw coal. Pingzhuang lignite has the highest swelling ratio (Q = 2.8); therefore its surface area decreases the most. The surface area of the coal particles is generally determined by the coal pore structure, so changes in the pore structure of coal during pyrolysis is influenced by two factors: structural differences and escape of volatiles. Two factors can cause the surface area and pore volume to decrease. One is destruction of a large number of micropores in the coal by the solvent during swelling, resulting in decreases in the surface area and pore volume. The second is removal of a significant amount of volatiles and breakage of crosslinks, resulting in ordering of the structural units and coalescence, causing the surface area and pore volume of the char to decrease. For coal with a low metamorphic degree, a crosslinked structure is formed by various functional groups because of the large number of side chains; its spatial structure is loose, and it has a large specific surface area. The crosslinking extent is reduced during swelling because the solvent breaks weak bonds, and the molecular arrangement tends to be regular, so the coking structure of swollen coal is similar to that of high metamorphic coal coke.

Pyrolysis Reactivity of Coal Before and After Swelling

The solvent destroys weak bonds in the coal during swelling and expands the macromolecular network, inevitably leading to reactivity changes. We used an IR spectroscopy–pyrolysis reactor, described in Chap. 4, Sect. 4.2, and dynamic calculations to study the changes in the functional groups in coal macromolecules after swelling, to investigate the effect of non-covalent forces on the coal structure and reactivity. As coal pyrolysis progresses, the functional groups in the coal decompose to CO, CO_2, CH_4, H_2O, and light hydrocarbons. This change is reflected in the IR spectrum, i.e., the intensities of the methylene group, aromatic ring, and ether bond absorptions change with increasing temperature. The activation energy can be obtained by calculating the changes in the intensities of the absorptions of the functional groups of the pyrolysis products under fast pyrolysis conditions. The activities of raw and swollen coal functional groups can be compared based on these activation energies. Pyrolysis is closely related to the coal structure and chemical properties [55–60]; therefore this method provides a description of the coal structure and reactivity from another perspective.

The activation energies obtained from the changes in the functional group absorption intensities are listed in Table 8.23; these are used to determine the coal macromolecular network activity. There are only slight differences between the raw coal and swollen coals, and no significant differences in the trends. The results

Table 8.23 Pyrolysis activation energies (J mol^{-1}) of functional groups in raw and swollen coals. Reprinted from ref. [53], Copyright 2000, with permission from Elsevier

Coal sample	Aromatic Hydrogen	Aliphatic Hydrogen	Diene	Ether bond
Pingzhuang lignite	22,115	10,805	16,416	10,929
Pingzhuang lignite, pyridine swollen	15,323	6651	14,097	7362
Pingzhuang lignite, NMP swollen	18,881	6659	14,994	7796
Fushun bituminous	20,957	14,982	13,195	9130
Fushun bituminous coal, pyridine swollen	16,707	11,966	12,290	9007
Fushun bituminous, NMP swollen	17,012	11,026	11,983	8852
Fuxin flame	21,368	17,598	18,007	10,571
Fuxin flame, pyridine swollen	21,266	10,772	14,314	9387
Fuxin flame, NMP swollen	19,264	10,819	14,350	9584
Pingshuo gas coal	19,067	23,158	14,626	9397
Pingshuo gas coal, pyridine swollen	17,432	15,034	15,149	10,063
Pingshuo gas coal, NMP swollen	17,902	15,527	14,137	9393
Zaozhuang coking coal	27,814	18,058	14,231	8726
Zaozhuang coking coal, pyridine swollen	27,478	11,457	13,865	9115
Zaozhuang coking coal, NMP swollen	26,352	10,465	14,097	8914

suggest that there is no significant difference between the pyrolysis reaction mechanisms of the swollen and raw coals, and that the main structure of swollen coal is nearly the same as that of raw coal.

References

1. van Krevelen DW (1961) Coal. Elsevier Scientific Publishing Company, Amsterdam 433
2. Sanada Y et al (1966) Fuel 45:295
3. Solomon PR et al (1994) Prepr Pap ACS Div Fuel Chem 39(1):68
4. Pande S et al (2001) Energy Fuels 15:1063
5. Nishioka M et al (2001) Energy Fuels 15:1270
6. Nishioka M et al (2002) Energy Fuels 16:1109
7. Giri CC et al (2000) Fuel 79:577
8. Peppas NA et al (1983) Chem Eng Commun 37:333
9. Brenner D et al (1985) ACS Div Fuel Chem Prepr 30(1):71
10. Alfrey T et al (1966) Polym Sci 2:249
11. van Krevelen DW (1946) Fuel 25:104
12. Solomon PR et al (1988) Energy Fuel 2:405
13. Wolf DW (1960) Fuel 39:25

14. Dryden IG (1951) Fuel 30:145
15. Meyers RA (1982) Coal structure. Academic Press, New York p254
16. Cody GD et al (1988) Energy Fuel 2:340
17. Solomon PR et al (1993) Fuel 72:587
18. Jurkiewicz A et al (1981) Fuel 60:1167
19. Jurkiewicz A et al (1982) Fuel 61:64
20. Brenner D (1983) Fuel 62:1347
21. Nishioka M (1993) ACS Div Fuel Chem Prepr 38(2):878
22. Brenner D (1985) Fuel 64:167
23. Szeliga J et al (1983) Fuel 62:1229
24. Hall PJ et al (1988) Fuel 67:863
25. Ndaji FE et al (1993) Fuel 72:1531
26. Aida T et al (1985) Am Chem Soc Div Fuel Chem Prepr 30(1):95
27. Ndaji FE et al (1995) Fuel 74:842
28. Brenner D (1984) Fuel 63:1225
29. Ndaji FE et al (1993) Fuel 74:932
30. Turpin M et al (1978) ICCS. Elsevier, New York, p 85
31. Larsen JW et al (1991) Energy Fuel 5:57
32. Nishioka M (1993) Fuel 72:1001
33. Suuberg EM et al (1993) Energy Fuel 7:384
34. Larsen JM et al (1989) Energy Fuel 3:557
35. Solomon PR (1990) Energy Fuel 4:42
36. Barton WA et al (1984) Fuel 63:1202
37. Kazuhiro M (1994) Energy Fuel 8:868
38. Miura K (1991) Energy Fuel 5:803
39. Solomon PR (1984) Fuel 63:1302
40. Suuberg EM (1985) Fuel 64:1668
41. Suuberg EM (1991) ACS Div Fuel Chem Prep 36:43
42. Yun YS et al (1993) Fuel 72:1245
43. Joseph JT (1991) Fuel 70(1):139
44. Suuberg EM et al (1991) Energy Fuel 11(6):1150
45. Fu XC et al (1990) Physical chemistry, 4th edn. Advanced Education Press, Beijing, p 942
46. He MJ (1991) Polymer physics. Fudan University Press, Shanghai, p 58
47. Kovac J (1978) Macromolecules 11:362
48. Barr-Howell BD et al (1985) PolymBull 13:91
49. Larsen JW et al (1990) Energy Fuel 4:107
50. Liotta R et al (1983) Fuel 62:781
51. Hombach HP (1983) Proc Abstr 78:19153x
52. Marzec A (1986) Fuel Process Technol 14:39
53. Xie KC, Li F et al (2000) Fuel Process Technol 64:241
54. Gether JS (1987) J Am Chem Soc Div Fuel Chem Prepr 32:239
55. van Niekerk D, Mathews JP (2010) Fuel 89:73
56. van Niekerk D et al (2008) Int J Coal Geol 76:290
57. Li X et al (2006) Fuel 85:1700
58. Yua J et al (2007) J Prog Energy Combust Sci 33:135
59. Maroto-Valer M et al (1998) Energy Fuel 12:833
60. Wan-taek Cho et al (2012) Korean J Chem Eng 29:190

Chapter 9
Coal Plasma Reactions

Abstract This chapter deals with the behavior of coal in a plasma medium. Pyrolysis behavior is discussed, and the reactions of coal with activated species (hydrogen) and acetylene formation are described. There then follows a discussion of the overall use of plasma technology in the coal industry. The behavior of the C–H–Ar–O system is described, followed by a discussion of the thermodynamic aspects of plasma operations. Experimental aspects are dealt with before returning to pyrolysis and the factors affecting that process in a plasma. The properties of pyrolysis residues are then covered, followed by a discussion of reactor dynamics and the application of various models to plasma conversion reactions and processes. Finally, coal coking behavior in an arc plasma is discussed.

9.1 Introduction to Plasma Coal Pyrolysis

In conventional coal thermal pyrolysis processes, the final products are volatile matter and solid residues. Volatile matter, consisting of tar and gaseous products, is generated through competing polycondensation and depolymerization reactions. Coal tar is mainly composed of aromatic compounds with a wide molecular-weight range; the product gases are mainly CH_4 and H_2, and are generally free of unsaturated gases such as C_2H_2 and C_2H_4. The solid residue is char or semi-coke or coke. An increase in the heating rate (10^2–10^4 K s^{-1}) significantly changes the quantity and composition distribution of the volatile matter. Desypris et al. [1] studied the fast pyrolysis of seven coal ranks, from lignite to anthracite. The results showed that the yield of volatile matter is higher than that predicted by the proximate analysis. Moreover, acetylene is generated in the product gases. Thermal decomposition intensifies with increasing pyrolysis speed and temperature, but thermal polycondensation is weakened. The yield of volatile matter produced through fast pyrolysis is therefore higher than that obtained under slow pyrolysis conditions. The product distribution tends toward low-molecular-weight compounds at higher pyrolysis speeds and temperatures. As the free energy of acetylene formation decreases at

K.-C. Xie, *Structure and Reactivity of Coal*, DOI 10.1007/978-3-662-47337-5_9

high temperature, the product gases of high-temperature rapid thermal decomposition contain a large proportion of acetylene. However, it is difficult to achieve temperatures that are sufficiently high for direct thermal decomposition of coal to acetylene in a laboratory by using conventional equipment. The development of plasma technology provides a method for the rapid one-step production of acetylene from coal because a thermal plasma has both high energy and high chemical reactivity. In this chapter, recent research work in this field is discussed.

9.1.1 Coal Pyrolysis and Other Reactions in Arc Plasma

Coal reactions under plasma conditions are diverse and complicated. The coal is subjected to the actions of high-temperature hydrogen atoms and other activated species when it comes into contact with the plasma jet. This produces three main types of product, i.e., CH species, char (semi-coke or coke), and volatiles. The char reacts with high-energy ions in the plasma and generates ionic carbon species, which then combine with hydrogen atoms, generating CH species. Some primary volatile compounds react with activated species in the plasma to yield CH species and other primary volatiles, and are converted to secondary products under the thermal reaction conditions. Some hydrocarbons (such as C_2H_2) in the secondary products can easily decompose to hydrogen and soot at high temperatures. Quenching therefore has a significant influence on the composition of the final product gases. After splitting decomposition, compound formation and polymerization, and other processes, the primary volatiles are converted to the final products, including pyrolysis residues, product gases, and charred materials. Figure 9.1 shows various coal transformation routes using plasma. These routes are complicated, and each plays a different role. This chapter focuses on the transformation routes and other factors that influence the final product during coal reactions in plasma.

The pyrolysis of coal using plasma is a new process; it is different from traditional coal reactions, and has attracted much attention. An arc plasma is most often used for coal pyrolysis, but there are various ways of generating the plasma. An arc plasma is a typical thermal plasma, and its most important characteristic is that it can provide extremely high temperatures. Furthermore, an arc plasma contains large numbers of various charged ions, neutral particles, electrons, and other activated species. An arc plasma is therefore considered to be an ideal and clean way to produce acetylene from coal. As the behavior of coal in an arc plasma is complicated, there are currently no generally accepted reaction mechanisms. Chakravartty et al. [2] suggested that coal reaches thermodynamic equilibrium in an arc plasma, and that the product distribution can be described by this equilibrium. Baumann et al. [3] proposed that the reactions of coal in a plasma are under the kinetic control of the coal pyrolysis process, the pyrolysis products, and activated species, and the product distribution does not correspond to that at thermodynamic equilibrium. Various studies have also indicated that hydrogen atoms can react directly with coal [4].

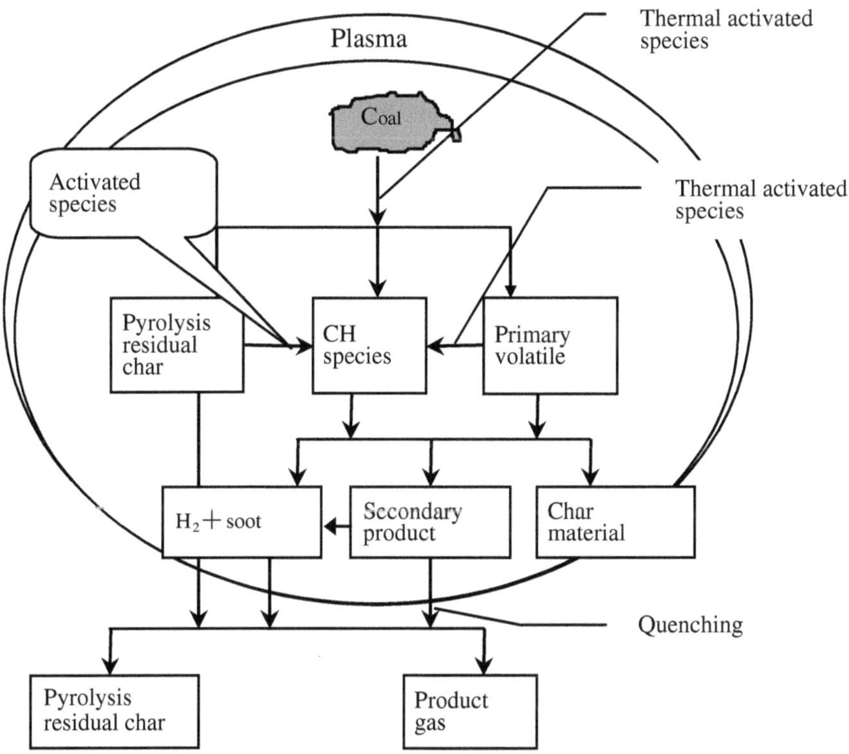

Fig. 9.1 Coal reaction routes in arc plasma

9.1.1.1 Coal Pyrolysis in Arc Plasma

Many experimental studies have shown that the main products of coal pyrolysis in an arc plasma are product gases, char, and soot. The volatiles are mainly composed of light gases (CH_4, C_2H_2, C_2H_4, H_2, CO, and CO_2). Char is the coal residue after devolatilization and soot is the carbon deposited by acetylene decomposition at high temperature. The pyrolysis is very fast because the temperature of the arc plasma is extremely high (ranging from thousands of degrees to tens of thousands of degrees). Thermal decomposition under such ultrashort pyrolysis conditions is violent, whereas thermal polycondensation occurs more slowly, and this is the main reason for the product gases being mostly light gases. Furthermore, acetylene decomposes easily to hydrogen and soot at high temperatures. Thus, there are many factors that influence the reactions of coal under plasma conditions [5–8].

Impact of Coal Properties on Pyrolysis

Under the same operating conditions, the yield of hydrocarbons from coal decreases with increasing coal rank [9–11]. The higher the volatile content of the coal, the higher the degree of coal conversion that can be obtained. This implies that the volatile content of coal is the main factor affecting coal pyrolysis in a plasma. The factors that influence the devolatilization process will influence the pyrolysis behavior in a plasma. It is important to note that the pyrolysis reactions of coal in a plasma are constrained by heat transfer and the coal pyrolysis kinetics. The thermodynamic parameters vary with pyrolysis temperature, and vary depending on coal type.

Impact of Pyrolysis Atmosphere

Bond et al. [9] performed coal pyrolysis experiments in an argon plasma at 4273 K. The pyrolysis gas consisted of hydrogen, acetylene, and carbon monoxide. When hydrogen was used as the plasma working gas, the main pyrolysis product was acetylene, and the coal conversion greatly increased. This is because hydrogen reacts with decomposed free radicals, and its presence depresses the decomposition of acetylene because acetylene decomposition is a volume-increasing reaction, based on thermodynamic analysis. The conventional hydrogenolysis of coal and thermal plasma pyrolysis of coal are therefore related to, but different from, each other in the same hydrogen atmosphere. A common feature is that all the free radicals produced during coal pyrolysis react with free radicals produced by hydrogen, so coal conversion is improved in both cases. However, in the conventional pyrolysis process, free hydrogen radicals diffuse into the coal particles and participate in the rearrangement and combination of various decomposed free radicals in a plastic mass; this depresses thermal polycondensation and increases coal conversion. Under plasma conditions, hydrogen atoms directly combine or exchange with free radical species produced from the coal, and depress acetylene and other labile hydrocarbon decomposition, so conversion is improved.

Chakravartty et al. [2] studied the molecular structure of coal and found that only aliphatic and alicyclic carbons are transformed into acetylene in an inert gas atmosphere, whereas aromatic carbons also react to form acetylene in a hydrogen atmosphere. Based on a summary of previous studies, Dixit et al. [12] showed that high acetylene yields are obtained when the C/H ratio is 0.080 for aliphatic structures, 0.195 for alicyclic structures, and 0.725 for aromatic structures. Under the same operating conditions, the acetylene yield increases with increasing C/H ratio. Dixit performed a theoretical calculation of the acetylene yield, based on Chakravartty's conclusions. It is assumed that all the oxygen in the coal is transformed into CO in the plasma jet, leading to consumption of a proportion of the carbon in the coal. Similarly, the nitrogen and sulfur in the coal also consume a proportion of carbon and hydrogen, generating HCN and H_2S. According to thermodynamic theory, at 4000 K, the free energy of acetylene formation is close to zero, allowing its formation.

The results of Dixit's theoretical calculations and the experimental findings differ under argon plasma conditions, but the two are comparable if only aliphatic and alicyclic structures are considered. Under hydrogen plasma conditions, the experimental values are close to the theoretical values. It is therefore concluded that only the aliphatic and alicyclic carbons are transformed into acetylene in an inert gas atmosphere, but aromatic carbon also participates in the reaction in a hydrogen atmosphere. The experimental value is always lower than the theoretical value because the experimental process is constrained by thermal transmission processes.

Thermodynamic Equilibrium

Baumann et al. [3] suggested that the gaseous component does not reach thermodynamic equilibrium when the coal reaction time is very short. However, Chakravartty et al. [2] proposed that coal transformation under high-energy and high-enthalpy conditions reaches thermodynamic equilibrium because the temperature in the high-temperature and high-enthalpy system exceeds the sublimation point of carbon. This means that all the carbon in the coal sublimes and reacts with hydrogen in the vapor phase, with the system reaching equilibrium. In conclusion, for coal pyrolysis in plasma to reach thermodynamic equilibrium, high-temperature, high-enthalpy conditions, or sufficient residence times in the reactor, are required.

9.1.1.2 Reactions of Coal with Activated Species in Plasma

An arc plasma is an ionic jet flow formed by a gas after the gas passes through an electric arc. An arc plasma not only has high temperature, but also contains a large number of activated ions. If hydrogen is introduced into the plasma working gas, a large number of activated hydrogen species are produced that enable bond breaking in the macromolecular coal structure, and coal conversion. A discussion of the functions of various activated species in coal transformation helps to identify another important coal transformation process in a plasma.

Reaction of Coal with Hydrogen Atoms in Non-plasma State

Hydrogen gas can be decomposed to hydrogen atoms under the action of a mercury catalyst [13]. The products of hydrogen and coal particles in the presence of mercury vapor are complex. The products contain a large amount of high-molecular-weight substances, and no gases such as CH_4 or C_2H_2. This indicates that hydrogen atoms in a non-plasma state can react with unstable bridging bonds among coal aromatic clusters. Hydrogen atoms generated by thermal decomposition of hydrogen gas can react with carbon to produce various hydrocarbons [14]; the products include methane and small amounts of ethane and propane. This indicates that hydrogen atoms can destroy various chemical bonds in the coal structure. The degree of destruction depends on the hydrogen atom energy.

Reaction of Coal with Hydrogen Atoms in Plasma State

Sanada used a high-voltage discharge to produce hydrogen atoms [15]. When pure carbon was added to the discharging area, methane was detected in the product gases. When the pure carbon left the discharge area along the discharge axis, the yield of methane was significantly reduced. When coal was used as the feedstock, CO, CO_2 and various hydrocarbons (C_2H_2, C_2H_6, C_3H_8, and C_4H_{10}) were produced. This indicated that C_mH_n is formed in the reaction between hydrogen and carbon in the coal structure, rather than by the reaction of gaseous hydrogen and carbon in the coal. This also indicated that hydrogen in coal is more prone to transfer. Vastola et al. [16] studied the reaction between hydrogen atoms with pure carbon and with coal, using a microwave hydrogen plasma. Almost no reaction occurred when the pure carbon was outside the discharging area (there are mainly hydrogen atoms outside the discharging area, and a few ions). This indicated that the reaction between pure carbon and hydrogen atoms is difficult. For carbon in the discharge area, the products were mainly methane and acetylene because this carbon was vaporized as a result of attack by, and collisions with, high-energy ions. The vaporized carbon combined with hydrogen atoms to form CH, CH_2, and other free radicals (the presence of CH and CH_2 has been proved using spectroscopic experiments). These free radicals formed intermediate species that generated hydrocarbon gases. Acetylene was the main product after quenching because of the high temperature in the discharge area. The reactions between hydrogen atoms and carbon are facilitated by the presence of high-energy ions in the plasma state. Methane, acetylene, and other hydrocarbons can be formed from coal even when outside the discharge area. In conclusion, hydrogen atoms have the capacity to destroy various bonds. The results are different for hydrogen atoms of different energies. Low-energy hydrogen atoms can only destroy weak bridging bonds in coal, and the products formed contain high-molecular-weight aromatic substances and no low-molecular-weight product gases. As the hydrogen atom energy increases, bonds in branched chain structures of coal are destroyed and the products contain low-molecular-weight hydrocarbons and acetylene. Some activated functional groups may be present on the coal surface, and the activation energy needed for hydrogen atoms to react with the functional groups and form hydrocarbons is relatively low.

Impact of Hydrogen on Reaction Products of Coal in Plasma

When hydrogen gas is used as the plasma working gas, both the coal conversion and the acetylene yields are significantly higher than when hydrogen is not used. Graphite was used for the reaction with hydrogen plasma in previous research, and acetylene and methane were generated. This indicated that under arc plasma conditions, the carbon in coal can react with hydrogen, resulting in coal conversion. Many free radicals are generated from coal in a plasma, and these readily combine with hydrogen atoms and generate hydrocarbon gases during quenching. A typical reaction is the formation reaction of acetylene:

$$C_2H^{\cdot} + H^{\cdot} \rightarrow C_2H_2$$

Acetylene is the main component of the gases produced in the reaction between coal and a thermal plasma, but acetylene is usually decomposed at high temperatures. The presence of hydrogen depresses decomposition of acetylene because acetylene decomposition is a volume-increasing reaction, based on thermodynamic analysis. Hydrogen therefore plays a "protective" role toward acetylene. Hydrogen has high thermal conductivity, so a hydrogen plasma has a high temperature and thermal capacity, based on thermodynamic analysis. A hydrogen plasma therefore not only accelerates pyrolysis, but also enables more carbon in the coal to participate in the reaction. However, a higher concentration of hydrogen does not give a higher coal conversion. Bond et al.'s [9] research indicated that the best acetylene yield from coal is obtained when the volume concentration of hydrogen is about 10 %. Mohammedi [17] found the highest concentration of hydrogen atoms in the plasma when the volume concentration of hydrogen was 10 %, in an Ar/H_2 plasma. It is assumed that the highest concentration of hydrogen atoms in the plasma leads to the highest acetylene yield from the coal. This trend indicates that hydrogen atoms are important in acetylene formation. However, other studies such as that by Garret et al. [18] showed that the maximum acetylene yield was obtained when Ar/H_2 (v/v) was 54:46.

9.1.1.3 Research on Acetylene Production from Plasma Pyrolysis of Coal

The volatiles in coal are decomposed to light hydrocarbons when coal is rapidly heated to above 1523 K. Acetylene is the main component of the product gases under appropriate operating conditions, including optimized reaction time, quenching rate, and coal type [19].

In earlier research, various energy sources were used for the rapid pyrolysis of coal. The first study using an electric arc was performed by James in Sheffield, UK, in 1962 [20]. He used an arc image furnace and discovered that the coal conversion exceeded the volatile components measured according to the British national standard, and 8 % of the carbon in the coal was transformed into acetylene.

In 1963, Bond et al. [21] used an anode drill method to supply coal powder to a plasma jet flow. The power of the plasma generator was 6 kW (pure argon) to 9 kW (argon/hydrogen mixed gas), the powder feed rate was 1 g min^{-1}, and the argon flow rate was 10 L min^{-1}. The experimental results indicated that the main hydrocarbon products in the product gases were methane, ethane, ethylene, acetylene, and propane. Only 20 % of the carbon in the coal was transformed into acetylene under an argon plasma, but the conversion rate increased to 40 % when 10 % hydrogen was added to the working gas. Later, Bond et al. [9] reported that for argon/hydrogen mixed gases, a hydrogen volume fraction of 10 % is the optimal concentration for the transformation of coal to acetylene. He found that the particle sizes of anthracite and some low-volatile-content feedstocks hardly changed after reaction with an arc jet flow. He deduced that the microparticles in the reaction residue were mainly produced by coal "bursting" and acetylene decomposition.

To increase the coal pyrolysis efficiency, Nicholson et al. [22] and Littlewood [23] used a hollow cathode axial feeding method to enable the feedstock to reach the hottest zone of the arc. Argon or a mixture of argon/hydrogen was used for the experiments, and the plasma generator power was 7–14 kW. The main components of the product gases were H_2 (46 %), C_2H_2 (31 %), and CO (23 %). The experiments showed that the total quantity of hydrogen and oxygen in the product gas was close to the total hydrogen and oxygen content in the raw coal; therefore the aliphatic and alicyclic carbon compounds in the coal were transformed. The highest acetylene yield was obtained with the shortest reaction time, and the reactor length had no effect on acetylene formation. Carbon conversion to acetylene increased and then decreased with increasing generator power, and the maximum acetylene yield was obtained at 7 kW. It is worth noting that when the hydrogen concentration in the argon/hydrogen mixed gas was 10, 74 % of the carbon in Barnborong coal was transformed into acetylene, which is the highest value so far reported in the literature. This is called a one-stage process because the coal pyrolysis and formation of an electric arc occur during the same stage. Reaction processes in which coal powders are fed into an arc jet flow are called two-stage processes. The one-stage process is an ideal way to improve the thermal efficiency of the overall process, but it has drawbacks: coke is generated at the cathode and anode during pyrolysis, preventing stable operation of the discharging system.

Chakravartty et al. [10] studied this subject at an early stage. Argon was used as the working gas, and coal powders were fed into the plasma jet flow with hydrogen as the carrier gas. The working power of the generator was 4–8 kW and Indian high-volatile-content coal was used as the feedstock. The results indicated that the transformation rate of coal to acetylene was related to the powder feed rate, particle distribution, volatile content of the coal, and other factors. The ash content of the coal, 20 %, had no adverse effect on the acetylene yield, but it increased the energy consumption. These conclusions are consistent with Nemets' experimental conclusions [24]. When hydrogen plasma was used, at a powder feed rate of 5–6 g min^{-1}, the acetylene yield increased from 10–20 % to 60–65 %, compared with an argon plasma [2].

Baumann et al. [3] and Beiers et al. [11] proposed that acetylene formation in coal pyrolysis in a plasma is affected by processes such as thermal conduction from the gas to the coal, the coal pyrolysis kinetics, and homogeneous gas-phase reactions. The coal is immediately heated when it is mixed with the plasma; the volatiles then escape and react with the activated species in the jet flow to form acetylene. The reaction is complete in 4 ms, and the coal is heated to the same temperature as that of the gas. It was predicted that coal will cause a bursting phenomenon in the arc plasma jet flow under rapid heating, in which the reaction surface area of the gaseous and solid phase are increased, thereby increasing the acetylene yield.

Bittner et al. [25] reported that in addition to the power, temperature, powder supply rate, and retention time, the chemical properties of the coal are a basic factor governing the transformation of coal to acetylene, i.e., the volatile content, heating value, and coking properties of the coal are the main factors that decide the distribution of acetylene, chars, and by-products. To study the impact of the coal properties on the acetylene yield, 13 types of coal, with volatile contents between 30 and 40 %, were used; the ash contents of some of the coals reached 35 %. The plasma generator power was 30 kW, and the coal powders were fed into the jet flow through four pipe channels, with hydrogen as the carrier gas. The average coal particle size was less than 250 μm, the particle sizes of five of the coal samples were less than 30 μm, and the system operating pressure was 40 kPa. The results indicated that oxygen and carbon in the coal combined to form CO, which was detrimental to acetylene formation. Coal samples rich in exinite formed acetylene more easily than did those poor in exinite. For coal samples with high and medium volatile contents, the acetylene yields increased with increasing coal rank. Although the volatile content was lower, the coal conversion increased. The coal conversion decreased with increasing total pore volume of the coal. It was also found that the acetylene yield was affected by the reactor length. Acetylene decomposed to soot and hydrogen if the reactor was too long. However, the coal did not react completely if the reactor was too short; this is related to the particle diameter distribution of the coal sample. The relationship between the acetylene yield and the quantity and composition of coal volatiles was investigated using methane, ethane, ethylene, and acetylene as the raw materials to simulate the light components of coal volatiles, and benzene, toluene, ethylbenzene, tetralin, hydrogen solution tar, anthracene oil, and cyclohexane were used to simulate the heavy components of coal volatiles. A 30-kW hydrogen plasma was used and the feedstock was injected into the jet flow at an initial temperature of 1773–2273 K and an angle of 35°. For an initial temperature of 1773 K, the order of the transformation rates of the light gases was $C_2H_6 > C_2H_4 > C_2H_2 > CH_4$, whereas the order for the heavy components was cyclohexane > hydrogen solution tar > tetralin > ethylbenzene > anthracene oil > benzene, showing a trend from high to low H/C ratios. Research has shown that acetylene formation is mainly related to the hydrogen atoms, molecules in excited states, atoms, and ions in the plasma jet flow. Once these activated species have been consumed, the reaction stops immediately. This process therefore cannot be described using traditional thermodynamic theory, and thermodynamic equilibrium is not reached for the millisecond reaction times.

Kulczycka [26] used various coals with different vitrinite contents to perform pyrolysis research in a mixed argon/hydrogen plasma (hydrogen content 33 %), with a generator power of 6.25–14.4 kW and powder feed rate of 0.1–1.3 g min^{-1}. The results showed that 45.5–52.8 % of the carbon in the vitrinite was transformed into acetylene.

Since the 1970s, the AVCO Corporation (Stratford, CT, USA) has been engaged in theoretical research on, and industrial development of, acetylene production from coal in arc plasmas. They used hydrogen as the quenching medium and achieved high acetylene yields. Deuterium and ^{13}C were added to show that complete atom exchange had occurred among acetylene molecules and other molecules in the quenching medium, resulting in a chain mechanism. In these experiments, a coal column was used as the consumable anode, and the acetylene concentration in the product gases was only 5.04 % (v/v). The experiments proved that decomposition of acetylene to carbon black and hydrogen continued without quenching; the results were consistent with the thermodynamic analysis. Cannon et al. [27] performed experiments using a plasma generator consisting of a tungsten cathode and a water-cooled copper anode with a power of 30 kW; the thermal efficiency of the plasma system was 75 %. The hydrogen flow rate was 3.5 $m^3 h^{-1}$, and the pressure in the reactor was 20–50 kPa. The maximum acetylene yield was 18 %, for Pittsburgh bituminous coal at a feed rate of 125–550 g min^{-1}. AVCO later developed a magnetic rotating DC reactor. The electric arcs rotated at high speed under the action of an external magnetic field, and the coal powders were injected into the hydrogen plasma arc area at 8000–15,000 K. The maximum acetylene yield was 33 %, the concentration was 16 %, and the coal conversion was 67 %. In 1980, AVCO performed semi-industrial-scale experiments and achieved an annual yield of 900 tons of acetylene, using a power of 1 MW and a coal feed rate of 450 kg h^{-1}.

Makino et al. [28] used three plasma generators with a power of 100 kW to study bituminous coal pyrolysis. The results showed that a large amount of acetylene was decomposed under the experimental conditions. The acetylene yield was higher if hydrogen was used as a quenching medium. The power consumption for acetylene production was 17 kWh kg^{-1}. The volatile content in the reaction residues was more than 15 %, and the ash content was also very high. The organic sulfur compounds in the coal were transformed into inorganic sulfur compounds during the reaction.

Fauchais et al. [29] reached a series of significant conclusions by comparing several processes: under the action of a magnetic field, the mixing efficiency of the arc and the feedstock increase; the hydrogen in the plasma plays an important role, not only because of its high thermal capacity and high thermal conductivity, but also because hydrogen can prevent decomposition of acetylene in the quenching process; operating the system at a negative pressure (0.5 atm) does not affect the thermal conduction process, and increases the acetylene yield; the waste heat can be completely used and more ethylene and acetylene are produced if light hydrocarbons are used as the quenching medium, so the energy consumption is reduced; acetylene may be partly oxidized if water is used as the quenching medium; and carbon black is formed on the reactor wall if the reactor wall is cooled.

9.1.2 Applications of Plasma Technology in Coal-Based Chemical Industry

9.1.2.1 Plasma Coal Gasification

Plasma coal gasification is the process in which syngas is generated from coal in an oxidic arc plasma atmosphere (such as oxygen, air, carbon dioxide, water vapor, or their mixture). As early as the 1960s, research on high-temperature gasification of coal using thermal plasmas such as hydrogen, hydrogen/carbon oxide mixtures, and water vapor was performed. To make full use of rich domestic peat resources, a plasma peat gasification process was developed in Canada. In this process, hot water vapor was injected into the plasma torch at 6273 K to produce syngas from the peat. Since the 1980s, a considerable amount of work has been done in the former Soviet Unit on plasma gasification of coal. The Republic of Tajikistan developed a coal gasification plant that can output 2.8×10^{10} m^3 of coal gas per year; the operating cost is 10 % lower than that of the conventional process, and the plant construction costs were only about 50 % of the equipment investment for the conventional process [30]. Russian and Polish scientists have also studied the factors influencing syngas production using this technology, and its advantages on an industrial scale.

The reactions in the plasma gasification process are complicated because various reactions take place at high temperatures. A number of particles that are not in thermodynamic equilibrium are present in a low-temperature water-vapor plasma. The following reactions take place when coal powders are injected into the plasma flow:

$$C + H_2O \rightarrow H_2 + CO$$
$$2C + O_2 \rightarrow 2CO.$$

As the activation energies of these two reactions are low, and the reactions are self-heat-supplying–heat-collecting reactions, the reaction proceeds smoothly. However, various factors affect the experimental results. As in the case of plasma pyrolysis, the composition of the plasma carrier gas, feed rate of the coal, residence time, type of coal, particle size, and other factors have significant impacts on the reaction. However, this process is clearly different from plasma pyrolysis of coal in terms of the reaction temperature and mechanism, selection of carrier gas, and other factors. Water vapor is added during the process, so the plasma is not only used as a thermal carrier for the reaction, but also participates in the reaction as the gasification medium. Most researchers believe that the oxygen participating in the reaction during gasification comes from the raw coal. The gasification speed clearly increases and the energy consumption of the entire process decreases if extra oxygen is added to the system. Under the same reaction conditions, therefore, coal with a high oxygen content consumes less energy. The energy consumption order during coal gasification is anthracite > bituminous coal > lignite; agreement on this point has basically been reached.

The first mathematical model of plasma gasification was developed by Kalinenko et al. [31], based on experimental thermodynamic data. These data were used to identify the reactions that could occur (there were 63 reactions if the intermediate reactions were counted). The experimental data from a single-plasma airflow reactor were used. The calculation results showed that the gasification reaction mainly occurred in the dynamic area.

A specific feature of plasma gasification is that there is no oxidation area; thermal energy is directly introduced into the reduction process. Georgiev et al. [32] divided up the reaction stages of coal plasma gasification, based on their experience. First, when coal particles exchange heat with the plasma, the coal particles are rapidly heated. At 900–1200 K, the volatiles are released by bursting. The released volatiles are gasified rapidly (the reaction is instantaneous), and the reaction rate depends on the temperatures in the reactor and the homogeneous phase, and the rates of heat and material exchange. In the final stage, semi-char coal particles are gasified by diffusion, chemisorption on activated ligand complexes, product desorption, and inverse diffusion. These reactions are notable in that when the volatiles are separated during bursting, the surface area and pore volume of the coal particles increase significantly, and a large number of chemical bonds are activated.

The method for the preliminary estimation of energy consumption has been improved because the energy consumption in the plasma gasification of coal is crucial for industrialization of the process [32]. The following factors were considered in calculating the energy consumption in the plasma gasification of 1 kg of coal: the thermal efficiency of the gasification reaction, the energy consumed in destroying the coal structure and releasing volatiles, the energy consumed in heating the plasma, the mineral content of the semi-char residues, and the final gas temperature. The calculation results indicated that this technology can be used for industrial production if the residual thermal energy lost after gasification can be recovered effectively. Qiu et al. [33] investigated the gasification of coal under steam and air plasma conditions at atmospheric pressure in a tube-type setup with the aim of producing synthesis gas. It was found that the content of H_2 and CO in the gas increases with increasing arc input power, and passes through a maximum with increasing current in the electromagnetic coil. Under the applied conditions, the content of H_2 + CO in the gas could reach 75 % in volume, with CO_2 representing less than 3.0 vol%. OES diagnosis reveals that the CO^+ ion and CH radical are probably the precursors or origins of CO species in the gas.

9.1.2.2 Microwave Plasma Coal-Processing Technology

Microwave plasmas provide a high-temperature thermal source for manufacturing acetylene through coal pyrolysis. Some researchers have used microwave plasmas to study coal gasification in water vapor atmospheres [34]. Thermodynamic analysis has shown that microwave plasmas are not suitable for continuous industrial production. The work on coal processing using microwave plasmas has mainly been performed in Japan. Kamei et al. [35], from the Chiba Technical Research Institute, injected lignitic coal into a methane microwave plasma. The results indicated that the carbon and hydrogen in the methane were mainly transformed into acetylene and hydrogen; the maximum acetylene concentration was 16 %. The carbon in the coal was mainly transformed into liquid products, the main components being C_{13}–C_{34} aliphatic hydrocarbons (H/C = 1.5–1.6), in a reaction time of 2 min. After reaction for 5 min, the carbon in the coal was transformed into CO_2.

Almost all the oxygen in the coal was finally transformed into CO, by depolymerization of carboxyl groups or from CO_2.

9.1.2.3 Coal Pyrolysis in Nitrogen Arc Plasma

A large amount of HCN is found after coal reacts with a nitrogen plasma jet. Like acetylene, HCN is a highly endothermic compound; therefore it is easily generated under high-temperature arc conditions. As early as the mid-1960s, Bond et al. [9] studied this, and achieved a coal conversion of 35 %, and production of some C_2H_2. Similar studies performed in China [36] showed that the main components of the gas generated under these conditions are hydrogen, methane, CO, acetylene, and cyanoacetylene; acetylene is considered to be the precursor of cyanoacetylene.

9.1.2.4 Low-Temperature Ashing of Coal

In low-temperature ashing, coal is added to a cold oxygen plasma and ashed. This technology has the advantage of leaving the inorganic mineral phases in the coal undamaged, so it is an ideal method for accurate measurement and investigation of the minerals in coal. Coal can be ashed in a cold oxygen plasma at 27.12 MHz and 373 K. The main mineral components in the coal samples tested were quartz and silicate; the secondary components included carbonates, sulfates, phosphates, and a small amount of chlorides.

9.1.2.5 Other Aspects of Plasma Applications

Research on the use of low-temperature plasmas for environmental protection has attracted the attention of scientists worldwide. Some examples include the removal of NO_x and SO_x from flue gases using pulse plasmas; the removal of NO, NO_2, NH_3, and other harmful gases from flue gases using microwave plasmas; and the removal of CO_2 and CO from flue gases using narrow-pulse corona discharge technology. The measurement of various elements in coal using inductively coupled plasma has become a mature technology, and is more convenient and accurate than other chemical measurement methods. Some research has focused on the composition of metal–carbon nanoscale catalysts in plasmas, and excellent catalytic activities have been achieved.

To improve the efficiency of solid fuel use and decrease fuel oil rates in the fuel balance of thermal power plants, plasma-assisted combustion technology is now popular, and is being disseminated around the world. In the framework of this concept, a portion of pulverized solid fuel is separated from the main pulverized solid flow and undergoes activation by an arc plasma in a special chamber with a plasmatron. The air plasma flame serves as a source of heat and additional oxidation, and provides a high-temperature medium enriched with radicals, where the fuel mixture is heated, volatile components of coal are extracted, and carbon is partially gasified. This active blended fuel can ignite the main pulverized solid flow supplied to the furnace. This technology provides boiler start-up and stabilization of pulverized solid flame and eliminates the necessity for additional highly reactive fuel [37, 38].

There is an increased interest in nanomaterials formed in plasmas, especially thermal plasmas, and the preparation of nanoscale carbon materials using coal as the raw material has attracted much attention. Pang et al. [39] successfully produced fullerenes from raw coal in arc plasmas, and reported that the formation of polycyclic hydrocarbons is related to the coal structure, based on a comparison of the $^{13}C/^{12}C$ abundances in the polycyclic hydrocarbons. Geldard et al. [40] added naphthalene to a hollow graphite anode under arc conditions to generate C_{60} and C_{70}. These studies prove that not only C_1 and C_2, but also C_{60} and C_{70}, can be transformed under arc conditions, and fullerenes can be produced from polycyclic hydrocarbons and macromolecular fragments. We also successfully synthesized carbon nanotubes by injecting coal powder into a plasma jet [41], and the yield of carbon nanotubes increased when a metal catalyst was introduced [42]. Not only carbon nanotubes, but also other fullerene materials such as carbon nanobeans can be generated under thermal plasma conditions [43].

9.2 Thermodynamic Equilibrium of Chemical Reactions in Multiphase Carbon–Hydrogen–Argon–Oxygen Systems

In this section, the use of modern chemical reaction equilibrium theories and calculation methods to perform a comprehensive examination of multiphase C–H–Ar–O systems under plasma conditions is described. The aim is to identify the optimal reaction conditions and increase the maximum acetylene yield.

Acetylene is the simplest hydrocarbon and contains a triple bond. The acetylene formation reaction, with a hydrocarbon as the raw material, is

$$C_nH_m \leftrightarrow \frac{n}{2}C_2H_2 + \frac{m-n}{2}H_2.$$

This reaction needs a high temperature and energy, based on the thermodynamic properties of hydrocarbons (enthalpy of formation $\Delta_f H_m$ and Gibbs free energy of formation $\Delta_f G_m$) with changing temperature, and the thermodynamic equilibrium in the chemical reaction between hydrocarbons.

Compared with other hydrocarbons, acetylene is highly unstable at low temperatures. The free energies of formation of most hydrocarbons increase with increasing temperature, but the free energy of formation of acetylene gradually decreases. When the temperature exceeds 1230 K, acetylene becomes more stable than other hydrocarbons. Hydrocarbons with shorter molecular chains have an intersection point at higher temperatures in the curves of the free energies of formation of other hydrocarbons and acetylene. The temperature needed for acetylene formation from methane is the highest. However, even at temperatures above 1230 K, acetylene remains unstable with respect to solid carbon and gaseous hydrogen. The decomposition of acetylene to carbon and hydrogen is always spontaneous below 4200 K, based on thermodynamic analysis.

The above descriptions only give a basic analysis of acetylene formation in simple situations. As a result of the difficulty of measuring the temperature of a thermal plasma in a chemical reaction, and the short retention time of the reactants in the plasma jet, a completely reliable and systematic method for accurately predicting the final products of chemical reactions in thermal plasmas has not yet been developed. Bittner et al. [44] calculated the equilibrium composition of a single-phase system, C:H:O:Ar = 1:13.39:0.073:0.605, under a pressure of 0.105 MPa, and considered a total of 14 gaseous components, i.e., H, H_2, C, C_2, C_3, CH_2, CH_3, CH_4, C_2H, C_2H_2, C_2H_3, C_2H_4, CO, and Ar. The results indicated that at 2773 K, the concentration of C_2H_2 is the highest. Dixit et al. [12] reported a simple method for estimating the highest acetylene yield generated from coal. In this method, the operating conditions of the thermal plasma are not considered, and only the ultimate coal analysis is used for data analysis. This method therefore lacks a strict theoretical basis, and the predicted results differ greatly from the experimental values. The predicted acetylene yield is 127.8 times larger than the experiment yield, suggesting that this method is not suitable for designing experiments.

Thermal chemical equilibrium analysis of a thermal plasma chemical reaction system is based on chemical thermodynamics. This analysis has a reliable theoretical basis, and can be used to predict the reaction direction and progress, and to identify the optimal temperature scope of the reaction.

9.2.1 Calculation Methods

According to basic thermodynamic principles, at a certain pressure and temperature, there are generally three methods for determining the chemical equilibrium stage of the system: the minimum free energy function method, the equilibrium constant method, and the forward–reverse reaction rate equality method.

The forward–reverse reaction rate equality method is a thermodynamic equilibrium method, and is an extreme version of the chemical thermodynamic method. In the calculation, each elementary reaction and its rate is considered. However, this method cannot obtain the simplification and stability achieved using the chemical equilibrium method. It needs hundreds of thermodynamic constants for the elementary reactions in a complicated system, and is therefore unrealistic and unfeasible in most situations. Researchers therefore usually do not use this calculation method.

The first two methods are thermodynamic static equilibrium methods, and are essentially equivalent to each other. The equilibrium constant method is a typical method for calculation of chemical equilibrium, based on the core concept of the equilibrium constant. In this method, first, the number of independent components, S, and the number of independent reactions, k, are defined, using the phase rules for the chemical reaction system at a given temperature T and pressure P. In the equilibrium constant method, if α types of element exist in the system, it is necessary to solve $k + \alpha$ non-linear equations to determine the quantity of each component. The calculation is very large for multiple components, especially for multiphase equilibrium systems. The most difficult and complicated points are that the process and details of the reactions should be considered.

In practical engineering calculations, only the first method, i.e., the free energy function minimum method, can dispense with the detailed mechanism of a complicated chemical reaction. In this method, the system is processed by the optimal mathematical method, using the basic concept of thermodynamic equilibrium to obtain a direct calculation process. This is the method usually used at present. The basis for the application of the minimum free energy function is the well-known chemical thermodynamic principle: for a chemical reaction system composed of a certain number of elements at an instant pressure and temperature, the system is in an equilibrium state, under the constraints that the atomic composition is conserved and non-negative when the free energy function of the system is minimum.

9.2.1.1 Mathematical Description of Thermodynamic Equilibrium

Total Free Energy of System

Generally, the total free energy of a system that contains N components can be expressed as follows:

$$G = \sum_{i=1}^{N} n_i \mu_i = \sum_{i=1}^{N} n_i \left(\mu_i^0 + RT \ln a_i \right),$$

where

G is the total free energy of the system (J);

n_i is the number of moles of component i;

μ_i is the chemical potential of component i (J mol^{-1});

μ_i^0 is the standard-state chemical potential of component i (J mol^{-1});

a_i is the activity of component i.

To facilitate the description, a system that contains M condensed phases and N_P mixed phases is considered. The activity of a condensed phase that only contains elementary substances is 1, and the Gibbs free energy description of this system is as follows:

$$G = \sum_{j=1}^{M} n_j \mu_j^0 + \sum_{j=1}^{N_P} \sum_{i=1}^{N_C} n_{ij} \left(\mu_{ij}^0 + RT \ln \frac{f_{ij}}{f_{ij}^0} \right),$$

where

M is the number of condensed phases;
N_P is the number of mixed phases;
N_C is the number of components in the mixed phases;
n_{ij} is the total number of moles of component i in phase j;
μ_{ij}^0 is the standard-state chemical potential of component i in phase j;
f_{ij} is fugacity of component i in phase j;
f_{ij}^0 is the standard-state fugacity of component i in phase j.

The above equation is a general form; for a specific system, it can be further simplified, according to the actual situation. When the mixed phase is a gaseous mixture,

$$\frac{f_{ij}}{f_{ij}^0} = \gamma_{ij} x_{ij} P,$$

$$x_{ij} = \frac{n_{ij}}{n_j},$$

where

x_{ij} is the mole faction of component i in phase j;
γ_{ij} is the fugacity coefficient of component i in phase j; for an ideal gas mixture, $\gamma_{ij} = 1$;
P is the system pressure.

The total free energy of the high temperature constant-pressure multiphase C–H–Ar–O system discussed in this section can therefore be expressed as follows:

$$G = \sum_{j=1}^{M} n_j \mu_j^0 + \sum_{j=1}^{N_P} \sum_{i=1}^{N_C} n_{ij} \left(\mu_{ij}^0 + RT \ln x_{ij} P \right).$$

Constraint Condition

When the system is simultaneously at chemical reaction and phase equilibria, the number of atoms of each element in the system is constant, according to the law of mass conservation. The total number of atoms of elements distributed in each phase and component is always equal to the number of atoms initially introduced into the system, i.e.,

$$\sum_{j=1}^{M} A_{kj}n_j + \sum_{j=1}^{N_P}\sum_{i=1}^{N_C} A_{kj}n_{ij} = b_k \quad k = 1, 2, \ldots N_E$$

where

A_{kj} is the number of atoms that include element k in component j;
n_j is the number of moles of component j in the condensed phase;
b_k is the total number of moles of atoms of element k in the system;
N_E is the number of element types contained in the system.

The mole number, n_{ij}, of each component must meet non-negative conditions:

$$n_{ij} \geq 0 \quad i = 1, 2, \ldots, N_C \quad j = 1, 2, \ldots, N_P$$
$$n_j \geq 0 \quad j = 1, 2, \ldots, M$$

Optimization of Non-linear Constraint

At constant temperature and constant pressure, the basis for determining whether the system reaches chemical reaction thermodynamic equilibrium is the minimum free energy value of the system:

$$(dG)_{T,P} = 0$$

Mathematically, this is expressed by the following optimization of the non-linear constraint:

$$\min G(\vec{n})$$

$$H_k\left(\vec{n}\right) = 0 \quad k = 1, 2, \ldots, N_E$$

$$Q_l\left(\vec{n}\right) \leq 0 \quad l = 1, 2, \ldots, N_C \times N_P + M$$

where

$\vec{n}$ is the mole vector quantity;
G is total free energy of the system, see section "Total Free Energy of System";
H_k is the quality constraint in section "Constraint Condition";
Q_l is the inequality constraint in section "Constraint Condition".

The existence and uniqueness of an optimal solution to the non-linear constraint optimization problem shown in the above equation have been proved [45]. A generalized reduced gradient algorithm was used to solve the equation [46]; this method has the advantages of excellent convergence and insensitivity to initial values.

9.2.2 Thermodynamic Data and Equilibria

Table 9.1 shows the thermodynamic data for all the components [47]. The constant-pressure thermal capacity C_p is fitted by the data for the C_p coefficient in the table by the equation:

$$C_P = A + B \times 10^{-3}T + C \times 10^5 T^{-2} + D \times 10^{-6} T^2,$$

where T_1 and T_2 in the table are the upper limit and lower limit, respectively, of the fitting interval. The standard states of all the components are 298.15 K and 10 1325 Pa (1 atm), as is the convention. The free energy of component i can be calculated at any temperature T using

$$\begin{aligned} H(T) &= H^0 + \int_{298.15}^{T} C_P \mathrm{d}T, \\ S(T) &= S^0 + \int_{298.15}^{T} C_p \mathrm{d}T, \\ G(T) &= H(T) - TS(T), \\ &= H^0 + TS^0 + aT\ln T + bT^2 + cT^{-1} + \mathrm{d}T^3. \end{aligned}$$

In the equation, the coefficients a, b, c, and d are calculated, using the integral-relation method, from the data for the coefficients A, B, C, and D of C_p.

9.2.2.1 Thermodynamic Calculation Results

C–H System

The C–H equilibrium system is the basis for establishing the process parameters for acetylene production using hydrocarbon feedstocks. The C–H equilibrium system is important, so the thermodynamic equilibrium of the C–H system has been calculated for various temperatures and H/C ratios. Holmen et al. [48] reported the mole fractions of 10 gaseous components, i.e., H, H_2, CH_4, CH_3, C_2H, C_2H_2, C_2H_4, C_2H_6, C_3H_4, and C_6H_6, of the C–H system, with $H/C = 4$ under 1 bar of pressure, at 773–2573 K. Dai [49] calculated the gaseous equilibrium of the C–H system with $H/C = 1$ and H/C = 4 at 2000–5000 K, and seven components, namely H, H_2, C_2H, C_2H_2, CH, C, and C_2, were considered. In these calculations, only the gaseous phase was considered, which means that the C–H system was considered as a single-phase system, and the important fact that solid carbon was present was neglected. The experimental results indicated that solid carbon is inevitably produced in the C–H system. Solid carbon is always a component with minimum free energy in the C–H system and is spontaneously generated at temperatures below 3500 K.

Table 9.1 Basic thermodynamic data

Component	H^0 (kJ mol^{-1})	S^0 (J mol^{-1} K^{-1})	C_p (J mol^{-1} K^{-1}) coefficient				T_1 (K)	T_2 (K)
			A	*B*	*C*	*D*		
CO(g)	−110.541	197.552	28.409	4.100	−0.460	0.000	298.15	6000.00
Ar(g)	0.000	154.846	20.790	0.000	0.000	0.000	298.15	6000.00
CH(g)	594.128	183.040	29.490	−2.630	0.000	5.050	298.15	600.00
			18.320	16.170	9.980	−2.900	600.00	2600.00
			48.630	−2.250	−235.380	0.240	2600.00	6000.00
CH_2(g)	397.480	181.167	25.263	27.372	−1.439	−6.004	298.15	6000.00
CH_3(g)	133.637	193.037	22.962	49.175	−0.025	−11.657	298.15	6000.00
CH_4(g)	−74.810	186.188	12.447	76.689	1.448	−18.004	298.15	6000.00
C_2H(g)	477.000	207.444	16.680	91.150	0.430	−81.410	298.15	400.00
			33.640	23.410	−2.680	−5.840	400.00	1100.00
			45.180	8.940	−22.600	−0.850	1100.00	3600.00
			74.190	−0.100	−832.400	−0.100	3600.00	6000.00
C_2H_2(g)	226.731	200.849	43.627	31.652	−7.506	−6.309	298.15	6000.00
C_2H_4(g)	52.467	219.225	32.635	59.831	0.000	0.000	298.15	6000.00
C(g)	716.677	157.988	20.133	0.490	0.569	0.008	298.15	6000.00
C_2(g)	837.737	199.271	32.012	3.481	8.994	−0.184	298.15	6000.00
C_3(g)	820.064	237.137	38.284	7.335	−3.057	−0.706	298.15	6000.00
H(g)	217.999	114.716	20.790	0.000	0.000	0.000	298.15	6000.00
H_2(g)	0.000	130.679	16.920	61.459	0.590	−79.559	298.15	400.00
			28.280	0.418	0.820	1.469	400.00	1600.00
			29.769	3.109	−40.120	−0.180	1600.00	6000.00
C(s)	0.000	5.740	0.109	38.940	−1.481	−17.385	298.15	1100.00
			24.439	0.435	−31.627	0.000	1100.00	6000.00

Correctly setting the equilibrium components of the system is necessary to define the system equilibrium composition in the free energy minimization method. The mole numbers or concentrations of components that do not actually exist in the system should be set at zero or very small values when the number of possible components in the system is greater than the number of components present. However, serious errors arise if the number of possible components in the system is less than the number of components present. It is therefore necessary to list all the possible components contained in the system, based on the temperature and pressure used in the calculation, and then delete the relevant components according to the calculation results and finally determine the components in the system. The components of the C–H equilibrium system were calculated at 300–5000 K, with $H/C = 1$, 2, and 4. For comparison, the C–H system was processed as a single-phase system and a multiphase system. The gaseous components were H, H_2, CH, CH_2, CH_3, CH_4, C_2H, C_2H_2, C_2H_4, C, C_2, and C_3, and the solid phase consisted of only monoplasmatic carbon.

Single-Phase C–H System

The calculation results for the equilibrium components of a single-phase C–H system with $H/C = 1$ are shown in Figs. 9.2 and 9.3. Below 2000 K, C_2H_2 was almost the only component in the system. The quantities of C_2H and H_2 increased gradually at higher temperatures, and reached a maximum at 3500 K; a large

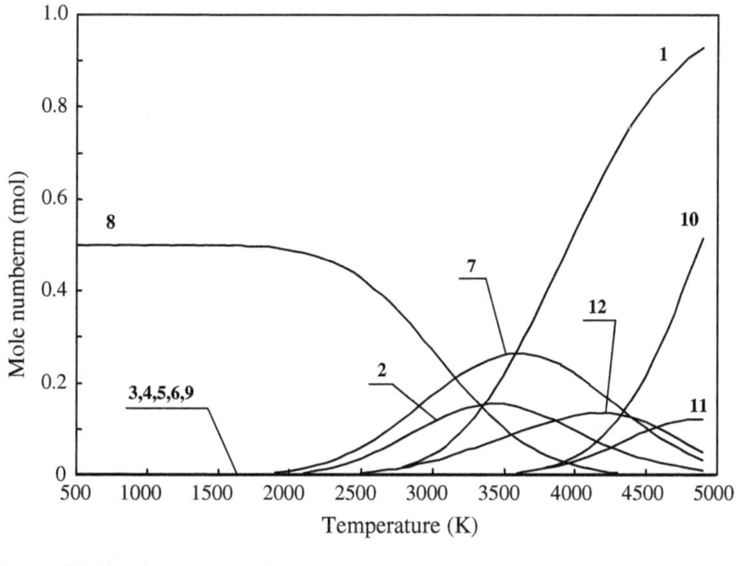

Fig. 9.2 Equilibrium composition of single-phase C–H system ($C{:}H$ = 1 mol:1 mol):1, H; 2, H_2; 3, CH; 4, CH_2; 5, CH_3; 6, CH_4; 7, C_2H; 8, C_2H_2; 9, C_2H_4; 10, C; 11, C_2; and 12, C_3

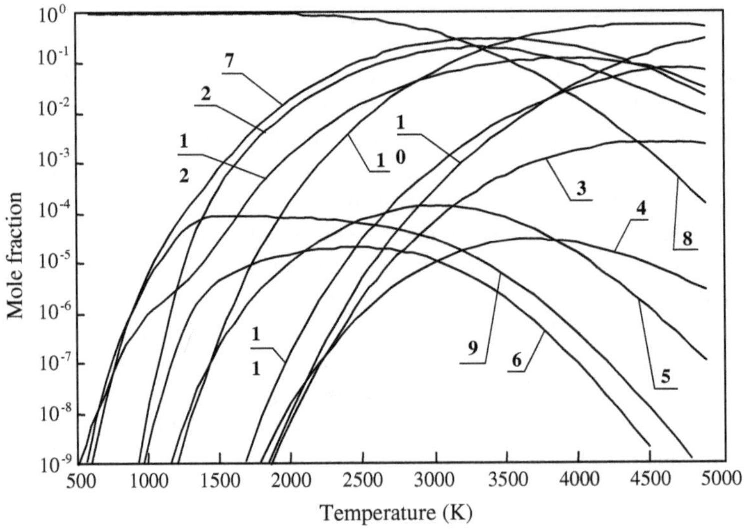

Fig. 9.3 Gaseous equilibrium concentrations of all components of single-phase C–H system for *H*/*C* = 1: 1, H; 2, H_2; 3, CH; 4, CH_2; 5, CH_3; 6, CH_4; 7, C_2H; 8, C_2H_2; 9, C_2H_4; 10, C; 11, C_2; and 12, C_3

number of H radicals were also present in the system. Further temperature increases caused C_2H, H_2, and other components to decompose to C and H radicals, especially above 4500 K. Figure 9.3 shows that the concentration of C_2H_2 decreased independently, and the concentrations of H_2, CH_2, CH_3, CH_4, C_2H_4, C_2H, and C_3 reached maximum values across the whole temperature range, but the concentrations of C, C_2, H, and CH increased independently.

When the *H*/*C* ratio was 2 or 4, the distribution of the equilibrium components of the single-phase C–H system was different from the distribution generated when *H*/*C* = 1. The equilibrium component distributions had similar curves to those in Figs. 9.2 and 9.3. The main characteristics are as follows.

When *H*/*C* = 2 and the temperature of the system was below 1000 K, the main components in the system were C_2H_4, C_2H_2, and CH_4. With further increases in temperature, the quantities of C_2H_4 and CH_4 gradually decreased, and the quantities of C_2H_2 and H_2 increased. In the range 1600–2500 K, the quantities of C_2H_4, CH_4, C_2H_2, and H_2 reached maximum values. When the temperature was increased further, the quantity of C_2H_2 decreased, whereas the quantity of CH_4 remained at the original level until the temperature reached 3500 K; the quantity of C_2H increased to the maximum value at the same time. At temperatures above 4500 K, large amounts of H and C radicals were present in the system, and the mole fraction of C_2H_4 decreased rapidly. The concentration of CH_4 increased when the temperature was below 1000 K, but decreased significantly above this temperature. C_2H_2 was the main component in the system, with a concentration of about 49 %, over a wide temperature range, i.e., 1200–3000 K. Other components such as CH_2

and CH_3 were produced with increasing temperature, and further decomposed to C and H radicals as the temperature continued to increase. The mole fractions of CH_2 and CH_3 were small.

For $H/C = 4$, CH_4 was the only component in the system at low temperatures (below 1000 K). The CH_4 decomposed, and the main components in the system were H_2 and C_2H_2, when the temperature reached 1600 K. In the range 1700–2500 K, CH_4 was converted to C_2H_2 and H_2. The C_2H_2 and H_2 decomposed if the temperature continued to rise. C_2H was present as a transitional intermediate product at 2600–4500 K, and the quantities of H and C radicals gradually increased.

Multiphase C–H System

The equilibrium composition of a multiphase C–H system is different from that of a single-phase C–H system. Below 1000 K, the main components in the system were CH_4, H_2, and C(s) rather than C_2H_2. At 1000–2500 K, H_2 was the dominant gaseous component in the system, and almost all the carbon species in the system were in the form of solid carbon. The amount of the solid component gradually decreased and H_2 decomposed to H with increasing temperature. At the same time, the quantities of C_2H_2 and C_2H gradually increased, and reached maximum values at 3500 and 3800 K, respectively. The maximum value for C_2H was about 4.11 times that for C_2H_2. The main components in the system were H, gaseous C, and C_2 above 4500 K; this is the same as in the single-phase system. The concentration of CH_4 decreased quickly with increasing temperature, whereas the H_2 mol fraction was above 0.92 at 1000–2700 K. The concentration of H continued to increase with increasing temperature. The mole fractions of C_2H_2 and C_2H reached maximum values of 0.0899 and 0.2590, at 3400 and 3800 K, respectively. The changes in the concentrations of other radical components such as CH_3, CH_2, C(g), C_3, and C_2H_4 were close to those in the single-phase C–H system, and only the mole fraction values were different.

For $H/C = 2$ and 4, the distributions of the equilibrium components of a multiphase C–H system were different from that generated when $H/C = 1$.

When $H/C = 2$ and the temperature of the system was below 1000 K, the main components in the system were CH_4, C(s), and H_2. The system contained only C(s) in the solid phase and H_2 in the gaseous phase at 1000–2500 K. The amounts of other components in the gaseous phase, such as H, C_2H_2, C_2H, CH_4, CH_3, and C_2H_4, were very low. With increasing temperature, H_2 decomposed to H, the amount of C(s) gradually decreased, and the amounts of C_2H_2 and C_2H reached maximum values at 3500 and 3700 K, respectively. The maximum amount of C_2H was 2.39 times that of C_2H_2. Among the gaseous components, the maximum mole fractions of C_2H_2 and C_2H were 0.0898 and 0.190, at 3400 and 3600 K, respectively.

When $H/C = 4$ and the temperature was below 1000 K, the main components in the multiphase C–H system were only CH_4, C(s), H_2, and an extremely small amount of C_2H_4. At 1500–2500 K, the quantity of CH_4 decreased to a small amount, the only solid component was C(s), and the gaseous component was mainly H_2. These results are very different from those for the single-phase system, which contained a large amount of C_2H_2. The quantities of C_2H_2 and C_2H reached maximum values at 3300 and 3800 K, respectively. Among the gaseous components, the maximum concentrations of C_2H_2 and C_2H were 0.0895 and 0.105, at 3300 and 3600 K, respectively.

By comparing the calculation results for the equilibrium compositions and gaseous equilibrium concentrations of single-phase and multiphase C–H systems, it is easily seen that a single-phase C–H system is in a metastable state, whereas a multiphase C–H system is in a thermodynamically stable state. The components of the single-phase system at low temperature depend on the *H*/*C* ratio, as a result of the atom conservation constraint. When $H/C = 1$, C_2H_2 is the main component, and when $H/C = 2$, C_2H_4 is the main component, but these results do not correspond to the actual situations. These calculation results are therefore unsuitable for selecting the parameters in the synthesis of C_2H_2 and C_2H_4. However, in the single-phase system in a medium-temperature range (1500–2500 K), C_2H_2 is one of the main components, and its quantity depends on the *H*/*C* ratio. When $H/C = 1$, the gaseous mole fraction of C_2H_2 reaches almost 100 %. When $H/C = 2$ and 4, the gaseous mole fractions of C_2H_2 are 0.50 and 0.25, respectively. However, this is impossible to achieve in real reaction processes; therefore only the calculation results for the multiphase C–H system can be used for designing actual reaction processes. No matter how the *H*/*C* ratio changes, the main components of the system at low temperatures are CH_4, C(s), and H_2. In the medium-temperature region (1500–2500 K), C(s) and H_2 are the main forms of C and H, and this is also the reason for the production of carbon black. It is worth noting that C(s) basically disappears when the temperature exceeds 3500 K, whereas the quantities of C_2H_2 and C_2H successively reach maximum values. Here, C_2H and C_2H_2 are considered to be the same because C_2H combines with H radicals to generate C_2H_2 in the quenching process.

The temperatures at which the mole fractions and gaseous phase concentrations of C_2H_2, C_2H, and $C_2H_2 + C_2H$ reach maximum values in the single-phase and multiphase C–H systems, for different *H*/*C* ratios, are listed in Tables 9.2 and 9.3, respectively. The data in the table show that the optimal temperatures for the formation of C_2H_2 and C_2H in the multiphase C–H system are not closely related to the *H*/*C* ratio. The C_2H_2 content always reaches the maximum value earlier than C_2H does, but the gaseous molar concentrations and mole numbers reach the maximum values synchronously; the optimal temperature range for acetylene production is 3400–3800 K.

Table 9.2 Optimal temperatures (K) for acetylene generation in single-phase C–H systems

Component	Corresponding mole number			Corresponding mole fraction		
	C_2H_2	C_2H	$C_2H_2 + C_2H$	C_2H_2	C_2H	$C_2H_2 + C_2H$
$H/C = 1$	<1500	3600	<1500	<1300	3400	<1500
$H/C = 2$	2000	3700	2400	1900	3500	2000
$H/C = 4$	2100	3800	2500	2100	3600	2200

Table 9.3 Optimal temperatures (K) for acetylene generation in multiphase C–H systems

Component	Corresponding mole number			Corresponding gaseous mole fraction		
	C_2H_2	C_2H	C_2H_2 + C_2H	C_2H_2	C_2H	C_2H_2 + C_2H
H/C = 1	3500	3800	3800	3400	3800	3700
H/C = 2	3500	3700	3600	3400	3600	3600
H/C = 4	3300	3800	3400	3300	3600	3400

C–H–Ar System

Based on the calculated equilibrium composition of the C–H system, we can investigate the impact of the presence of an inert component, Ar, on the generation of C_2H_2 by calculating the thermodynamic equilibrium of the multiphase C–H–Ar system.

The calculation results show that at 5000 K, Ar is not ionized to a significant extent, and the Ar mole number remains constant across the whole temperature range. The changes in the mole numbers of the other components in the system are almost the same as those in the C–H system without Ar. As a result of the presence of Ar, the mole numbers of all the components in the system change; for example, the maximum content of C_2H_2 decreased from 0.0617 to 0.047 mol and that of C_2H decreased from 0.253 to 0.240 mol, when the multiphase equilibrium conditions were changed from H:C = 1:1 to H:C:Ar = 1:1:1. The C_2H_2 and C_2H concentrations decreased by about 22.8 and 5.1 %, respectively. Correspondingly, the maximum concentrations of C_2H_2 and C_2H also decreased by 69.4 and 54.1 %, respectively. These results indicate that the presence of Ar adversely affects acetylene production.

When Ar:C:H = 1:1:2 and Ar:C:H = 1:1:4, the system behaviors were similar. The main difference was that the extent to which the maximum concentrations of C_2H_2 and C_2H decreased was less pronounced with increasing amounts of hydrogen species in the system. This is because the proportion of Ar in the system was reduced and the adverse impact caused by Ar was also weaker.

When liquid petroleum gas is cracked in a thermal hydrogen/argon plasma, the typical gas composition of the reaction system is C:H:Ar = 1:13.39:0.6. A large amount of H_2 is present in the system across the whole temperature range, so hydrogen species dominate the reactions of the system. The proportion of the inert component, Ar, is small, which results in a reduction in the optimal reaction temperature for acetylene production. The calculation results show that at 3000 K, the mole number and molar concentration of C_2H_2 simultaneously reach maximum values, 0.377 mol and 0.0478, and the mole number and molar concentration of C_2H_2 + C_2H also simultaneously reach maximum values, 0.495 mol and 0.0628. The optimal reaction temperature range for acetylene production is 2800–3200 K.

Coal contains oxygen, and when coal decomposes in a thermal hydrogen/argon plasma, a C–H–Ar–O system is generated. Oxygen is an impurity in acetylene production. Based on the calculated equilibrium composition of a multiphase C–H–Ar system, combined with the pyrolysis process of coal in a thermal hydrogen/argon plasma, we calculated the thermodynamic equilibrium of a typical C–H–Ar–O system under normal pressure, with the composition C:H:O:Ar = 1.000:8.638:0.160:3.339, to investigate the impact of the presence of oxygen as an impurity on acetylene production.

The oxygen in the C–H–O–Ar system was mainly present in the form of CO, and the quantity of CO did not vary with temperature. The changes in the mole numbers of the other components in the system were consistent with those in the C–H–Ar system. The maximum contents of C_2H_2 and C_2H were only 0.260 and 0.274 mol, respectively, as a result of the presence of oxygen. The maximum concentrations of C_2H_2 and C_2H were 0.0307 and 0.0276, respectively. The presence of oxygen therefore caused a decrease in the equilibrium production of acetylene. The maximum concentration of C_2H_2 was achieved at 3100 K, which was 300 K higher than the value calculated by Bittner et al. [44]. Based on all the above results, the optimal temperature range for acetylene production through coal pyrolysis in a plasma is 3000–3200 K.

9.3 Acetylene Production Experiments Involving Coal Pyrolysis in a Plasma

This section introduces the main experimental devices and methods used in this research on acetylene production by coal pyrolysis in a plasma.

9.3.1 Experimental Devices

The experimental devices were a plasma power control system, plasma generator, coal powder injector, coal powder distributor, reactor, spray quenching device, sampling system, cooling system, product analysis system, and other servo systems for water, electricity, and gas.

9.3.1.1 Reactor

The reactor is the place where the chemical reaction takes place, and is therefore the core of the apparatus. A downer reactor was used in the experiments. The plasma temperature is very high, so a water cooling jacket was placed around the outer wall of the reactor, and the inner wall of the reactor was lined with graphite. The inner diameter of the reactor was 20 mm and the length was 310 mm. The actual length of the operating part of the reactor was about 350 mm plus the length of the section connecting the reactor with the generator. The positions of the feed holes and the numbers and positions of the sampling openings are shown in Table 9.4.

Table 9.4 Feedstock inlets and sample outlets

Feedstock inlets* (mm)	Positions of sample outlets** (mm)				
	1#	2#	3#	4#	5#
−15	−22	−135	−200	−265	−365

*With the greater anode low-end plane as the criterion plane
**With the feedstock inlet as the criterion plane

To study coking changes, the graphite liner in the reactor was divided into four parts, marked A, B, C, and D, from the top to the bottom. The char sediments on the different parts were called Char A, Char B, Char C, and Char D. The lengths of parts A, B, C, and D were 60, 70, 90, and 80 mm, respectively. The bottom of D was also the bottom of the reactor. After the reaction was finished, the chars developed on all parts were extracted, weighed, and stored for characterization. Gaseous samples for gas chromatographic analysis were obtained from the sample outlets through a sampling system.

9.3.1.2 Quenching Unit

Quenching was used to prevent decomposition of the produced acetylene. Running water was used as the quenching medium; it was sprayed into the reactor tube through a quenching unit and was in direct contact with the hot gas. To ensure full contact between the quenching medium and the product, a split-ring-type structure was used in the quenching operation. The position of the quenching system was 10 mm from the bottom of the reactor, and the product gas outlet was about 160 mm below the quenching position. The lower part of the quenching unit was connected to a separating unit, which was a cask with a hollow cavity. The main function of this was to allow separation of gas from liquids and solids. The gas was extracted from the top of the separating unit, and the quenching water was mixed with the solid residues and drained from the bottom of the separating unit. The drained mixture was passed through filtering equipment to separate the water and solid residues.

9.3.1.3 System Assembly

The assembly order of the system was from the plasma generator, to the feed distributor, reactor, quenching device, separator, and finally the filter; Fig. 9.4 shows the assembly order. Note that to "freeze" the intermediate coking phase by water quenching, the quenching unit was assembled above the reactor in some experiments. Full details of the intermediate coking phase and the experimental design will be given in the following sections.

Typical experimental conditions are listed in Table 9.5.

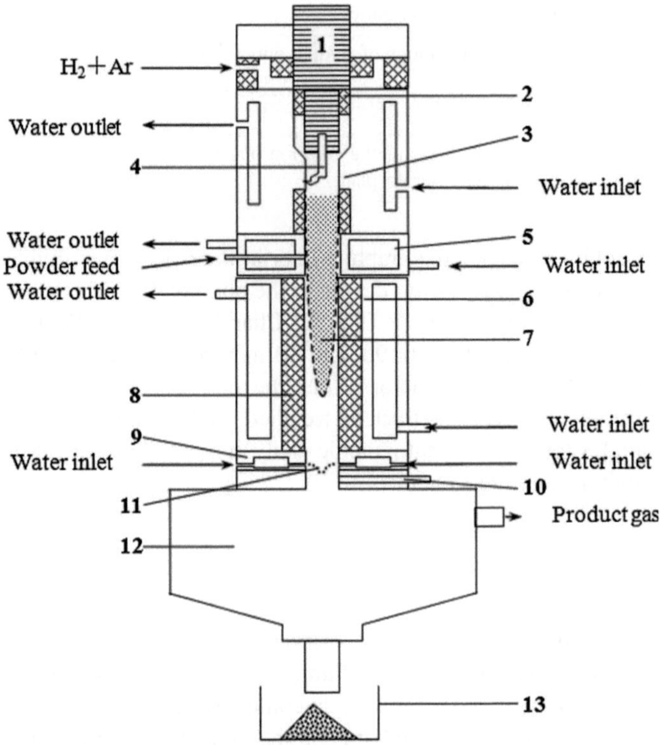

Fig. 9.4 Structural diagram of plasma generator and reactor system: *1* cathode; *2* insulator; *3* anode; *4* electric arc; *5* injector; *6* reactor; *7* plasma torch; *8* reactor wall; *9* sprayer; *10* gas-sampling pipe; *11* water curtain; *12* separator; and *13* filter

Table 9.5 Typical experimental conditions

	Condition
Generator power (kW)	42 ± 0.2
Flow rate of working gas ($m^3\ h^{-1}$)	Ar: 2.2 ± 0.1; H_2: 5.2 ± 0.5
Flow rate of carrier gas (Ar; $m^3\ h^{-1}$)	1.7 ± 0.1
Size of coal particles (μm)	5–25
Powder supply speed ($g\ s^{-1}$)	0.5–4.0
Reactor pressure (atm)	1
Flow rate of generator cooling water ($m^3\ h^{-1}$)	4.0
Flow rate of reactor cooling water ($m^3\ h^{-1}$)	4.2
Flow rate of quenching water ($m^3\ h^{-1}$)	0.5

9.3.2 Conversion Calculations

The coal conversion value is the quantity of product gases obtained from coal as a percentage, based on the amount of raw coal; the calculation methods are different under different experimental conditions.

9.3.2.1 Mass Loss Calculation Method (x_L)

The mass loss calculation method is one of the original methods used for calculating the coal conversion x_L, which is defined as the ratio of the mass m_1 of coal after the reaction and the mass m_0 of coal that participates in the reaction. The calculation results may change if different criteria are used. The calculation is expressed as follows:

$$x_L = 100 - \frac{m_1}{m_0} \times 100\,\%.$$

This method is not rigorous, but the actual operation is very complicated, especially in the large experimental systems discussed in this chapter. Many dead zones were identified in the experiments; therefore the calculations had large errors because much of the reaction residues could not be retrieved. Additionally, the residues collected contained soot decomposed by the product gases. This is therefore not a suitable method.

9.3.2.2 Ash-Content-Tracing Method (x_A)

The ash-content-tracing method is commonly used; it is an equilibrium method and is performed by measuring the ash content $A_{d,1}$ before the reaction and the ash content $A_{d,2}$ after the reaction, based on an invariable absolute magnitude in the solid residue before and after the reaction. The calculation is expressed as follows:

$$x_A = 100 - \frac{A_{d,1}}{A_{d,2}} \times 100\,\%.$$

This method is simple and convenient, and it is only necessary to compare the ash contents in the raw coal and the solid residue after the reaction. However, this method also has some drawbacks. First, under the action of a high-temperature plasma, some low-boiling-point ash substances will evaporate, invalidating one of the assumptions that the method is based on. The magnitude of the resulting error in the calculation has not yet been defined. Even if the impact can be neglected, for coal with a low ash content, errors frequently occur. The main reason is that this method has an amplification effect on coal with a low ash content, i.e., a slight error in measurement of the ash content in the sample has a significant impact on the calculation results.

9.3.2.3 Volatile-Tracing Method (x_V)

The theoretical basis of this method is that only the volatiles in the coal are transformed in the reaction process. This is an equilibrium method and performed by measuring the volatile content $V_{daf,1}$ (daf: dry ash-free basis) before the reaction and the volatile content $V_{daf,2}$ after the reaction. The calculation is expressed as follows:

$$x_V = V_{daf,1} - V_{daf,2}\left(1 - \frac{V_{daf,1}}{100}\right).$$

The amplification effect in this method for coal with a high volatile content is weak, and the calculation error is small, but the defect is that if the non-volatile components react, this method is not applicable.

9.3.2.4 Product Addition Method (x_P)

If we calculate the mass Σm_i of all the product gases, then we can calculate the conversion by dividing Σm_i by the mass m_0 of coal that participates in the reaction. This method of calculating the conversion is called the product addition method. The calculation is expressed as follows:

$$x_P = \frac{\Sigma m_i}{m_0} \times 100\,\%.$$

For the system discussed in this chapter, i includes CH_4, C_2H_4, C_2H_6, C_2H_2, CO, and CO_2.

This method is not affected by other factors and only depends on the accuracy and flexibility of the instruments used for the measurements, and is therefore a reasonable calculation method. Conversion using solid feed materials, as in the examples in this chapter, is calculated according to this method; however, for gaseous raw materials, this method is clumsy.

9.3.3 *Coal Pyrolysis in Arc Plasma*

Coal pyrolysis in an arc plasma is an ultrashort reaction occurring at high temperatures, and is a complicated process. The coal reaction mechanisms under such rapid and unique arc plasma conditions are different from those in conventional pyrolysis processes. It has been shown that the main products of coal after reaction in an arc plasma can be divided into gaseous and solid products. The gaseous products are light gases, with the main components being acetylene and carbon monoxide, and the solid products are mainly pyrolysis residues. All the products are directly produced by the reaction between coal and the plasma jet flow; therefore this section mainly focuses on the various properties of and changes in the gaseous and solid products, and summarizes the reaction mechanisms of coal in an arc plasma. Acetylene production requires high-temperature, high-enthalpy conditions, which are provided by an arc plasma. The various factors and changes that affect acetylene formation are discussed in detail.

9.3.3.1 Factors Influencing Acetylene Production

Impact of Coal Type

Yangcheng anthracite, Jingjiao coking coal, Xinzhi fat coal, and Baode long-flame coal were selected as representative coal types for the experiments. Table 9.6 shows the proximate and ultimate analyses for these four coal types.

Figure 9.5 shows the experimental results for plasma pyrolysis of different coal samples, with a powder supply rate of 1.5 g s^{-1}. The coal conversion and acetylene yield both improved with increasing volatile content of the coal. These results are consistent with many results reported in the literature [49]. The volatiles in coal are

Table 9.6 Proximate and ultimate analyses (%) of coals used in experiments

Coal	Proximate analysis			Ultimate analysis				
	M_{ad}	A_d	V_{daf}	C_{daf}	H_{daf}	N_{daf}	S_{daf}	O_{daf}
Baode coal	4.43	3.29	38.92	75.82	5.36	1.86	0.74	16.22
Xinzhi coal	1.00	9.20	34.90	82.84	4.84	1.43	0.49	10.40
Jingjiao coal	0.80	10.10	17.50	89.12	4.37	1.36	0.51	4.64
Yangcheng coal	1.55	9.67	7.69	92.69	3.23	1.10	1.63	1.35

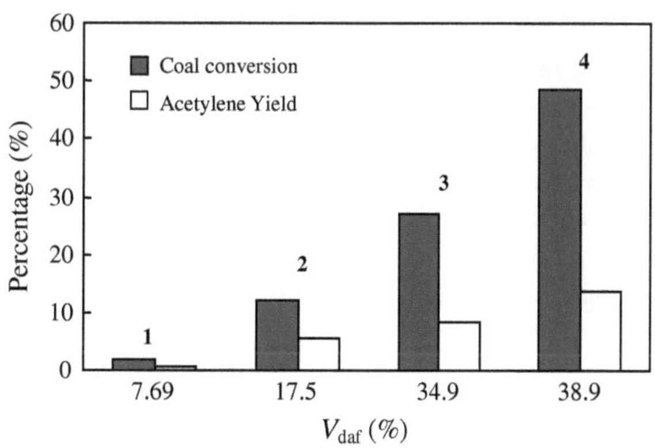

Fig. 9.5 Conversions of different coal types in arc plasma, and acetylene productivity: *1* Yangcheng coal; *2* Jingjiao coal; *3* Xinzhi coal; and *4* Baode coal

therefore considered to play important roles in coal conversion and acetylene production.

Only in the case of Baode coal did the total conversion exceed the volatile content of the coal. This indicates that the volatile content is not the only factor affecting coal conversion, although it plays an important role in coal conversion and acetylene production. It is the coal structure that determines the coal reactivity. For young coal, because the structure is rich in aliphatic side-chains and there is a small amount of crosslinks among aromatic microcrystalline components in the coal, structural changes in the coal occur very easily under plasma conditions.

The coking ability of coal is an important factor affecting coal feed during the experiments. For fat coal or coking coal with high volatile and low oxygen contents, a large amount of plastic mass is produced under high-temperature hydrogenous conditions. The plastic mass very easily becomes attached to the nozzle outlet and the reactor wall, blocking the nozzle and the reactor. Baode coal is a young coal with poor caking and coking properties, making it suitable for plasma pyrolysis. Anthracite is unsuitable for use as the feedstock for production of acetylene or other gaseous products because it has a low volatile content and stable structure. The feedstock selected for experiments should meet the following requirements: high volatile content, active coal structure, high H/C ratio, and poor coking and caking properties. The experiments indicated that only Baode coal meets the above requirements and is an ideal reaction feedstock. Baode coal was therefore selected as the feedstock in this research on acetylene production from coal in a plasma.

A number of studies have shown that activated particles can react with coal and cause coal conversion, but the reactivities shown by different carbonaceous materials and coals with different chemical structures and compositions differ significantly. For comparison, experiments using graphite powder as the feedstock, at a powder feed rate of 1.50 g s^{-1}, were performed; Table 9.7 shows the results. It was found that under same reaction conditions, the conversion of graphite powder was much lower than that of coal. The low conversion of graphite powder was mainly caused by the activated species in the plasma jet flow, and also demonstrated the high inertness of the graphite structure. The structure of Baode coal is much more easily activated than that of graphite, so the conversion of Baode coal is even higher than expected from its volatile content. The properties of coal are usually determined by its metamorphic degree. As the metamorphic degree increases, the content of aromatic structures in the coal gradually increases, whereas that of aliphatic structures decreases. Graphite is therefore the highest stage of coal metamorphosis. Under the same reaction conditions, the younger the coal is, the more favorable it is for coal conversion; this is shown in Fig. 9.5.

Table 9.7 Carbon conversions (%) of graphite and Baode coal

Sample	CH_4	$C_2H_6 + C_2H_4$	C_2H_2	CO	Other	Total
Graphite	0.009	0.005	0.032	1.834	–	1.880
Baode coal	1.230	0.830	12.640	11.550	–	26.250

Impact of Powder Feed Rate on Conversion

Figure 9.6 shows the relationship between coal conversion, on a dry, ash-free basis, and powder feed rate. The coal conversion decreased with increasing powder feed rate. The coal conversion even exceeded the volatile content of the coal (shown by the dotted line) under some conditions. Research on conventional rapid coal pyrolysis indicates that the total weight loss rate exceeds the volatile content when coal pyrolysis is accelerated [1]. Azhakesan et al. [50] reported that the weight loss of raw coal measured at a heating rate of 5000 K s^{-1} increased by 5 % compared with that of the volatile content. The usual explanation is that rapid pyrolysis reduces the polycondensation degree, and leads to more volatiles being released than under normal conditions. Based on the thermal pyrolysis reaction alone, the maximum coal conversion should be 45.35 % (air-dried basis), according to Azhakesan's research, although the heating rate is much faster under plasma jet flow conditions; this is less than the actual conversion of about 55 % (air-dried basis). Large amount of activated particles such as H^+, H^-, and Ar^+ and of electrons and other active species are present in an arc plasma jet flow [17], and can directly cause coal conversion when they strike the coal surface [15]. When coal enters the plasma jet flow, it is not only pyrolysis that causes coal conversion. The coal conversion exceeds the volatile content of the coal when the powder feed rate is low, indicating that devolatilization and the reactions between activated particles and coal play important roles in coal conversion. As the powder feed rate increased, the coal conversion gradually decreased and eventually became lower than the volatile content of the coal. It is necessary to further investigate which of the two main processes is dominant under such conditions.

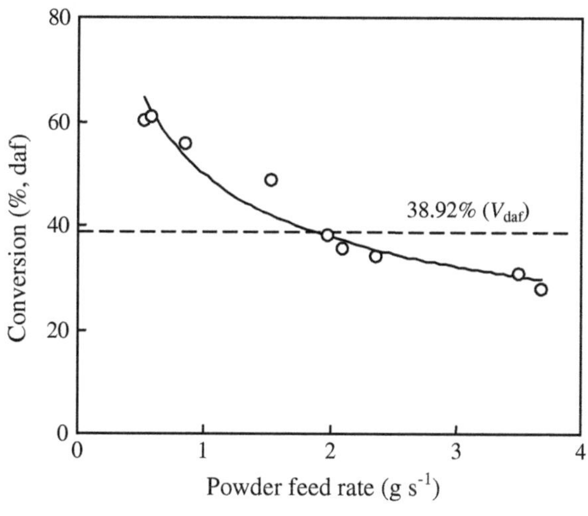

Fig. 9.6 Impact of powder feed rate on coal conversion. Reprinted from Ref. [5], Copyright 2001, with permission from American Chemical Society

Impact on Types of Reaction Product

When the rate of coal powder supply increases, the temperature of the system is lowered as a result of increased heat absorption by the coal powder. The secondary decomposition of volatiles is therefore incomplete. A comparison of gas-phase chromatograms of the product gases at powder supply rates of 0.58 and 3.49 g s^{-1} showed that only light products, i.e., methane, ethylene, and acetylene, were generated in the product gases at the lower powder feed rate, whereas as many as 11 types of hydrocarbon gases and their derivatives were generated at the higher powder feed rate; gas chromatography-mass spectrometry identified these products as C_3–C_5 hydrocarbons and their derivatives. With increasing powder feed rate, the distribution of product gases became complicated: not only the number of product types, but also the number of carbons in the products increased.

Impact on Acetylene Yield

Figure 9.7 shows the curve of acetylene yield versus powder feed rate. The acetylene yield decreased with increasing powder feed rate. The changes in acetylene production and coal conversion were consistent with the changes in the powder feed rate. The primary product in coal conversion, especially the primary volatile, is considered to be the precursor for acetylene formation. The acetylene concentration in the product gas increased, although the acetylene yield decreased, with increasing powder feed rate, as shown in Fig. 9.8. The acetylene output increased because the working gas flow rate and electric power were kept constant in the experiments. The use of the same electric power in the experiments suggests that the energy consumption in acetylene formation might decrease at high powder feed rates. Figure 9.9 shows the relationship between acetylene energy consumption and

Fig. 9.7 Impact of powder feed rate on acetylene yield

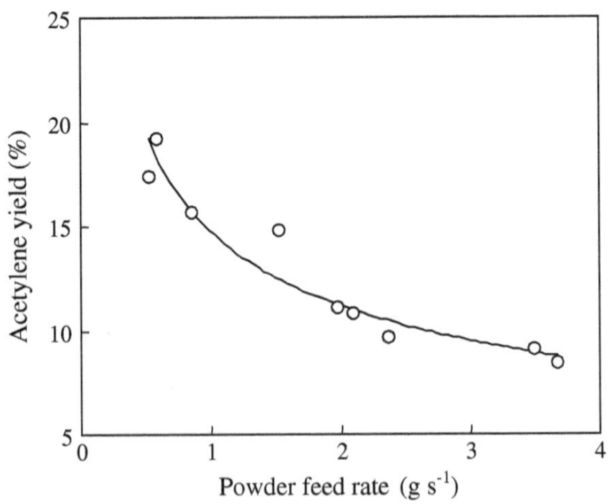

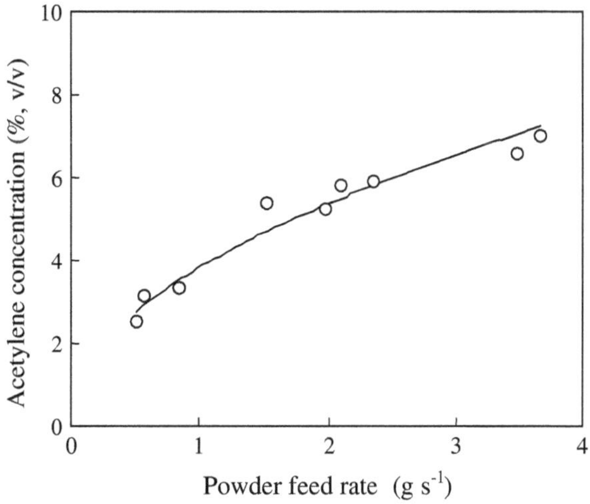

Fig. 9.8 Impact of powder feed rate on acetylene concentration in product gases. Reprinted from Ref. [5], Copyright 2001, with permission from American Chemical Society

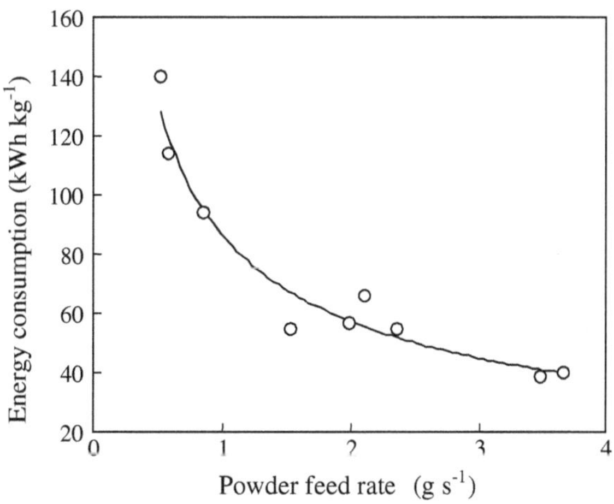

Fig. 9.9 Impact of powder feed rate on energy consumption in acetylene production

powder feed rate. The energy consumption decreased steadily, but the reduction slope decreased gradually and non-linearly with increasing powder feed rate. In conclusion, increasing the powder feed rate led to reductions in the coal conversion and acetylene yield, but the energy consumption for acetylene formation improved with increasing powder feed rate. For industrial production, the optimal point among these conflicting factors should be identified, depending on the process conditions.

Impact on Other Product Yields

In addition to acetylene, the final pyrolysis product gases also contain CH_4, C_2H_4, C_2H_6, CO, and a small amount of CO_2 and C_3–C_5 hydrocarbons. Figure 9.10 shows

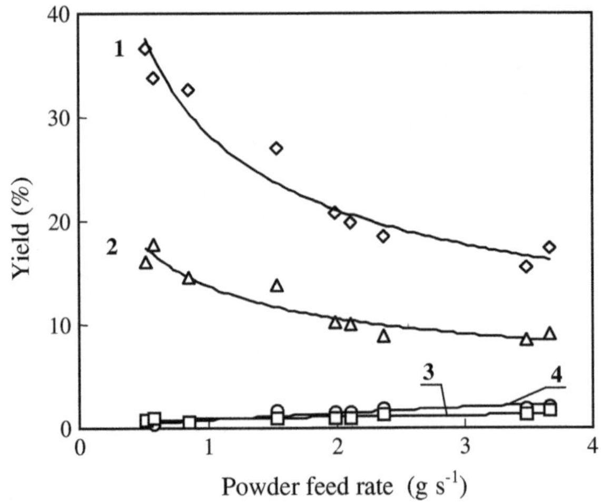

Fig. 9.10 Impact of powder feed rate on product yields: *1* CO; *2* C_2H_2; *3* $C_2H_4 + C_2H_6$; and *4* CH_4

the changes in the yields of some of the products with changes in the powder feed rate. The figure shows that CO is the main product, along with acetylene, and the CO yield is higher than that of acetylene under most conditions. CO is mainly generated as a result of the large amount of oxygen and adsorbed air and water in the coal. Experiments have shown that the oxygen contained in the working gas can also react with coal to generate CO. Generally, the yields of CO and C_2H_2 decrease at high powder feed rates, whereas the yields of CH_4 and $C_2H_4 + C_2H_6$ are basically constant or increase slightly. As in the case of acetylene, the outputs of CH_4 and $C_2H_4 + C_2H_6$ increase with increasing powder feed rate.

Impact on Competition Between Acetylene and Carbon Monoxide

Because CO is one of main components of the product gases, and accounts for a certain amount of the source carbon, it has an adverse effect on acetylene production. Experiments have shown that the production of C_2H_2 and CO compete with each other. At each sample outlet, when the C_2H_2 concentration was high, the concentration of CO was low, and vice versa. The quantity of conservable active carbon in the coal under certain conditions is constant; therefore the quantity of C_2H_2 decreases as the quantity of CO increases. Figure 9.11 shows the ratios of conservable active carbon allocated to CO and C_2H_2, at different sample outlets under the same reaction conditions. The results clearly reflect competition between the two products. As shown in Fig. 9.11, the ratio of conservable active carbon present as C_2H_2 changed from high to low, and finally increased again over the entire reaction process. The increase at position 5# was large because the quenching operation occurred before this position. This indicates that quenching can prevent CO production and promote C_2H_2 production, but the total amount of carbon

Fig. 9.11 Distribution of converted carbon in CO and C_2H_2 at different sampling positions: *1* C_2H_2 and *2* CO

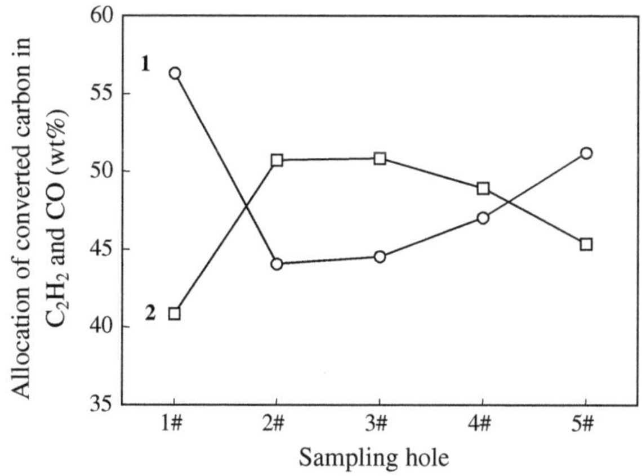

converted to these two products was about 92 % of the entire converted carbon, and did not change significantly with changes in the sample outlet positions.

The differences among the ratios of converted carbon present as CO and C_2H_2 calculated at sample outlets 1#–5# were used to calculate the standard deviation and to determine the dispersion degrees of the different feed rates; the greater the deviation was, the greater the dispersion was, which indicated more intense competition. Figure 9.12 shows that as the powder feed rate increased, the standard deviation gradually decreased, indicating that competition between formations of the two products decreased at higher powder feed rates. Figure 9.13 shows the selectivity changes for production of CO and C_2H_2 with changes in the powder feed rate. As the powder feed rate increased, the selectivity for C_2H_2 was basically unchanged, whereas the selectivity for CO decreased slightly, which is favorable for acetylene production.

Fig. 9.12 Impact of powder feed rate on competition between formation of CO and C_2H_2

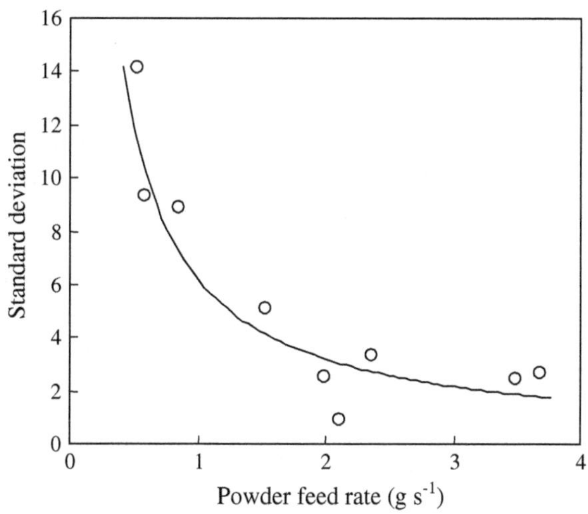

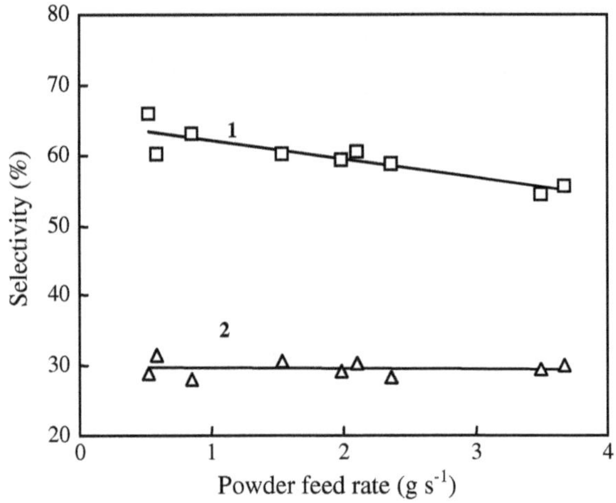

Fig. 9.13 Impact of powder feed rate on selectivities for production of CO and C_2H_2: *1* CO and *2* C_2H_2

Impact of Coal Specific Enthalpy on Coal Conversion and Acetylene Yield

The coal specific enthalpy is defined as the energy that can be distributed in the coal powder per unit mass under the operating conditions, and it reflects the average reaction temperature of the coal powder in the plasma jet flow. Assuming that the coal is evenly mixed with the plasma jet flow, and that the heat absorbed by the device wall can be neglected, the coal specific enthalpy (J g^{-1}) can be obtained from the effective power divided by the powder feed rate:

$$\mathrm{CSE} = \frac{\eta \cdot P}{v_c},$$

where
η is the efficiency coefficient of the plasma generator; $\eta = 0.6$ in this experiment;
P is the power of the plasma generator (W);
v_c is powder supply rate (g s^{-1}).

Under given power and working gas flow conditions, the coal specific enthalpy reflects not only the average temperature of the plasma jet flow and the coal powder reaction, but also the coal energy consumption per unit mass of coal powder. The coal specific enthalpy provides a high average temperature for the reaction, and adequate energy. Figure 9.14 shows the increase in coal conversion with increasing coal specific enthalpy. For a low coal specific enthalpy, the coal conversion quickly increased by about 30 %, which was close to the proximate analysis volatile value. The reduction in the powder feed rate was consistent with the increase in the generator power, for calculation of the coal specific enthalpy, so the impact of the generator power is not discussed in this chapter.

Fig. 9.14 Impact of coal specific enthalpy (CSE) on coal conversion

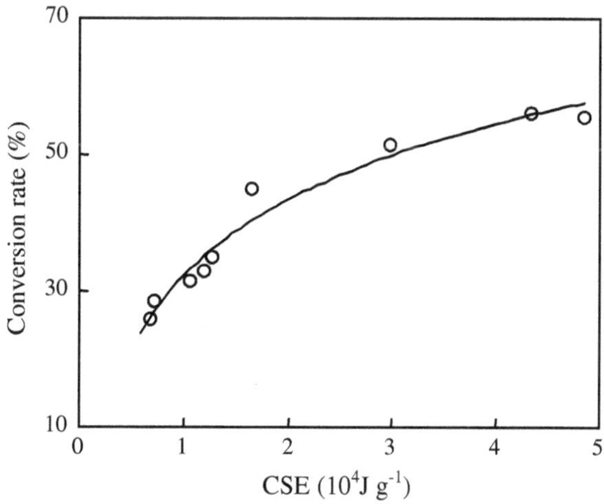

Impact of Different Positions in Reactor on Acetylene Production

To study the changes in acetylene formation with changes in reactor length, samples were obtained from five outlets at different positions in the reactor; sample outlet 5# was below the quenching device. Figure 9.15 shows the acetylene concentrations measured at different sample outlets for different powder feed rates. Because of the impact of quenching on acetylene formation, the acetylene concentration at position 5# is not marked in the figure. When the coal enters the jet flow, it moves toward the bottom of the reactor, driven by the jet flow, and reacts with the plasma during this movement. Figure 9.16 shows the changes in coal

Fig. 9.15 Acetylene concentration at different sample outlets under different powder feed rates: *1* 3.49 g s^{-1}; *2* 1.53 g s^{-1}; and *3* 0.52 g s^{-1}

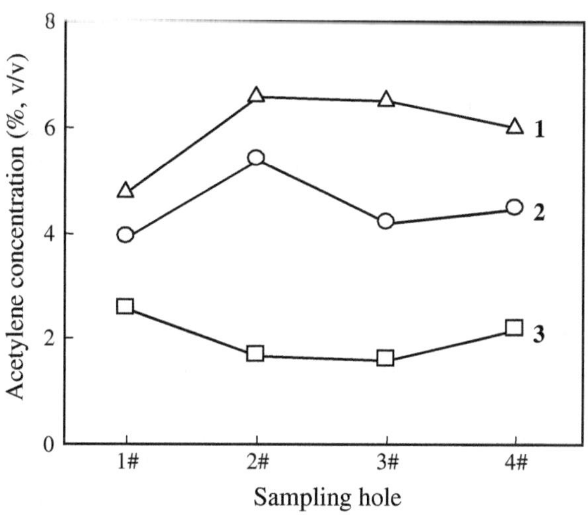

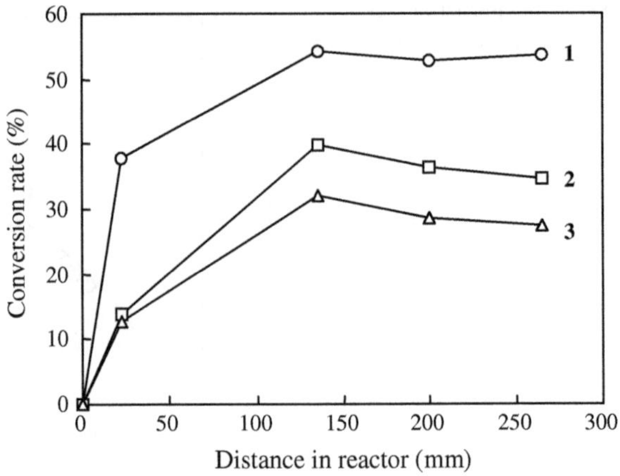

Fig. 9.16 Coal conversions at different sample outlets at different powder feed rates: *1* 0.58 g s^{-1}; *2* 1.98 g s^{-1}; and *3* 3.49 g s^{-1}. Reprinted from Ref. [5], Copyright 2001, with permission from American Chemical Society

conversion with distance traveled in the reactor. At 100 mm after the coal powder enters the jet flow, the coal conversion process is almost complete, and the conversion process slows.

Impact of Quenching on Acetylene Formation

The quenching operation is an important factor affecting the acetylene yield. In the experiments, sample outlets 4# and 5# were located in front of and behind the quenching unit, respectively. The acetylene concentration in the quenched product gases was higher than that before quenching under most conditions. The protective role of the quenching operation toward acetylene was more obvious when the powder feed rate was low. The quenching mechanism has two effects. First, quenching can prevent acetylene from decomposing and the high-temperature product gases can be "frozen" quickly to fix the components, thus protecting the acetylene. Second, quenching can promote the recombination of various radicals in the thermally decomposed gas, and the acetylene-forming reactions can increase the acetylene concentration in the product gases. The acetylene yield can reach 59.9 % under ideal conditions, according to the radical recombination theory [49].

9.3.3.2 Pyrolysis Residue Properties

As well as gaseous products, the pyrolysis products of coal powder in a plasma jet flow contain a large amount of solid products; these are mainly pyrolysis residues and carbon deposits on the reactor wall (also called coking products). Research on changes in the properties of the reaction residues is helpful in understanding the mechanism of the coal pyrolysis process.

Table 9.8 Proximate and ultimate analyses (%) of raw coal and reaction residues

Powder supply rate (g s^{-1})	Proximate analysis			Ultimate analysis	
	M_{ad}	A_d	V_{daf}	C_{daf}	H_{daf}
0.52	1.25	5.84	5.87	–	–
0.58	0.70	9.84	6.52	95.29	0.54
0.85	0.65	6.88	6.57	–	–
1.98	1.24	6.12	10.39	92.02	2.00
3.49	0.51	4.53	22.41	80.89	3.24
Raw coal	4.43	3.29	38.92	75.82	5.36

Proximate and Ultimate Analyses

Table 9.8 shows the proximate and ultimate analyses of Baode raw coal and the reaction residues produced at different powder feed rates. In the ultimate analysis, the changes in carbon and hydrogen reflect the main trends in the changes; therefore only the analytical data for these elements are given in the table. With increasing powder feed rate, the ash content in the reaction residues gradually decreased, the volatile content gradually increased, the carbon content decreased, and the hydrogen content increased. The data became consistent with those for raw coal with increasing powder feed rate. As the powder feed rate approached infinity, all the values became the same as those for raw coal. These trends prove that the coal conversion degree decreases with increasing powder feed rate, which is consistent with previous experimental results.

Differential Thermal Gravimetric (DTG) Analysis

The purpose of the DTG experiments was to investigate the pyrolysis weight losses and gasification reactivities of residues before gasification. The experimental con ditions were a heating rate of 5 °C min^{-1}, from room temperature to 1200 °C, in a CO_2 atmosphere. To prevent ash affecting the residue sample, all the samples for the thermogravimetric experiments were pretreated with hydrochloric and hydrofluoric acid to wash out the ash. Figure 9.17 shows the DTG curves of the reaction residues and raw coal at different powder feed rates. As shown in Fig. 9.17, the gasification reaction with CO_2 in all the samples started at about 800 °C, and pyrolysis mainly took place before gasification. The curve for raw coal shows two sharp weight loss peaks before gasification: the first at 400–500 °C and the second at 600–700 °C, suggesting that the activation energies for the pyrolysis of volatiles contained in the coal were distributed in different ways. With decreasing powder feed rate, the peak between 400 and 500 °C disappeared; the peak between 600 and 700 °C disappeared when the feed rate was 0.52 g s^{-1}. These results indicate that for coal pyrolysis in a plasma jet flow, the volatiles in coal with low pyrolysis activation energies were most easily separated. This is another factor that should be

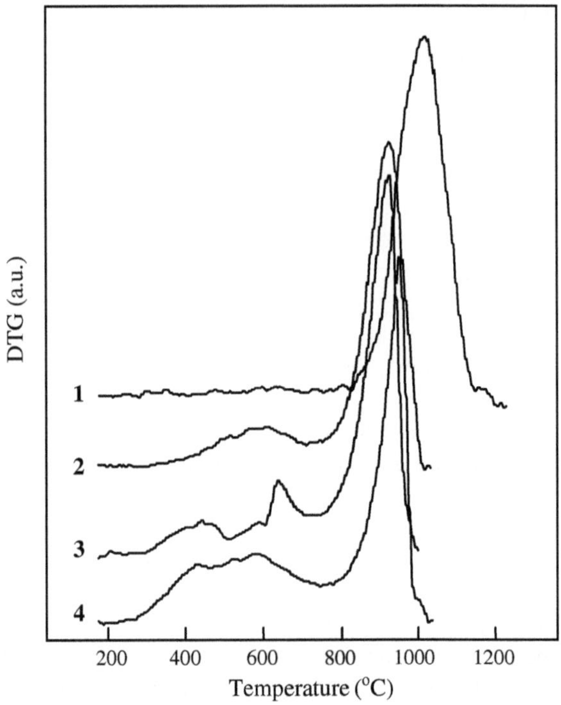

Fig. 9.17 DTG curves for residues and raw coal under CO_2 atmosphere: *1* 0.52 g s^{-1}; *2* 1.98 g s^{-1}; *3* 3.49 g s^{-1}; and *4* raw coal

considered when selecting the type of coal feedstock. When the coal specific enthalpy was high, the average temperature of the reaction system was high; therefore volatiles with high activation energies were separated during pyrolysis. If the temperature for the fastest gasification reaction is used to judge the gasification reaction activity, different residue samples show different gasification activities under a CO_2 atmosphere. The residue samples from a feed rate of 0.52 g s^{-1} had the maximum gasification activity, followed by raw coal, and the other two activity values were almost equal.

Specific Surface Area and Pore Volume

Tests showed that the specific surface area and pore volume of Baode raw coal are 3.01 m^2 g^{-1} and 0.012 mL g^{-1}, respectively, and that these values changed after reaction. Figures 9.18 and 9.19 show that the maximum specific surface area of the reaction residues was about 50 m^2 g^{-1} and the maximum pore volume was about 0.25 mL g^{-1}. At high powder feed rates, both the specific surface area and the pore volume decreased. At higher powder feed rates, the degree of coal conversion decreases as devolatilization is one of the main coal conversion processes. It is easy to understand that coal surface parameters and conversion processes are closely related, with high conversion associated with a large specific surface area and pore volume.

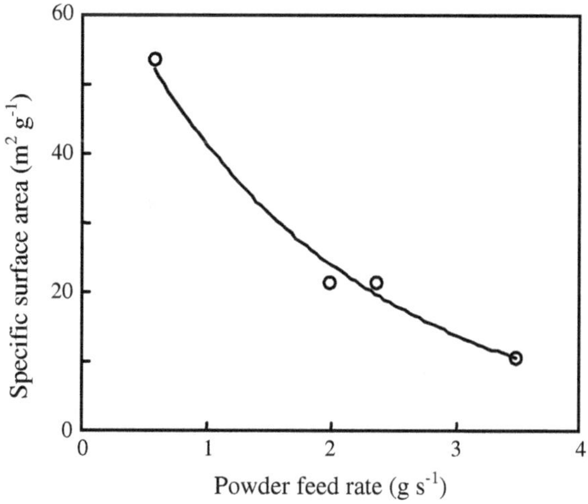

Fig. 9.18 Changes in Brunauer–Emmett–Teller specific surface area with powder feed rate

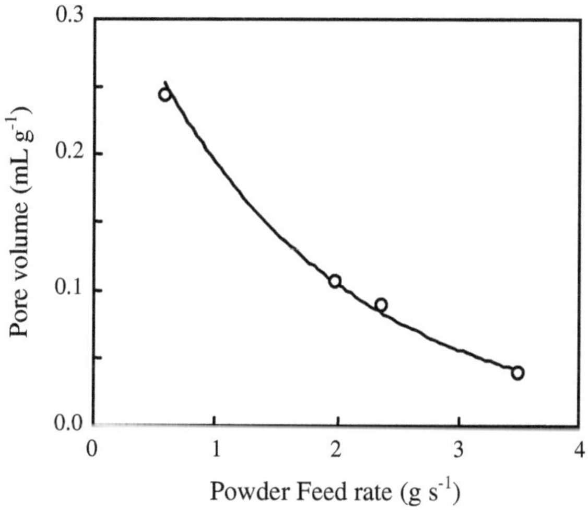

Fig. 9.19 Changes in pore volume with powder feed rate

X-ray Diffraction (XRD) Analysis

In addition to the peak for graphite, the XRD patterns of the reaction residues also contained peaks for inorganic salts, derived from the raw coal. To avoid the effects of inorganic salt ash, the reaction samples were pretreated with hydrofluoric acid and hydrochloric acid to wash out the ash. Figure 9.20 shows the patterns before and after pretreatment; the patterns show that the peak from inorganic salt disappeared after washing.

The crystal parameters obtained using XRD for the aromatic microcrystals can be used to calculate the interlayer spacing d of the microcrystals in the reaction

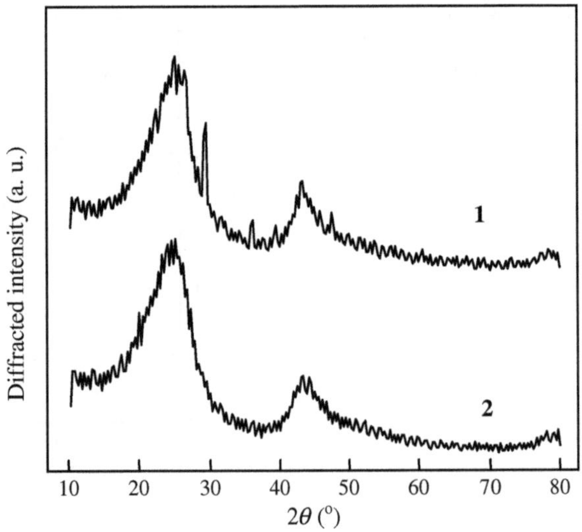

Fig. 9.20 XRD patterns before and after pretreatment: *1* before washing and *2* after washing

residue, and the sizes L_a and deposition heights L_c at different powder feed rates. Table 9.9 shows the calculation results. The *d* value of the (002) crystal surface of the reaction residue at the minimum powder feed rate was minimum, whereas the cell size was maximum. With increasing powder feed rate, the *d* value gradually increased; the microcrystal cell size gradually decreased, the graphitization degree decreased, and the crystal parameters became closer to those of the raw coal. The gasification activation was maximum for the residues produced at a feed rate of 0.52 g s^{-1} because of the high graphitization degree. Although the graphitization degree of the raw coal was the lowest, the coal porosity was undeveloped, so it showed high gasification activity. The XRD and surface analysis results provided a reasonable explanation for the gasification activation degrees observed in the DTG analysis.

Table 9.9 Microcrystalline structures of reaction residues at different powder feed rates

Powder feed rate (g s^{-1})	Parameters of microcrystalline structure (Å)		
	d	L_c	L_a
0.52	3.4809	14.97	40.19
0.85	3.5757	12.62	35.59
1.53	3.5701	12.17	29.60
1.98	3.5743	12.06	28.70
3.49	3.5729	9.60	27.88
Raw coal	3.5814	8.89	17.88

Scanning Electron and Transmission Electron Microscopies

After the coal particles reacted in the jet flow, many of their properties had changed. Scanning electron microscopy (SEM) was used to observe the changes in the particle shapes. The SEM images of the raw coal showed clear edges, a dense texture, a particle distribution of 2–25 μm, and no particle aggregation. The coal particles had no clear edges after reaction with the jet flow, but a large number of pores of size less than 1 μm were produced; this was in agreement with the specific surface area results. A small number of particles with shapes similar to those of the raw coal were found. This may indicate that not all the coal particles reacted with the plasma, and the mixture of the coal powder and the plasma jet flow was not homogeneous. No free small particles were found in the residue, and a large number of small particles were aggregated; this may be caused by aggregation of decomposed carbon black particles.

Transmission electron microscopy showed that the reaction residues contained 10–20 nm carbon black particles; these particles were ball shaped and connected together in chains or balls to form large aggregates. These results proved the decomposition of acetylene or other hydrocarbons, and supported the SEM results.

9.3.3.3 Reaction Dynamics and Reaction Mechanism Analysis of Coal in Plasma

Changes in Acetylene Concentration

A thermal plasma jet is a high-temperature flow with a high-enthalpy content. After the coal enters the plasma jet flow, the primary pyrolysis reaction is production of volatiles and coal chars. The gaseous volatiles produced in the pyrolysis process decompose further to light hydrocarbon gases in a mixed system; acetylene and carbon monoxide are the main components. The acetylene further decomposes to soot and hydrogen in such an environment. These three processes are described by the following equations:

$$\text{Coal} \xrightarrow{k_1} \text{Volatile} + \text{Char},$$
$$\text{Volatile} \xrightarrow{k_2} C_2H_2 + CO + CH_4 + C_2H_4 + H_2,$$
$$C_2H_2 \underset{k_4}{\overset{k_3}{\rightleftarrows}} 2C + H_2.$$

As the data in Table 9.8 show, certain amounts of volatiles were still detected in the reaction residues; therefore coal pyrolysis in a plasma can be considered as a single-direction slow reaction. The thermal reaction of volatiles in the jet flow is a gaseous process, and no coal tar was detected in any of the experiments. The reaction is therefore assumed to be a single-direction irreversible process. The experiments indicated that $k_2 \gg k_1$, i.e., the secondary pyrolysis rate of the volatiles is far greater than that of coal pyrolysis. Bond et al. [9] used an argon arc plasma to carry out coal pyrolysis experiments and found that the acetylene yields in the product gases were low. When hydrogen was added to the argon, the acetylene yields increased significantly, indicating that hydrogen suppresses acetylene decomposition. The decomposition reaction of acetylene can therefore be considered to be reversible. In the following discussion of the dynamic reaction, we assume that all the reactions are first order, the reaction system is homogeneous with an even temperature throughout and no temperature gradient in the coal particles, and the activated components will react with the plasma in the coal pyrolysis reaction. If the amount of activated components in the coal is ξ, then the concentration of the coal powder can be expressed by ξ, which may vary among different coal types.

Based on the above assumptions and reaction mechanism, we can derive the following equations:

$$-\frac{\mathrm{d}c_\xi}{\mathrm{d}t} = k_1 c_\xi,$$

$$\frac{\mathrm{d}c_\mathrm{V}}{\mathrm{d}t} = k_1 c_\xi - k_2 c_\mathrm{V},$$

$$\frac{\mathrm{d}c_{\mathrm{C_2H_2}}}{\mathrm{d}t} = k_2 c_\mathrm{V} - k_3 c_{\mathrm{C_2H_2}} + k_4 c_\mathrm{C}^2 c_{\mathrm{H_2}},$$

$$\frac{1}{2}\frac{\mathrm{d}c_\mathrm{C}}{\mathrm{d}t} = k_3 c_{\mathrm{C_2H_2}} - k_4 c_\mathrm{C}^2 c_{\mathrm{H_2}}.$$

If we assume a steady state, i.e., $\frac{\mathrm{d}c_\mathrm{V}}{\mathrm{d}_t} = 0$, and $k_2 \gg k_1$, we get:

$$\frac{\mathrm{d}c_{\mathrm{C_2H_2}}}{\mathrm{d}t} = k_1 c_\xi - k_3 c_{\mathrm{C_2H_2}} + k_4 c_\mathrm{C}^2 c_{\mathrm{H_2}}.$$

In the initial stage, the concentration of the activated component is high and the acetylene concentration is low. The acetylene concentration increases rapidly by acetylene generation, whereas the acetylene concentration decreases slowly by acetylene decomposition; the net result is that the acetylene concentration increases. However, as the reaction progresses, the concentration of activated components in the coal gradually decreases, and the acetylene concentration increases. Acetylene generation becomes slower than its decomposition, which results in the acetylene concentration gradually dropping after reaching a maximum value. As the reaction continues, the hydrogen concentration in the system gradually increases as a result of coal pyrolysis and acetylene decomposition, and the third term in the right-hand side of the equation gradually increases, so the acetylene concentration increases again, after reaching a minimum value. These changes in acetylene concentration are in agreement with the changes in acetylene concentration with reactor position shown in Fig. 9.15. As the temperature increases, the reaction rate increases, and the time needed for acetylene to reach the maximum concentration decreases.

Effect of Plasma Jet Flow on Coal

Based on the rapid pyrolysis theory, we assumed 45.35 % to be the maximum theoretical value of coal conversion under coal pyrolysis conditions, and then calculated the coal conversion using the ash-content-tracing method, volatile-tracing method, and product addition method introduced in 3.2; Table 9.10 shows the calculation results. The results showed a large difference between x_V and x_P at powder feed rates below 1.98 g s^{-1}. According to the concept of pyrolysis, coal conversion will not exceed the theoretical value. The conversion higher than the theoretical volatile content could be caused by the actions of large amounts of various activated species in the plasma jet flow; x_V and x_P were clearly consistent after the powder feed rate reached 1.98 g s^{-1}. These results suggest that pyrolysis and activated species play important roles in coal conversion at a low powder feed rate. With increasing powder feed rate, the effects of activated particles in the jet flow gradually decreased, and when the powder feed rate exceeded 2.0 g s^{-1}, the effects of activated species were no longer observed, and pyrolysis dominated.

As the data in Table 9.10 show, the x_A values were inconsistent with x_P and showed irregular fluctuations. XRD analysis of the reaction residues also indicated that the samples were contaminated with inorganic salts after they were collected. Such contamination might cause scale formation when the quenching water reacted with the high-temperature pyrolysis gas. The lower the powder feed rate is, the higher the temperature of the reaction system, and therefore the more pronounced scale formation becomes. The ash content of Baode coal is low, so the deviations in measurements in the ash-content-tracing method have an obvious amplification effect.

Based on the above experimental results and discussion, we can describe the coal reaction mechanism in an arc plasma jet flow as follows. When the coal enters the plasma jet flow, it is heated for an ultrashort time under high-temperature and mainly undergoes pyrolysis. This reaction takes place inside the coal particles under thermal action. At the same time, a large amount of activated species contained in the plasma jet flow collide with the coal surface, causing coal conversion. When the powder feed rate is relatively low, such effects are obvious, but with increasing powder feed rate, the effects become less obvious. When the powder feed rate is more than 2.0 g s^{-1}, the effects are no longer observed. At this time, the coal is mainly converted by pyrolysis. The primary volatiles produced in the pyrolysis reaction cause secondary reactions in the plasma jet, and produce the main products, i.e., acetylene and carbon monoxide. The produced acetylene can decompose to hydrogen and carbon in the plasma environment. The hydrogen in the system

Table 9.10 Coal conversions calculated using different calculation methods

Powder feed rate (g s^{-1})		0.52	0.58	0.85	1.98	3.49	3.67
Conversion (%)	x_A	44.94	67.43	53.62	47.63	30.03	50.48
	x_V	38.77	38.56	38.24	34.64	26.63	25.86
	x_P	55.58	56.13	51.55	35.08	28.52	25.83

depresses acetylene decomposition and plays an important role in protecting acetylene. In the quenching process, hydrogen can combine with other radicals and form acetylene. When the volatiles are separated from the coal particles, many pores are produced in the coal particles; therefore the specific surface area and pore volume of the reaction residue are higher than those of raw coal. The greater the degree of conversion, the greater the pore volume and specific surface area are. The clearly defined forms of the coal particles also change to shapeless forms with multiple pores. As the coal particles pass through the high-temperature zone, the crystal structure is gradually graphitized. The lower the powder feed rate, the higher the graphitization degree, and the greater the cell size is. Many of the properties of the reaction residues are different to those of coal, but all these properties gradually approach those of coal as the strength of the reaction decreases.

A small amount of carbon is deposited on the reactor wall and forms coking products, although this has little impact on the conversion because the quantity is small. However, the formation of these coking products is one of the secondary reactions of the primary volatiles, a procedure that cannot be neglected in the overall reaction of coal in an arc plasma jet flow. Aspects of this will be discussed further.

9.4 Simulation of Coal Devolatilization in Plasma

9.4.1 Devolatilization Model

When coal is heated, unstable chemical bonds in the coal break, and decomposition to light gases and heavy molecular fragments occurs. These products can evaporate and form char or remain within the coal as metaplast. As some of the raw coal is converted to char, the chemical model of the devolatilization process is an important component in the description of any thermal process involving coal.

The coal devolatilization model was developed from simple weight loss expressions in the early stages to complex chemical and physical processes [51–53]. In coal pyrolysis, the labile bonds in aromatic clusters break and produce fragments of finite sizes. The light fragments are separated from the coal particles in the form of coal tar, as a result of the high vapor pressure, but the heavy fragments remain in the coal and are finally reconnected to the lattice because the vapor pressure is low under standard devolatilization conditions. These heavy components plus the residual lattice are considered to be metaplast, and the quality and quantity of metaplast and the subsequent crosslinking reactions decide the softening behavior of coal particles.

The number of broken labile bonds and the mass of finite-sized molecular fragments separated from the infinite coal lattice are highly non-linear, which indicates that coal pyrolysis is not a simple evaporation process. In a series of experiments performed under the same heating conditions, Freihant et al. [54] collected coal tars using a heated screen reactor and then measured the evaporation temperatures of these coal tars. The secondary evaporation temperature was found to be significantly lower than the initial temperature at which the coal tars were released from the coal. These results indicate that coal pyrolysis is not only an evaporation process, and it may be necessary to describe the coal pyrolysis reaction with reference to the lattice networks.

Pitt [55] was the first to suggest that the bond breaking had different types of activation energies. He considered coal as an assembly of a large number of species capable of simultaneous first-order decomposition reactions. Anthony et al. [56] used a similar concept to develop the distributed activation energy model. Kobayashi et al. [57] introduced chemical diversity into the devolatilization process through a set of double competing reactions, giving priority to char production at low temperatures. A large amount of activation energy data for different chemical bonds supports this viewpoint, i.e., the breaking of chemical bonds in coal is constrained by the distribution of chemical bond types in the coal. If the average activation energy in a certain range is used, a set of differential equations can be obtained for simulation through calculation of the standard deviation and corresponding Gaussian distribution.

Distributed-energy chain statistics (DISCHAIN) [58, 59] is used to predict monomer species formation. The monomer species are the sources of volatile tar, and of char by polymerization. Niksa [60] first used the Γ distribution function to introduce the molecular weights of tar products into the devolatilization flash model, which introduced a multicomponent gas–liquid equilibrium mechanism for the treatment of tar evaporation. The results indicated that the diffusion of volatiles in the metaplast did not obviously influence the pyrolysis behavior. The flash model and the DISCHAIN model were combined in a new model called FLASHCHAIN [61], which shows the mass distribution of each molecular-weight group from the general equilibrium. These molecular-weight groups are constructed based on the chain statistics, flash process, and coking mechanism. The FLASHCHAIN vapor pressure function originates from an empirical equation, and is inconsistent with the vapor pressure data for single packages. The coordination number is uniformly set at 2 in the chain equation. The molecular weight of bridges in the model is higher than that of aromatic components in the cluster, but the predicted results are anastomotic with many experimental data.

Solomon [51, 62] produced detailed chemical models of tar and light gases released in the devolatilization process and provided 19 first-order energy distribution rate expressions based on the release of various light gases. Such a set of differential equations might be useful in all coal devolatilization lattice models.

Fu et al. [63] reported a general pyrolysis model for coal particles (the Fu–Zhang model). He assumed that the activation energy E and reaction rate parameter k of coal pyrolysis are not dependent on the coal type, and are only related to the final temperature T_∞ of the coal particles and the heating rate; he obtained the relationship between E, k, and T_∞, and thereby solved the problem of the mathematical simulation of the pyrolysis of coal particles.

In conclusion, most of the above models are experiential, although the simulation results are related to the experimental results under many conditions; the scientific validity of methods that depend on adjustment of the input parameters is doubtful. The input parameters used in the chemical percolation devolatilization (CPD) model are coal structure parameters that have been determined experimentally, and is strong theoretical based on theory rather than experience. The CPD model was therefore selected for simulation research on coal devolatilization in plasmas. Because the maximum CPD heating rate reported in the literature is only 10^4 K s^{-1}, ultrashort pyrolysis under arc plasma jet flow conditions is still in the primary stages of investigation.

9.4.2 *Chemical Percolation Devolatilization (CPD) Model*

9.4.2.1 Principles of CPD Model

In the CPD model, coal is considered as a heavy matrix with various aromatic clusters of different sizes as connecting sites. These aromatic clusters are connected through various bridging bonds; some labile bonds are easy to break in the pyrolysis process, whereas other bridging bonds are stable below a given temperature. The bonds that remain intact in the pyrolysis process are char bridges. Obviously, labile and char bridges are defined relatively, and this mainly depends on the pyrolysis temperature and the kinetics for a given bridge. The branched chains connected to aromatic clusters include fat, carbonyl, and other functional groups, which are the precursors of light gases. The fragments separated from the coal matrix are composed of one aromatic cluster or several aromatic clusters, formed by connection of a labile or char bridge. A cluster is therefore composed of several fused aromatic clusters with ancillary substances, whereas a fragment is composed of several internally connected clusters. In coal, a small number of clusters are not connected with the infinite matrix and can be extracted using an appropriate solvent under conditions that do not destroy the covalent bonds. Many assumptions have been made with regard to the heavy structure of coal. In this research, we used the simplest appropriate assumptions and broadly defined the coal structure as clusters, bridges, side chains, and rings, as shown in Fig. 9.21. The pyrolysis products of coal include light gases, tar, and char.

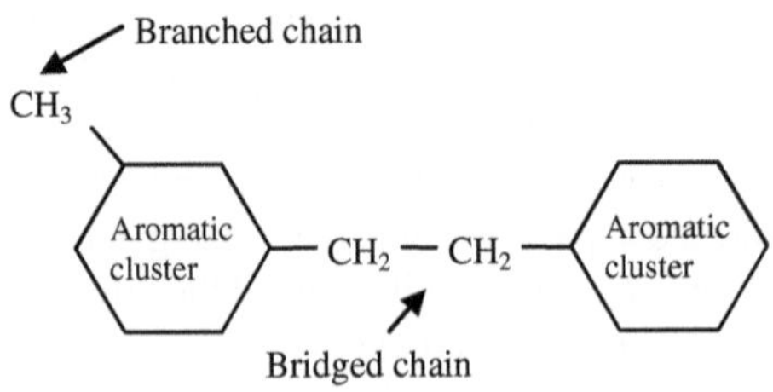

Fig. 9.21 Representation of coal structure in CPD model

In this model, the distributed activation energy is generally used, but it is sometimes necessary to redefine the characteristics of the activated bonds to take account of significantly different heating methods. Based on the CPD model, the input parameters are selected in combination with the final gas output and structural parameters obtained using nuclear magnetic resonance (NMR) spectroscopy. The definitions of the fitting parameters in the kinetic differential equation are also important because these parameters constrain the breaking and crossing of bridging chains. The methods used in the CPD model have the following characteristics. The input parameters are defined based on chemical methods; some of these parameters are obtained from NMR spectroscopic data, and the parameters reflect the chemical differences among coals of different types and ranks. Lattice statistics mathematical functions with clear connotations are used. The distribution of tar molecular clusters of different sizes and the distributions of certain species in the infinite matrix are directly given by the analytical expression in percolation theory. The data on the activation energy for tar formation come from Solomon's [52] research results. The average activation energies and the pre-exponential factors of light gases are obtained as a set of complete chemical reaction parameter weighted averages reported by Solomon et al. The distributed activation energies for light gases and tar are calculated using a simplified method.

To introduce chemical factors into the devolatilization model, it is necessary to describe the model using analytical data and input parameters, and then describe the devolatilization control process using differential equations. Karr [64] selected chemical analysis techniques that are currently used to characterize coal structures, including Fourier-transform infrared spectroscopy, pyrolysis-mass spectrometry, and solid-state NMR spectroscopy. The NMR method can directly determine the relative numbers of carbon atoms in coals with different structures, and provide data on different carbon structures. The numbers of hydrogen and oxygen atoms can be obtained by elemental analysis, so the structural characteristics can be indirectly deduced from the NMR data. The different chemical compositions in the coal can be determined using solid state NMR spectroscopy, and these serve as necessary chemical information in the CPD model; for example, the number of aromatic carbons in a typical fused aromatic cluster can be calculated from the number of aromatic bridgehead carbons [65, 66]. From these data and the number of carbon atoms included in the side chains or bridges surrounding each cluster, the average molecular weight of a certain cluster and the branch points of each cluster can be roughly estimated. All these parameters are of great significance in the characterization of the lattice statistics and selection of input parameters in the chemical devolatilization model.

9.4.2.2 Chemical Reaction Route

The CPD model assumes that the order of the pyrolysis reactions is as follows. A highly reactive bridge intermediate species is formed by breaking a labile bridge, and is then quickly consumed by one of two competing processes: the reactive bridge material is either released in the form of light gases with concurrent relinking of the two associated sites within the reaction cage to give a stable or char bridge, or the bridge materials are stabilized by hydrogen radicals to form side chains, which can finally be converted to light gases through a subsequent slow reaction.

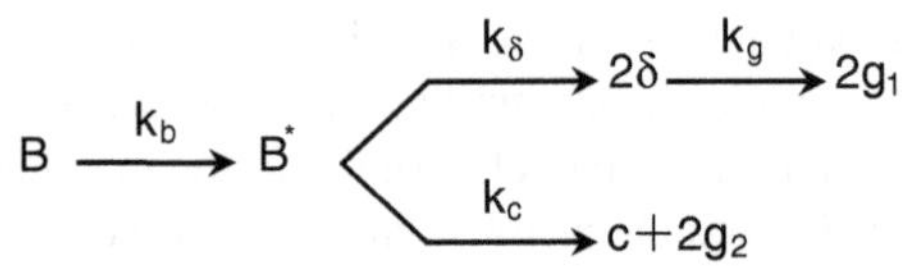

Activated bridged intermediates B* are formed from labile bridges B at a relatively slow rate, with a rate constant k_b, and are soon consumed by one of the competing reactions. In one competing process, the activated bridged intermediate B* cleaves two half-side-chains δ, with a rate constant k_δ, and these side chains are still connected to their own aromatic clusters. When an adequate number of bridging chains are cleaved from the finite fragments, they evaporate to form tar when the molecular weight is low enough. The side chains δ eventually undergo bond-breaking reactions and form g_1. In the other competing process, the activated intermediate B* forms stable cross-bridging bonds, and releases the light gas g_2, with a rate constant k_c. In this model, all the mass connected in the infinite lattice is combined into char, and the metaplast remains in the finite coal fragment in the condensed phase. The coal can be divided into light gases, tar, metaplast, and char. The char also includes the coal that did not react, so the infinite lattice not only includes labile bridging bonds, but also stable cross-bridging bonds formed in the pyrolysis process. The percolation statistics determine the populations of finite fragments (i.e., tar plus metaplast) as a function of the ratio of intact to broken bridges.

The competing actions of the activated intermediate B* are controlled by the rate of formation of the side chains and the cross-bridging bonds, and a group of dynamic equations can be obtained based on the supposed pyrolysis reaction mechanism:

$$\begin{aligned}
\frac{\mathrm{d}B}{\mathrm{d}t} &= -k_b B,\\
\frac{\mathrm{d}B^*}{\mathrm{d}t} &= k_b B - (k_\delta + k_c)B^*,\\
\frac{\mathrm{d}c}{\mathrm{d}t} &= k_c B^* = \frac{k_b B}{(\rho+1)},\\
\frac{\mathrm{d}\delta}{\mathrm{d}t} &= 2k_\delta B^* - k_g\delta = \frac{2\rho k_b B}{(\rho+1)} - k_g\delta,\\
\frac{\mathrm{d}g_1}{\mathrm{d}t} &= k_g\delta,\\
\frac{\mathrm{d}g_2}{\mathrm{d}t} &= \frac{2\mathrm{d}c}{\mathrm{d}t} = \frac{2k_b B}{(\rho+1)},
\end{aligned}$$

where ρ is the composite constant, defined as k_δ/k_c.

The steady-state approximation of B* is

$$\frac{dB^*}{dt} = 0,$$

$$B^* = \frac{k_b B}{k_\delta + k_c}.$$

The fraction p of intact bridging bonds and the proportion f of broken bridges can be calculated from the population parameters B and c of the bridging bonds:

$$p = B + c,$$

$$f = 1 - p.$$

The dynamic variable p and the theoretical percolation variable (σ + 1) can be combined using these equations. Percolation theory places no constraint on the types of bridging bonds used for characterization of the system, as long as the bridging bonds can be divided into broken and non-broken bonds. In addition to the dynamic variables for the populations of the bridging bonds, it is also necessary to control two gaseous dynamic variables, g_1 and g_2, and the dynamic variable δ of the metastable side chains, which can be calculated from the mass of broken bridging bonds. As the dynamic variables are expressed with a half bridge as the criterion, a factor 2 is needed to relate the dynamic variables to the population of the bridging bonds.

9.4.2.3 Mass Conversion and Initial Conditions

The mass conversion relation forms the constraint equations for the dynamic variables:

$$g = g_1 + g_2,$$

$$g_1 = 2f - g,$$

$$g_2 = 2(c - c_0).$$

The initial conditions for the gaseous dynamic variables are given by

$$c(0) = c_0,$$

$$B(0) = B_0,$$

$$\delta(0) = 2f_0 = 2(1 - c_0 - B_0),$$

$$g(0) = g_1(0) = g_2(0) = 0.$$

It is important to note that all the initial conditions are expressed by c_0 and B_0.

9.4.2.4 Reaction Dynamic Parameters

The reaction rate equations for the bridging-bond-breaking and gas-releasing steps are given by the Arrhenius equation, with the distribution variable V:

$$k_c = A_c \exp\left(-\frac{E_c \pm V_c}{RT}\right),$$
$$k_\delta = A_\delta \exp\left(-\frac{E_\delta \pm V_\delta}{RT}\right).$$

As the competing process is only related to the ratio of the rate constants, $\rho = k_\delta/k_c$, these two equations can be combined as follows:

$$\rho = k_\delta/k_c = A_\rho \exp\left(-\frac{E_\rho \pm V_\rho}{RT}\right),$$
$$A_\rho = A_\delta/A_c,$$
$$E_\rho = E_\delta - E_c.$$

9.4.2.5 Weight Ratios of Light Gases, Tar, and Soot

In the CPD model, the population parameter of the bridges is normalized by the total number of possible bridging bonds in the intact lattice. The site is defined as the fused ring portion of the aromatic cluster, and the cluster is defined as the site plus any portion of attachments connected to the site. The finite fragments formed by broken bridging bonds can be composed of an aromatic cluster (monomer), two clusters connected by a labile or char bridge (dimer), or n clusters (fragment size n) connected by $n - 1$ bridging bonds. The dynamic parameters for the bridging bonds given in the above differential equation can be related to the mass of independent clusters and the mass of the bridges; the total mass m_{total} of each cluster is

$$m_{\text{total}}(t) = m_a + m_b(1 - c_0)(\sigma + 1)/2,$$

where

m_a is the average mass of fused ring sites (g);
m_b is the mass of bridges (g).

As some bridges might be stable from the beginning of the reaction, m_b is corrected using $(1 - c_0)$. The term $(\sigma + 1)/2$ is the ratio of bridges to sites, and converts the bridge parameter $(1 - c_0)$ to a per-cluster quantity, f.

The mass m_{gas} of gas released in time t, expressed on a single cluster basis, is

$$m_{gas}(t) = \frac{m_b g(\sigma + 1)}{4}.$$

The fraction of bridging bonds released as gas can be converted to the mass of each cluster using $(\sigma + 1)/2$, and another factor, 0.5, is used to convert m_b to the mass of half bridges to calculate the average quantity of side chains and released light gases.

Taking time as the function, the mass of finite fragments of size n, $m_{f,n}$, formed by breaking labile bonds, can be calculated from the bridge population parameters B and p, according to the following equation:

$$m_{f,n} = nm_a + (n - 1)m_b \frac{B}{p} + \frac{\tau \delta m_b}{4(1 - p)}.$$

In the equation, the first term represents the molecular weight of n clusters in one fragment (if $n = 1$, the cluster is a monomer such as benzene, toluene, or naphthalene; if $n = 2$, it is a dimer such as two benzenes connected by an aliphatic bond). The second term represents the molecular weight of the labile bridging bonds, which is calculated from m_b multiplied by the fraction of intact labile bridging bonds. The third term represents the average molecular weight of the side chains released from the gas, which is calculated by multiplying the fraction of side chains by the number of broken bonds τ, and then multiplying by the mass of each side chain, $m_\delta = m_b/2$.

The total mass $m_{fin,n}$ of all the separated fragments of size n is obtained from the mass of the fragments multiplied by the population of the fragments:

$$m_{fin,n} = m_{f,n} \cdot Q_n(p).$$

The total mass m_{fin} of all the separated fragments (assumed to be the mass of tar in the CPD model) can be calculated by adding the masses of the various fragments:

$$m_{fin} = \sum_{n=1}^{\infty} m_{fin,n} = \Phi m_a F(p) + \Omega m_b K(p),$$

$$\Phi = 1 + r\left[\frac{B}{P} + \frac{(\sigma - 1) \cdot \delta}{4(1 - P)}\right],$$

$$\Omega = \frac{\delta}{2(1 - P)} - \frac{B}{P},$$

$$r = m_b/m_a.$$

The mass fractions f_g, f_{fin}, and f_c of the gas, finite fragments, and char, respectively, can be calculated using the following equations:

$$f_g = \frac{m_g}{m_t},$$
$$f_{fin} = \frac{m_{fin}}{m_t},$$
$$f_c = 1 - f_g - f_{fin}.$$

As both B and δ become zero at infinite time, the final amount of light gases produced can be calculated as

$$g(\infty) = 2(1 - c_0).$$

From the calculation of the final amount of light gases produced combined with the equations for calculating σ and c_0 from f_g, we get

$$r = \frac{2f_g(\infty)}{(1 - c_0)(\sigma + 1)\left[1 - f_g(\infty)\right]}.$$

The equation for the calculation of f_g and f_{fin} with the conditions $t = \infty$, $\Omega = 0$, and $\Phi = 1$, give the final amount of finite fragments produced, f_{tar}:

$$f_{tar}(\infty) = \left[1 - f_g(\infty)\right]F(P)|_{t=\infty}.$$

9.4.3 Applications of CPD Model

9.4.3.1 Coal Particle Track Calculation

If the speed of the coal particles is known, the thermal conductivity process from the high-temperature arc plasma to the moving coal particles can be determined. The speed of the coal particles and the arc plasma depend on the flow rate of the working gas and the effective working power of the plasma generator. If the coal particles collide with the reactor wall in a completely elastic mode, the path of the motion of the coal particles can be calculated using the following differential equations [67]:

$$\frac{dv_r}{dt} = -\frac{3}{4}c_D\frac{\rho_g}{\rho_p d_p}\left[v_r^2 + (u - v_z)^2\right]^{\frac{1}{2}}v_r,$$
$$\frac{dv_z}{dt} = \frac{3}{4}c_D\frac{\rho_g}{\rho_p d_p}\left[v_r^2 + (u - v_z)^2\right]^{\frac{1}{2}}(u - v_z) - g.$$

where

v_r and v_z are the radial and axial speeds of the coal particle (m s^{-1});
c_D is the pulling coefficient;
d_p is the diameter of the coal particle (m);
u is the speed of the plasma jet flow (m s^{-1});
g is the gravitational constant (m s^{-2});
ρ_g is the plasma density (g m^{-3});
ρ_p is the coal particle density (g m^{-3}).

9.4.3.2 Determination of Model Parameters

For the CPD model, the most important input parameters are the structural data obtained from the ^{13}C NMR spectra of the raw coal. Genetti et al. [68] used the NMR results for 30 types of coal, performed regression treatment with the proximate and ultimate analytical data of the raw coal and the independent variables, and obtained the following general correction relationship:

$$M_\delta = c_1 + c_2C + c_3 10^{(c_4H)} + c_5O + c_6O^2 + c_7V + c_8V^2,$$
$$Y = c_1 + c_2C + c_3C^2 + c_4H + c_5H^2 + c_6O + c_7O^2 + c_8N + c_9N^2 + c_{10}V + c_{11}V^2,$$

where

$c_{0\text{–}11}$ are correction coefficients (see Table 9.11 for the values);
M_δ is the molecular weight of the side chain;
Y is M_{clust} (molecular weight of cluster), p_0 (fraction of initial intact bridging bonds), or $\sigma + 1$ (coordination number).

Table 9.11 Correction calculation coefficients for ^{13}C NMR chemical structural parameters

Correction calculation coefficient	M_δ	M_{clust}	p_0	$\sigma + 1$
c_1	162.8622	934.235	5.388075	−0.76947
c_2	−1.638111	16.84401	−0.1539078	0.2435051
c_3	5.27×10^7	−0.167003	0.001100592	−0.001919523
c_4	−5.076032	−464.6925	0.1450201	−0.795627
c_5	−0.1605678	51.17301	−0.02162992	0.0519163
c_6	0.002824456	−1.873122	0.0364868	0.04458839
c_7	0.115939	0.03017347	−0.0007587561	−0.003810489
c_8	−0.002498698	207.1326	−0.05401453	0.5653824
c_9	0.0	−41.07443	0.01861638	−0.03658786
c_{10}	0.0	0.2054622	0.0	0.01970864
c_{11}	0.0	−0.02503995	0.0	−0.0001384398

The coal structural parameters can be conveniently calculated from a combination of the correction relationship and chemical analysis of the coal. Introducing the data in Table 9.11 into the above correction relationship, in combination with the proximate and ultimate analytical data for Baode coal, we get the ^{13}C NMR chemical structural parameters for Baode coal, shown in Table 9.12.

According to the CPD theory, the dynamic parameters that control the devolatilization process do not change with changing coal type; therefore the parameters reported in the literature (Table 9.13) can be used directly [69].

9.4.3.3 Energy Equilibrium

No coal tar products were observed in the experiments under the given conditions. The total production of light gases and tar calculated using the CPD simulation model is therefore considered to represent the coal conversion. The material used in the experiments was air dried. The water content compared with that of young Baode coal therefore cannot be neglected, and it is necessary to consider the evaporation of water in the coal in the calculation. When the coal enters the plasma jet flow, heat is transmitted to the coal by convection and radiation, coal devolatilization and water evaporation take place, the volatiles undergo secondary decomposition, and other chemical reactions take place successively. The energy equilibrium equation for the coal particles is expressed as follows:

Table 9.12 Chemical structural parameters of Baode coal

Structural parameter	M_δ	M_{clust}	p_0	$\sigma + 1$
Calculation result	38.23	421.52	0.558	5.088

Table 9.13 General dynamic parameters in CPD model

Dynamic parameter	Value
E_b (kJ mol^{-1})	232.7
A_b (s^{-1})	2.6×10^5
σ_b (kJ mol^{-1})	4.3
E_g (kJ mol^{-1})	289.8
A_g (s^{-1})	3.0×10^{15}
σ_g (kJ mol^{-1})	34.0
ρ	0.9

$$v_p m_p c_p \frac{dT}{dz} = hA_p(T_g - T_p)\frac{B}{e^B - 1} - \sigma\varepsilon_p A_p\left(T_p^4 - T_w^4\right) - v_p\frac{dm_p}{dz}\Delta H,$$
$$h = \frac{Nu \cdot k_g}{d_p},$$
$$\theta = \frac{B}{e^B - 1},$$
$$B = \frac{c_{pg}}{2\pi d_p k_g} \cdot \frac{dm_p}{dt},$$
$$Nu = 2 + 0.6Re^{0.5}Pr^{1/3}.$$

where

m_p is the coal particle mass (kg);
c_p is the heat of melting of the coal particle (J kg^{-1} K^{-1});
A_p is the outer surface area of the coal particle (m^2);
T_g is the temperature of the plasma flow (K);
T_p is the temperature of the coal particle (K);
ε_p is the thermal radiation coefficient (J m^{-2} K^{-4});
T_w is the temperature of the reactor wall (K);
ΔH is the reaction enthalpy (J kg^{-1});
k_g is the thermal conductivity coefficient of the gas (J s^{-1} m^{-1} K^{-1});
d_p is the diameter of the coal particle (m);
c_{pg} is the specific heat of the plasma flow (J kg^{-1} K^{-1});
Re is the Reynolds dimensionless number;
Pr is the Prandtl dimensionless number.

The temperature for the calculation of the thermal conductivity coefficient of the gas is the temperature of the film on the surface of the coal particle, i.e., $(T_p + T_g)/2$. The coal for the experiments had a high water content; therefore it is necessary to add a water evaporation model to the simulation to reflect the temperature change process of the coal particle completely, and accordingly determine the reaction rate constant correctly. In fast-heating experiments because the mass transfer rate is very fast, the water evaporation rate can be expressed as follows:

$$W = \theta_w k_m \pi d_p^2\left(\frac{x_{w,0} - x_{w,\infty}}{1 - x_{w,0}}\right),$$
$$\theta_w = \frac{B_w}{e^{B_w} - 1},$$
$$B_w = \frac{W}{2\pi d_p D_w \rho_g},$$
$$x_{w,0} = \frac{P_w}{P_t}.$$

where

B_w is the transfer coefficient of the vapor;
$x_{w,0}$ is the mole fraction of the vapor on the coal particle surface;
$x_{w,\infty}$ is the mole fraction of the vapor in the plasma flow;
k_m is the mass diffusion coefficient (g m^{-2} s^{-1});
W is the evaporation rate of water (g s^{-1});
D_w is the diffusion coefficient of the vapor (m^2 s^{-1});
ρ_g is the density of the plasma flow (g m^{-3});
P_w is the vapor pressure (Pa);
P_t is the total pressure (Pa).

The P_w is calculated from the Antoine vapor pressure correction:

$$\ln P_w = A - \frac{B}{T + C},$$

where the coefficients A, B, and C, respectively, have the values 18.3036 mm Hg, 3816.44 mm Hg, and 46.13 K.

Taking the impacts of water, secondary decomposition, and other reactions into consideration, the energy equilibrium equation can be expressed as

$$m_p c_p \frac{dT}{dt} = \theta h A_p (T_g - T_p) \frac{B}{e^B - 1} - \sigma \varepsilon_p A_p \left(T_p^4 - T_w^4\right) - \sum_i r_i \Delta H_i.$$

In the equation, i includes water evaporation, devolatilization, secondary decomposition of volatiles, and thermal effect processes such as formation of acetylene, ethylene, and carbon monoxide.

9.4.3.4 Calculation Process

In the actual calculation, it is first necessary to input the initial experimental conditions to calculate the initial reaction temperature $T_{g,0}$ and the initial speed $v_{g,0}$ of the plasma jet flow. At the same time, the initial speed $v_{p,0}$ of the coal particles is calculated. Room temperature is considered to be the initial temperature $T_{p,0}$ of the coal particles. The ^{13}C NMR data for the coal is input and the time step of the calculation is set at 1×10^{-6} s. The temperatures of the gas and coal particles are calculated according to the rules of mass and heat transfer, and the changes in the structural parameters of the coal are calculated, using the CPD dynamic equation, at a temperature of $(T_p + T_g)/2$. The coal conversion is calculated by the statistical method used in CPD (in the calculation, the coal conversion is the total amount of light gases and tar). The axial speed of the coal particle is multiplied by the time step to give the axial displacement of the coal particle in the time step, and the displacements are accumulated. The gas temperature, gas speed, coal particle temperature, and coal particle speed at the end of the i time step and the structural parameters in the CPD model are taken as the initial values, and the above calculation process is repeated until the axial displacement of the coal particles is greater than the length of the reactor; the calculation is then stopped immediately. In the calculation process, it is necessary to record the calculation results for each time step; therefore the coal conversion and temperatures of the gas, coal particles, and other species in the reaction process can be obtained after the calculation has been completed. The time step in the calculation should not be too large, or convergence is not achieved in the calculation. The procedure codes were programmed using Fortran77. Figure 9.22 shows the detailed calculation process.

9.4.3.5 Calculation Results and Discussion

Simulation calculations were performed at powder feed rates of 0.58, 1.98, and 3.49 g s^{-1}. It was observed that the coal conversion reached the maximum value at a short distance and then did not change significantly along the length of the reactor; this indicated that under the conditions used, the coal conversion process in the plasma jet flow was completed quickly and over a short distance. The remaining reactor length and retention time had no obvious effects on the coal conversion. Additionally, increasing the powder feed rate reduced the coal conversion. Figure 9.23 shows the CPD model simulation results and the experimental results. When the powder feed rate was 0.58 g s^{-1}, the experimental results were significantly different from the simulation results. The simulation values remained constant after reaching about 45 % in a short time, and did not reach 55 %.

Under conventional rapid heating conditions, the coal weight loss increases by about 5 % (relative to raw coal) [50], i.e., the maximum conversion of Baode coal is about 45 %. The CPD simulation result of 45 % is therefore reasonable. According to the CPD theory, the main variables that control coal conversion are p_0 and c_0, and ($p_0 - c_0$), which represents the active bridging bonds in the coal structure. The coal conversion process under heating involves breaking active bridging bonds. Once the active bridging bonds have all been broken, coal pyrolysis stops immediately. According to the CPD theory, when the active bridging bonds in the coal have been consumed, the simulation calculation value no longer changes. When the powder feed rate was 0.58 g s^{-1}, the simulation results were inconsistent with the experimental results; this reflects the limitations of the CPD theory. The fact that the experimental value was higher than the simulation value indicated that under plasma jet flow conditions, not only the active bridging bonds but also other stable bridging bonds were broken, improving the coal conversion. An arc plasma has a high temperature and high enthalpy, as well as a large number of activated ions and particles. Activated high-energy species such as H^+, H^-, and Ar^+ can destroy stable coal structures that cannot be converted under common high-temperature conditions, allowing for improved coal conversion. Research [70] has shown that activated species can cause coal conversion by colliding with the coal surface, forming light compounds, suggesting that the active species in the jet flow may accelerate coal conversion at low powder feed rates.

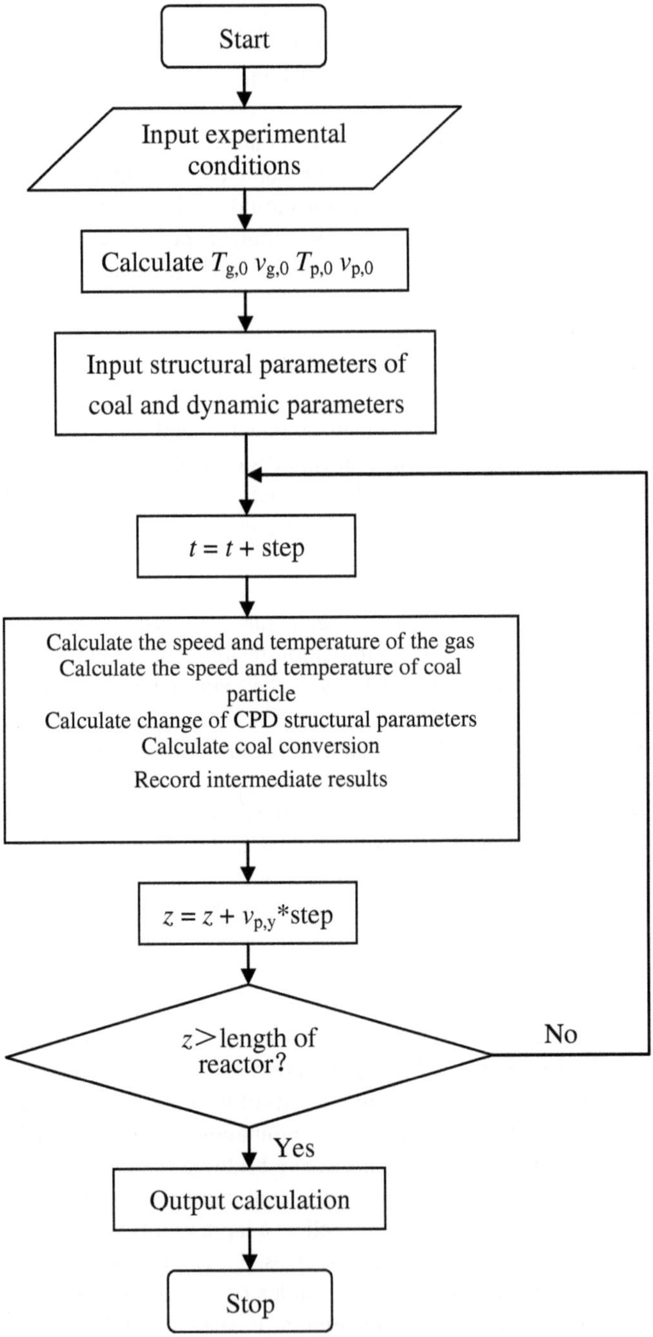

Fig. 9.22 Flow diagram of simulation of pyrolysis conversion of coal in plasma

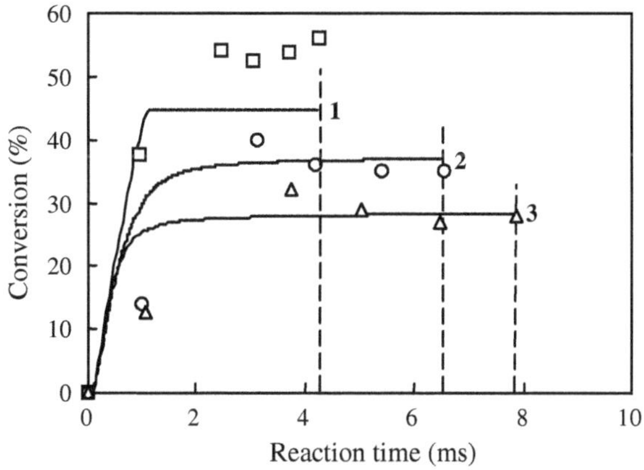

Fig. 9.23 Changes in coal conversion with time at different powder feed rates: *1* 0.58 g s^{-1} (*white square*); 2, 1.98 g s^{-1} (*circle*); and 3, 3.49 g s^{-1} (*triangle*); the full lines represent simulation results and the dashed lines represent experimental results. Reprinted from Ref. [5], Copyright 2001, with permission from American Chemical Society

As shown in Fig. 9.23, at powder feed rates of 1.98 and 3.49 g s^{-1}, the simulation results corresponded to the experiment results. When large amounts of coal particles enter the jet flow, the heat absorption and reaction cause the temperature to decrease dramatically. As the lifetimes of activated species at low temperatures are short, and the reactions of the activated species are suppressed, whereas pyrolysis reactions increase. It is reasonable to describe the whole conversion process using CPD theory. For the first sample outlet, the simulation result was significantly different from the experimental value because of uneven mixing of the coal particles and jet flow in the initial reaction stage, and obstacles to thermal diffusion and mass transfer of the gaseous and solid phases. In fact, it is very difficult for the coal particles and jet flow to be mixed evenly at the beginning of the experimental process, which is why the simulation value is larger than the experimental value.

The temperature change processes of the gaseous and solid phases can be calculated by CPD simulation. As shown in Fig. 9.24, in the initial 1.5 ms, the temperatures of the gaseous and solid phases both changed dramatically. The temperature of the gaseous phase decreased quickly, whereas the solid phase was heated. At later stages, the temperatures of the two phases gradually became closer to each other, and decreased slightly. The chemical reaction of the coal in the plasma mainly took place in the initial 1.5 ms; therefore it is predicted that the reduction in the gas-phase temperature is mainly caused by the reaction of, and heat absorption by, coal particles. Figure 9.23 shows the retention times of coal particles at different powder feed rates (the horizontal coordinate values are indicated by vertical dashed lines). Changes in the powder feed rates directly affect the system temperature, and the temperature decides the extent to which the gas volume increases, and also affects the airflow speed, which decides the retention time of the coal particles in the reactor. At a high powder feed rate, the retention time of the

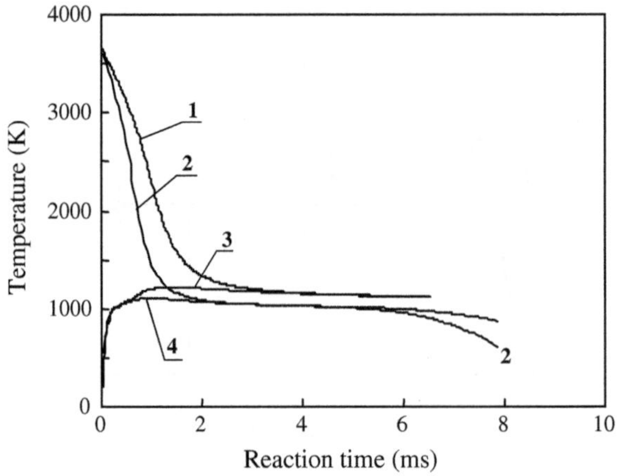

Fig. 9.24 Changes in average temperatures of coal particles and plasma with time at different powder feed rates: *1* plasma temperature, 1.98 g s^{-1}; *2* plasma temperature, 3.49 g s^{-1}; *3* coal particle temperature, 1.98 g s^{-1}; and *4* coal particle temperature, 3.49 g s^{-1}. Reprinted from Ref. [5], Copyright 2001, with permission from American Chemical Society

coal particles in the reactor is longer than that at a low powder feed rate. However, regardless of whether the powder feed rate was high or low, the calculation and experimental results both indicated that the reaction of the coal particles was complete in about 1.5 ms. This result is inconsistent with the conclusions of some research, which showed that a longer retention time led to higher coal conversion. It is suggested that under certain experimental conditions, increasing the reactor length does little to improve the coal conversion.

However, simulation of coal devolatilization in a plasma reactor is not easy, and computational fluid dynamics should be used to obtain more realistic devolatilization processes in the future [71].

9.5 Coke Formation Mechanism in Arc Plasma

Research on solid pyrolysis products was performed as early as the late 1960s [20], but mainly focused on the reaction residues. In the mid-1970s, Chakravartty et al. [10] observed many hard coking products at the anode outlet, and selected four different types of coal to carry out XRD analysis on the coking products. Many researchers have reported methods for removing these coking materials. VEB Kolloidchemie Leipzig (Germany) periodically introduced a certain amount of oxygen in a continuous working process to make the plasma oxygenous, so the coking materials were constantly oxidized and removed. This method guarantees continuous operation, with simple control. This technique enables alternating implementation of the decoking and pyrolysis processes, but a major drawback is that it cannot be used for continuous operation. The AVCO Corporation used a periodic vapor-spraying method to remove coking materials on the reactor walls, and added nitrogen to the plasma to convert free carbons to HCN, thereby preventing the formation of coking materials to a certain extent. These examples show that the formation of coking materials is a problem in the industrialization of plasma pyrolysis processes, but there is a lack of systematic research on the formation mechanism of coking materials. Research on the mechanism of coke formation would increase our understanding of coking, enabling development of methods for preventing or reducing coking.

In this section, the properties of carbon deposits are explored by investigating the impact of operating conditions on carbon deposits on reactor walls, and deposit characterization, and these properties are compared with those formed under plasma conditions from methane gas, liquefied petroleum gas, and biomass. The corresponding mechanisms are discussed, providing a basis for preventing or decreasing carbon deposits from coal.

9.5.1 Factors Affecting Coking

To study whether the coke formation mechanisms are the same along the axial direction, the graphite liner of the reactor was divided into four parts, namely parts A, B, C, and D; the relevant experiments are described in Sect. 9.3.1.1. After the experiments, the coking materials were removed, and designated as cokes A, B, C, and D. Two regimes were used to investigate the coking mechanism. In one regime, referred to as regime 1, a quenching device was installed below the reactor. In the other regime, regime 2, the quenching device was placed above the reactor. Under regime 1, coking materials formed on the reactor walls, whereas under regime 2, no coking materials were formed on the reactor walls below the quenching device.

The experimental feedstocks were Baode coal, methane gas, liquefied petroleum gas, and biomass (sawdust powder). The purity of the methane gas was above 99.9 %; Table 9.14 shows the composition of the liquefied petroleum gas, and Table 9.15 shows the proximate and ultimate analyses of the sawdust powder.

Table 9.14 Chemical composition (%) of liquefied petroleum gas

CH_4	$C_2H_6 + C_2H_4$	C_3H_8	C_3H_6	C_4H_{10}	C_4H_8	C_5H_{12}
0.4	0.82	17.9	48.2	13.0	19.5	0.18

Table 9.15 Proximate and ultimate analyses (%) of sawdust powder

Proximate analysis			Ultimate analysis				
M_{ad}	A_d	V_{daf}	C_{daf}	H_{daf}	N_{daf}	S_{daf}	O_{daf}
5.65	2.79	78.62	44.39	6.93	3.25	0.097	45.33

Provided by CAS Guangzhou Energy Institute

The experiments indicated that feedstocks with different carbon contents showed different coking patterns in the arc plasma jet flow. The coking patterns were different for different raw materials, and the cokes formed at different positions were also different. Research on the impact of different materials and reaction conditions on coking and the properties of the coking materials formed at different positions is the basis for exploration of the coking mechanism.

9.5.1.1 Impact of Coal Type on Coking

Different coal types have different coking and caking properties, and the difficulty of forming deposits under plasma jet flow conditions varies. The degree of difficulty in depositing a quantity C_{ck} per unit mass and unit time is called the coking capacity:

$$C_{ck} = \frac{m_{ck}}{m_0 \cdot t},$$

where

m_{ck} is the mass of coking material (g);

m_0 is the mass of reaction material (g);

t is the reaction time (s).

Unlike the coking ability, the coking capacity is a quantitative concept, and is related not only to the properties of the raw material, but also to the operating conditions. Figure 9.25 shows the coking capacities of coal under a plasma jet flow. The results indicate that Xinzhi coal shows the highest coking capacity, followed by Jingjiao coal and then Baode coal. (Yangcheng coal did not show any coking capacity and therefore is not included in the figure.) Graphite and anthracite did not

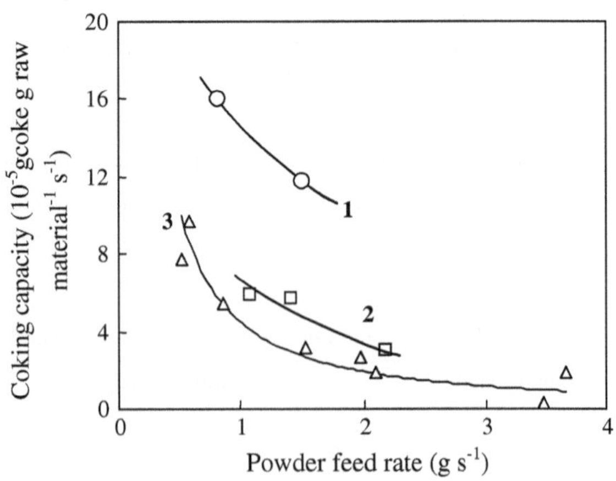

Fig. 9.25 Coking capacities of different coal types in plasma: *1* Xinzhi coal; *2* Jingjiao coal; and *3* Baode coal

show coking capacities. The coking ability and caking property of coal, reflected by the coking process therefore have a large impact on the coking capacity under a plasma jet flow. The coking capacity decreases with increasing powder feed rate. The previous discussion indicated that coal conversion decreases with increasing powder feed rate. The changes in coking capacity are therefore consistent with the conversion changes, and the number and quality of the primary volatiles in coal conversion determine the coking capacity. The volatile content of coal is one of the crucial factors that affect coking, and with increasing powder feed rate, the amount of volatiles separated from the coal powder per unit mass per unit time decreases because of the decreased conversion. The coking capacity also decreases. The coal conversion degree decides the coking degree, and volatiles can only form coke when they have been separated. Gaseous organic matter makes an important contribution to coke formation.

9.5.1.2 Impact of Other Carbonaceous Materials on Coking

Methane, liquefied petroleum gas, and biomass are carbonaceous materials, and their reactions in a plasma jet flow are similar in some sub-processes. However, they have different coking capacities. Table 9.16 lists their maximum coking capacities; the coking capacity of liquefied petroleum gas is the highest, and biomass does not form coke. The coking capacity order is liquefied petroleum gas > Baode coal > methane > biomass.

Liquefied petroleum gas and methane are hydrocarbon gases that contain only carbon and hydrogen. The coking capacity of methane is lower than that of liquefied petroleum gas; therefore heavy hydrocarbons form coke more easily than light hydrocarbons do. For coal and biomass, the molecular weights of the primary volatiles are higher than that of the liquefied petroleum gas primary volatile. However, the coking capacities of coal and biomass are lower than that of liquefied petroleum gas. The effect of molecular weight on the coking capacity is therefore unclear. The ultimate analysis shows that the biomass contains about 45 % oxygen, and oxygen radicals have high oxidative activity. The experimental results shown in Sect. 9.3 indicate that oxygen combines easily with free carbons to produce carbon monoxide. Biomass therefore does not form coke because it contains a large amount of oxygen, which can consume all the carbon that might form coke. Baode coal has a higher coking ability, and a lower oxygen content than biomass. This suggests that the coking ability of a material is affected by its elemental composition

Table 9.16 Coking capacities of different carbonaceous materials in arc plasma jet flow

Raw material	Maximum coking capacity (g soot g raw material $^{-1}$ s^{-1})
Liquefied petroleum gas	0.15
Baode coal	9.71×10^{-5}
Methane gas	5.23×10^{-6}
Biomass	0.0

as well as its molecular weight. In the case of bituminite, the molecular weight of the primary volatiles is high and the primary volatiles contain oxygen. When the oxygen content is low, bituminite shows a high coking capacity (for example, Xinzhi coal and Jingjiao coal); conversely, the coking ability is low when the oxygen content is high. As the coking capacities are different, the coking speeds are different. The coking speed, r_{ck}, is defined as the quantity of coking materials formed in unit time:

$$r_{ck} = \frac{m_{ck}}{t}.$$

Figures 9.26 and 9.27 show the changes in the coking speeds of coal and liquefied petroleum gas, respectively, with changes in the feedstock feed rate. The

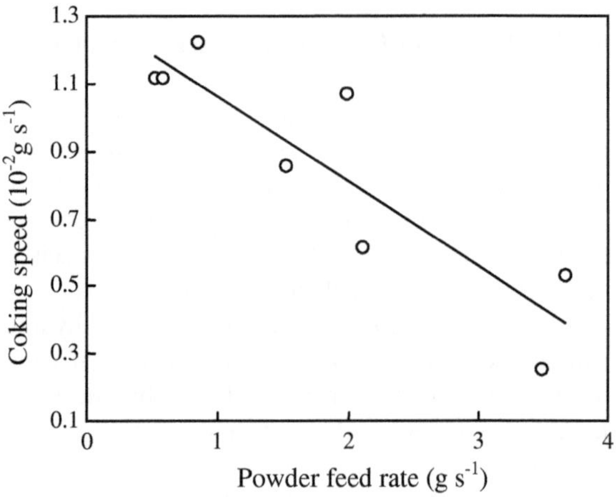

Fig. 9.26 Changes in coking speed of Baode coal with changes in powder feed rate

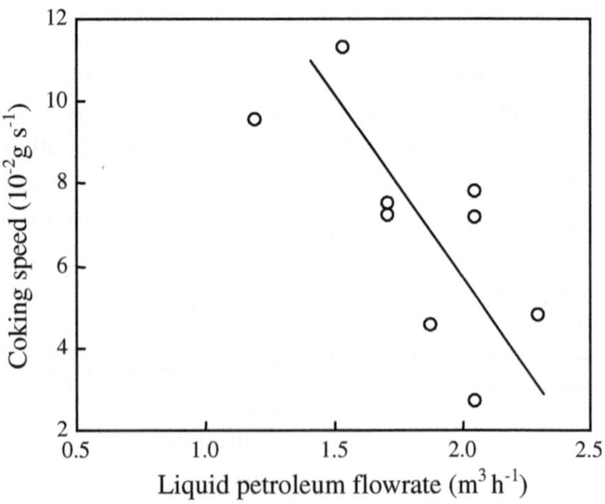

Fig. 9.27 Changes in coking speed of liquefied petroleum gas with changes in gas flow rate

coking speed of the liquefied petroleum gas was higher than that of Baode coal. The quality and quantity of volatiles in the coal are the intrinsic factors that decide the coking capacity, and the reaction conditions and environmental changes are the external causes that affect the coking capacity. When the liquefied petroleum gas enters the jet flow, it directly undergoes various reactions; the reaction between the liquefied petroleum gas and the plasma is homogeneous, whereas the reaction between the coal and the plasma is a gas–solid process. The coking of liquefied petroleum gas therefore better reflects the coking of hydrocarbons in a jet flow. The coking of liquefied petroleum gas provides an important reference for research on the coking mechanism of coal in a plasma. From Figs. 9.26 and 9.27, it can be seen that coking speeds generally decrease with increasing feedstock feed rate, which is consistent with the conversion changes. Therefore, regardless of whether the material is a gas or a solid, the conversion affects the coking, i.e., conversion is the basis of the coking process and all the causes of coking are based on the conversion. The higher the conversion degree is, the more serious the coking phenomenon is.

Coking materials reduce the inner diameter of the reactor and increase the gas flow resistance; therefore, as the amount of coke increases, the flow resistance of the feedstock entering the system increases. Based on the pressure before the working argon entered the system as the criterion, the time from the beginning of the experiment to the moment when the pressure reached 0.08 MPa was recorded; the experimental results are shown in Fig. 9.28. The first change seen in the figure is caused by the sudden increases in the temperature as a result of ignition of the electric arc and the pressure jump caused by volume expansion. The pressure then increases quickly after a relatively slow stage. If the coking speed is constant, when the inner wall of the reactor is covered by coking material, the inner diameter of the reactor is reduced, and the pressure increases accordingly. The time needed to reach 0.08 MPa gradually increases with increasing liquefied petroleum gas feed rate,

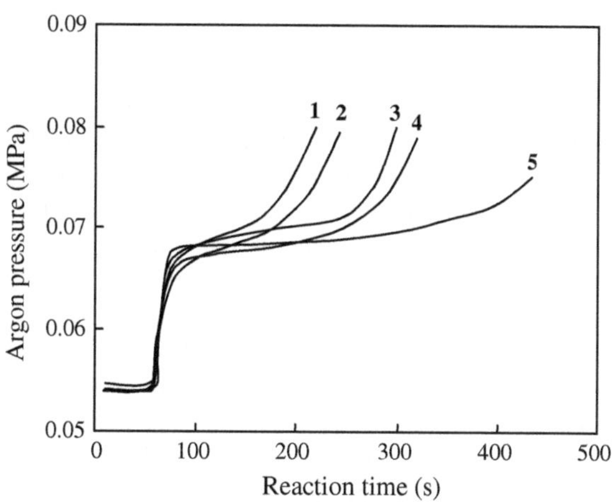

Fig. 9.28 Changes in argon pressure with reaction time at different liquefied petroleum gas flow rates: *1* 1.36 $m^3\ h^{-1}$; *2* 1.02 $m^3\ h^{-1}$; *3* 1.70 $m^3\ h^{-1}$; *4* 2.04 $m^3\ h^{-1}$; and *5* 2.38 $m^3\ h^{-1}$

which indicates that the coking speed decreases. The experiment indicated that the necessary time is shortest at a liquefied petroleum gas flow rate of 1.36 m^3 h^{-1}, showing that this feed rate reduces coking.

9.5.2 Properties of Coking Material

9.5.2.1 XRD Analysis

The coke formation temperatures differ from the top to the bottom of the reactor. This affects the crystal structures of the coking materials. From the top to the bottom of the reactor, the intensities of the (002) and (100) peaks for the deposited materials gradually decreased, reflecting the gradual reduction in the reaction temperature.

The peak (002) changed significantly with variation in the powder feed rate, whereas at positions B and C, the intensities gradually decreased. The (100) peaks showed similar changes, but they were less obvious than those of the (002) peak. The crystal structures of coking materials deposited under the same conditions approached that of graphite from the bottom to the top of the reactor. As the powder feed rate decreased, the crystal structure of the coking material at a given position gradually changed to that of graphite, reflecting changes in the formation temperature of the coking material.

9.5.2.2 Morphologies of Coking Materials

A large amount of information on the coking materials was obtained using SEM. SEM low-magnification images of sections of coking materials at all positions, at a powder feed rate of 0.58 g s^{-1}, show that the coking materials gradually changed from dense to loose, from positions A to D. At position D, the particles simply formed a pile of loose coking material with low mechanical strength. The low-magnification SEM images show that the coking materials at all positions were denser on the side next to the graphite liner than they were on the outer side. All the hard coking materials formed had similar characteristics.

At a low powder feed rate, i.e., 0.58 g s^{-1}, the SEM images show that at all positions, the coking materials consisted mainly of piles of small particles. The average particle size at position A was about 0.5–1.0 μm, and that at position B was about 2–5 μm. At position C, a large amount of filaments was mixed with particles and formed a three-dimensional network. At position D, large particles of size 5–10 μm were loosely piled, with no obvious connections among the particles. The particle sizes in these samples gradually increased from position A to position D; the number of connections between the particles gradually decreased and the structures gradually became looser; this corresponds to the observation that the mechanical strength changed from strong to weak. The morphologies and other properties of the coking materials at different parts also varied, suggesting different formation mechanisms at each position.

The morphologies of the coking materials formed at all positions also changed with the powder feed rate. The morphology of the coking material at position A at a powder feed rate of 1.98 g s^{-1} was similar to that of the material at position B at a powder feed rate of 0.58 g s^{-1}. The morphology of the coking material at position B at a powder feed rate of 1.98 g s^{-1} was similar to that of the material at position C at a powder feed rate of 0.58 g s^{-1}. The morphology of the coking material at position C at a powder feed rate of 1.98 g s^{-1} was similar to that of the material at position D at a powder feed rate of 0.58 g s^{-1}. At a powder feed rate of 3.49 g s^{-1}, the morphology of the coking material at position A was similar to that at position B when the powder feed rate was 1.98 g s^{-1}. The morphology of the coking material at the higher position, with a feed rate of 3.49 g s^{-1}, was similar to that observed at the lower position with a feed rate of 1.98 g s^{-1}. The results indicate that as the powder feed rate increased, the phenomena observed at the bottom of the reactor gradually occurred further up. The XRD results indicate that the formation of coking materials was closely related to the formation temperature. Therefore, at a powder feed rate of 3.49 g s^{-1}, the temperature at position A was close to the temperature at position B at a powder feed rate of 1.98 g s^{-1}, and the temperature at position C at a powder feed rate of 0.58 g s^{-1}; this shows that the temperature changes in the reaction system with increasing powder feed rate. The previous practice of relating the average temperature of the reaction system to the coal specific enthalpy is therefore reasonable.

Liquefied petroleum gas has a high coking ability and does not contain any inorganic impurities. Research on the formation of coking materials from liquefied petroleum gas is therefore important for understanding coking mechanisms. From the SEM images, it is obvious that the structure gradually became less interconnected from positions A to D, similar to the situation for coal. All the coking materials at A, B, and C consisted of large particles composed of spherical particles of size about 1 μm, with irregular pore structures between the large particles. The coking material at D formed a porous three-dimensional network structure. This morphology is similar to that of the coking material formed in the middle part of the reactor with coal as the raw material, but the pores were smaller, and the coking material was denser. Many balls and filaments composed of carbonaceous materials were observed.

The adoption of regime 2 in the experiments prevented coking. The pyrolysis residues collected in regime 2 contained many fluid substances, indicating that new phases were produced, i.e., coking precursors.

9.5.2.3 Investigation of Coking

Coking at Different Positions in Reactor

The above research refers to all the coking materials on the reactor wall. However, experiments have indicated that the coking materials are not evenly deposited on the reactor wall, and the coking materials at different positions have different physical properties. Taking the coking materials from Baode coal as an example, the hardness differed from positions A to D. Coke A could not be crushed without a breaker, whereas coke D could be broken easily. This indicated that the formation mechanisms of coke at different positions in the reactor were different.

Figure 9.29 shows the changes in coking speeds for Baode coal and liquefied petroleum gas with changes in feed rates. The coking speed for Baode coal on all parts of the reactor decreased with increasing powder feed rate; the order was A > B > C > D. The coking speeds for liquefied petroleum gas on all parts of the reactor also decreased (or were basically unchanged) with increasing gas flow rate, but the order was B > A > C > D. For a solid material (coal), increasing the powder feed rate increases the concentration of powder particles in the reaction system. This increase in the concentration of the solid phase hinders contact between the coking precursor and the reactor wall. Also, increasing the powder feed rate reduces the system temperature, so conversion is reduced, and the coking capacity is also reduced. These are the main reasons for the decreased coking speed. For the liquefied petroleum gas, the solid substance was the reaction product; therefore the "obstacle" function of the solid product was hysteretic. A comparison of the coking speed changes for Baode coal and liquefied petroleum gas at all positions in the reactor shows that the coking speed decreased most quickly at the parts with the highest coking speeds (parts A and B) with increasing feed rate, whereas the coking speeds at parts C and D decreased slowly, or basically remained unchanged. Although the total coking speed was decreasing, the relative amounts of coking at each position were changing.

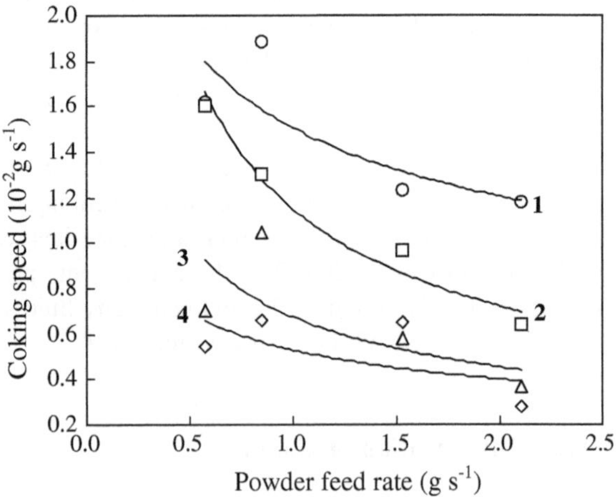

Fig. 9.29 Changes in coking speed with changes in powder feed rate: *1* coke A; *2* coke B; *3* coke C; and *4* coke D

Coking Precursors

For coal, the coking capacity gradually decreased at higher powder feed rates, and the liquefied petroleum gas showed the same general trend. The increase in the material feed rate means that the average temperature of the reaction system and the conversion both decreased. It is therefore deduced that the primary volatiles from material conversion are the main causes of coking. In fact, coking precursors must be present for coke formation to occur; the coking precursors do not consist of a single species or phase. In this book, all the species that can form coking materials are called coking precursors. The coking precursors collide with the reactor wall during flow through the reactor, and are deposited on the wall, forming a coking material. Because the temperatures at different positions in the reactor are different, the reactions are also different. When intense high-temperature decomposition occurs and large numbers of free carbons are generated, the typical reaction equation is as follows:

$$C_nH_m \rightarrow nC + \frac{m}{2}H_2.$$

Under such conditions, free carbons are the main coking precursors. At low temperatures, heavy products can be produced by polymerization, radical combination, and other reactions:

$$C_nH_m + C_nH_m \rightarrow C_{2n}H_{2(m-1)} + H_2,$$
$$C_nH_m + C_nH_m \rightarrow C_{2n}H_{2m}.$$

When the vapor pressure of these heavy organic products is lower than the system pressure, the materials exist as small liquid drops when recombination is complete. When the viscous liquid drops collide with the reactor wall, they become attached, and are rapidly carbonized at high temperature to form a coking material. These composite or polymeric heavy materials are coking precursors. Many of the properties of coking materials and pyrolysis residues at the bottom of the reactor are similar. The pyrolysis residues can be considered as another form of coking on the reactor wall, i.e., pyrolysis residues are coking precursors. Of course, other types of coking materials are formed in the reaction; nanoscale carbon black particles are evidence of the pyrolysis of free carbon precursors:

$$nC \rightarrow C_n.$$

A liquefied petroleum gas flow rate of 1.36 $m^3\ h^{-1}$ was favorable for coking; Table 9.17 shows the proximate and ultimate analyses of the pyrolysis residues formed under regimes 1 and 2 at this flow rate. Regime 2 is used to ensure that the coking precursor can be directly "frozen", by using a large amount of water before coking occurs. A comparison of the results indicates that under regime 2, the hydrogen and volatile content of residues both clearly increased; this indirectly indicates the presence of coking precursors. The low-melting point ash in the coal

Table 9.17 Carbon, Hydrogen, and volatiles (%) in pyrolysis residues of liquefied petroleum gas under regimes 1 and 2

Experimental condition	V_{daf}	C_{daf}	H_{daf}
Regime 1	13.88	95.46	1.54
Regime 2	28.99	91.69	2.01

The liquefied petroleum gas flow rate under both regimes was 1.36 m^3 h^{-1}

fuses and aggregates at high temperatures. This material then mixes easily with carbon and adheres to the reactor wall, thereby increasing the amount of carbon deposited on the reactor wall.

Coking Process

Many factors cause coking, and the coking mechanisms at different temperatures differ. This has a great impact on the formation of coking materials. When coal particles enter the plasma jet flow, they first react with the plasma and produce primary volatiles. Some primary volatiles decompose to free carbons at high temperatures, and collide with the reactor walls under the action of thermophoretic and centrifugal forces (because the jet flow is rotating). The coking effect of methane shows that the coking capacity of pure free carbons is poor, but the fusible ash in the coal might promote the formation of coking materials at this point. The fusible ash has a high viscosity and can attach to the reactor wall and capture free carbons and other particles. The system temperature soon decreases significantly because of the reaction heat absorption and heat absorption by coal particles. At this point, coking precursors are mainly formed by recombination and polymerization reactions (although other reactions also occur). Under such conditions, coking materials are mainly produced from heavy organic materials, but this does not eliminate the possibility of other coking mechanisms. As the temperature decreases, coking materials are mainly formed by the pyrolysis residues, which simply form piles. If the powder feed rate increases, the reaction temperature of the system decreases. At a low powder feed rate, some phenomena occur at the bottom of the reactor that occur in the upper part of the reactor at higher powder feed rates. This indicates that the coking temperature affects the formation of coking precursors, and therefore the coking mechanisms are different. In fact, under many conditions, various coking precursors coexist and simultaneously act as precursors, but they have different advantages under different conditions. There are various coking reactions, and coking mechanisms are complicated. More research is therefore needed to improve our understanding.

During the entire reaction process, other external factors affect the formation of coking materials; for example, if the coal contains oxygen and nitrogen, which can consume free carbons, they can reduce the coking speed, resins in coal can accelerate coking, and solid substances can increase the diffusion of coking precursors toward the reactor wall. Many factors can cause coking, and coking constraint or removal can only be achieved when these factors are weakened or removed.

Table 9.18 Coking capacity and coking speed of coal–sawdust mixture

Sample	Coking capacity (g coke raw material^{-1} s^{-1})	Coking speed (g s^{-1})
Coal (75 %) + sawdust powder (25 %)	1.86×10^{-5}	1.09×10^{-2}
Coal	2.63×10^{-5}	4.28×10^{-2}

Coking Prevention and Constraint

The above discussion of coking mechanisms shows that preventing or effectively constraining the deposition of coking materials on the reactor wall can be achieved by either controlling the formation of coking precursors chemically or by physically preventing the precursors from colliding with the reactor wall.

To introduce more oxygen, a mixture of Baode coal and sawdust powder was used as the reaction raw material. The weight ratio of raw coal to sawdust powder was 3:1, and the particle size of the sawdust powder was 40–80 μm. For comparison, an experiment with raw coal was performed using the same feed rate (1.98 g s^{-1}). The experimental results are shown in Table 9.18. The data in the table show that the coking capacity and the coking speed decreased after the addition of sawdust powder. This was mainly because the large amount of oxygen in the sawdust prevents the formation of coking intermediates. It is important to note that too much oxygen reduces the amount of acetylene produced. Under regime 2, the use of a large amount of cold water prevented deposition of the coking intermediate phase on the reactor wall. This proved that the intermediate phase helps to prevent the formation of coking materials. However, this method decreases the temperature of the reaction system and adversely affects conversion. In conclusion, controlling the formation of an intermediate phase and preventing collisions between the coking intermediate phase and the reactor wall are effective means of preventing or constraining coking.

References

1. Desypris J et al (1982) Fuel 61:807
2. Chakravartty SC et al (1984) Indian J Technol 22:146
3. Baumann H et al (1988) Fuel 67:1120
4. Wong CL et al (1986) Fuel 65:1483
5. Tian YJ et al (2001) Energy Fuels 15(6):1354
6. Wang BJ et al (2007) Energy Sources Part A 29:1195
7. Bao WR et al (2006) Trans IChemE Part B 84(B3):222
8. Bao WR et al (2007) Energy Fuels 21:2082
9. Bond RL et al (1966) Fuel 45(5):381
10. Chakravartty SC et al (1976) Fuel 55:43
11. Beiers HG et al (1988) Fuel 67:1012
12. Dixit LP et al (1982) Fuel Process Technol 6:85
13. Sundaram MS (1979) PhD thesis, Oklahoma State University

14. Gill PS et al (1967) Carbon 5:43
15. Sanada Y et al (1969) Fuel 48:375
16. Vastola FJ et al (1963) Carbon 1:11
17. Mohammedi MN (1996) Plasma Chem Plasma Process 16(1):191
18. Garret MT et al (1969) Int Symp Coal Sci Technol, India C1:1
19. Fan Y (1983) Physical properties of low-temperature plasma. Tsinghua University Press, Beijing p5
20. James AH (1962) Research project report, Sheffield University
21. Bond RL et al (1963) Nature 200:1313
22. Nicholson R et al (1972) Nature 236:397
23. Littlewood K (1984) Mater Res Symp 30:127
24. Nemets Y (1980) Solid Fuel Chem 14(5):88
25. Bittner D et al (1990) Fuel Process Technol 24:311
26. Kulczycka J (1978) Solid Fuel Chem 12(3):89
27. Cannon RE et al (1970) Ind Chem Res Dev 9(3):343
28. Makino M et al (1983) J Fuel Soc Jpn 62:1013
29. Fauchais P et al (1980) Int Chem Eng 20(2):289
30. Li DS et al (1999) Coal Convers 22(2):12
31. Kalinenko RA et al (1987) Kinet Katal. USSR XXVIII:723
32. Georgiev IB et al (1987) USSR Technol Sci 4:83
33. Qiu JS et al (2004) Fuel Process Technol 85:969
34. Djebara D et al (1991) Fuel 70(12):1473
35. Kamei O et al (1999) Fuel 77(13):1503
36. Qiu J et al (1999) J Chem Eng 50(5):586
37. Messerle VE et al (2013) Fuel Process Technol 107:93
38. Messerle VE et al (2014) Fuel 126:294
39. Pang LS et al (1995) Coal Sci Technol 24:1184
40. Geldard L et al (1998) Fuel 77:15
41. Tian YJ et al (2001) Chem J Chine Univ 22(9):1456
42. Tian YJ et al (2004) Catal Today 89:233
43. Tian YJ et al (2004) Carbon 42:2597
44. Bittner D et al (1981) Erdöl Kohle 34(6):237
45. Jiang Y (1995) SIAM J Optim 5(4):813
46. Tang H, Qin X (1994) Optimized method. University of Technology Press, Dalian
47. Chase MW et al (1985) J Phys Chem Ref Data 14(Supplement No 1)
48. Holmen A et al (1995) Fuel Process Technol 42:249
49. Dai B (1999) Chem Eng Sci 54:957
50. Azhakesan M et al (1991) Fuel 70(3):322
51. Schlosberg RH (1985) Chemistry of coal conversion. Plenum, New York, p 121
52. Serio MA et al (1987) Energy Fuels 1:138
53. Grant DM et al (1989) Energy Fuels 3:175
54. Freihant JD et al (1990) 7th annual international Pittsburgh coal conference, Pittsburgh, Pennsylvania
55. Pitt GT (1962) Fuel 41:267
56. Anthony DB et al (1975) 15th symposium (international) on combustion, Pittsburgh, p 1303
57. Kobayashi H et al (1977) 15th symposium (international) in combustion, Pittsburgh, p 411
58. Niksa S (1986) Combust Flame 66:111
59. Niksa S et al (1986) Combust Flame 66:95
60. Niksa S (1988) AIChE J 34:190
61. Niksa S et al (1991) Energy Fuels 5:647
62. Pugmire RJ et al (1991) Fuel 70:414
63. Fu W et al (1988) Sci China A 12:1283
64. Karr C (1979) Analytical methods for coal and coal products, vol 3. Academic Press, New York

65. Sethi NK et al (1987) ACS Div Fuel Chem Preprints 32:155
66. Solum M et al (1987) ACS Div Fuel Chem Preprints 32:273
67. Honda T et al (1985) J Chem Eng Jpn 18(5):414
68. Genetti DB et al (1995) 8th international conference on coal science, Spain
69. Fletcher TH et al (1992) Energy Fuels 6:414
70. Sharkey AG et al (1964) Nature 202:988
71. Yan BH et al (2012) Fuel Process Technol 100:1

MIX
Papier aus verantwortungsvollen Quellen
Paper from responsible sources
FSC® C105338
FSC
www.fsc.org

If you have any concerns about our products,
you can contact us on
ProductSafety@springernature.com

In case Publisher is established outside the EU,
the EU authorized representative is:
Springer Nature Customer Service Center GmbH
Europaplatz 3, 69115 Heidelberg, Germany

Printed by Libri Plureos GmbH
in Hamburg, Germany